Understanding Real Traffic

Boris S. Kerner

Understanding Real Traffic

Paradigm Shift in Transportation Science

≙ Springer

Boris S. Kerner
Physics of Transport and Traffic
University of Duisburg-Essen
Duisburg, Nordrhein-Westfalen, Germany

ISBN 978-3-030-79604-4   ISBN 978-3-030-79602-0 (eBook)
https://doi.org/10.1007/978-3-030-79602-0

© The Editor(s) (if applicable) and The Author(s), under exclusive license to Springer Nature Switzerland AG 2021
This work is subject to copyright. All rights are solely and exclusively licensed by the Publisher, whether the whole or part of the material is concerned, specifically the rights of translation, reprinting, reuse of illustrations, recitation, broadcasting, reproduction on microfilms or in any other physical way, and transmission or information storage and retrieval, electronic adaptation, computer software, or by similar or dissimilar methodology now known or hereafter developed.
The use of general descriptive names, registered names, trademarks, service marks, etc. in this publication does not imply, even in the absence of a specific statement, that such names are exempt from the relevant protective laws and regulations and therefore free for general use.
The publisher, the authors and the editors are safe to assume that the advice and information in this book are believed to be true and accurate at the date of publication. Neither the publisher nor the authors or the editors give a warranty, expressed or implied, with respect to the material contained herein or for any errors or omissions that may have been made. The publisher remains neutral with regard to jurisdictional claims in published maps and institutional affiliations.

This Springer imprint is published by the registered company Springer Nature Switzerland AG
The registered company address is: Gewerbestrasse 11, 6330 Cham, Switzerland

Foreword

One of the few really new ideas in traffic science is due to the author. Careful observation and study of empirical data collected on autobahns in Germany and other countries led him to develop the three-phase theory discussed in this book. The phases of traffic were identified as free flow, synchronous flow and wide moving jams. The author describes these phases by examining patterns of congestion in traffic data. He emphasizes how phases can only be fully understood by looking at patterns of congestion as they evolve over time. The nucleation of congested phases is explained and examples found in real traffic phenomena are given. The presentation is so well done that a reader can get a good grasp of the fundamentals by merely looking at the instructive diagrams and figures.

Unlike the three previous, more technical books by the author on the same topics, this book is written so that nonspecialists can qualitatively understand all the basic features of traffic dynamics. However, even specialists (*e.g.,* experienced traffic researchers) can learn from the intuition and points of view of the author. As in all his books, the author's encyclopedic knowledge of the history and literature of the field is evident. The books serve as an excellent resource in this regard.

Three-phase theory challenges the long-standing beliefs of traffic researchers and, consequently, has only been slowly accepted. This new book could help the traffic community understand the principles of the theory and the author's insights better.

Plymouth, Michigan, USA   Craig Davis
May 2021

Preface

Already in 1992 when the author began working in the field of traffic and transportation science at a Research Institute of the Daimler Company, many practical traffic engineers knew that none of the application approaches developed in the framework of standard traffic and transportation theories could reliably perform in real installations on traffic network roads. While giving a talk "traffic jam without obvious reason" in 1995 at the traffic engineering consulting firm Heusch/Boesefeldt GmbH in Aachen (Germany), the failure of traffic applications of standard vehicular traffic theories in real traffic installations was openly deplored by the engineers: "Since 25 years we tried at the Heusch/Boesefeldt GmbH to apply control and management approaches developed by the traffic research community based on the standard traffic theories, however, nothing works in real installations".

While working later during many years in many engineering projects in the field "traffic" at the Daimler Company, the author recognized that indeed no noticeable progress was done in the field of applications of standard traffic and transportation theories for the reduction of traffic congestion in real traffic and transportation networks.

The above-mentioned encounter with the engineers of the Heusch/Boesefeldt GmbH has affected the author to begin with a study of real (empirical) field traffic data. The data have been measured through sequences of road detectors installed over many highway sections in Germany and the Netherlands. The understanding of the empirical traffic data led to three-phase traffic theory introduced by the author at the end of 1990s. The three-phase traffic theory is incommensurable with any standard traffic theory.

> The *understanding real traffic* has resulted in the *paradigm shift in traffic and transportation science*.

Reviews of the three-phase traffic theory and traffic control based on this theory as well as the criticism of standard traffic theories and models have already been done in the books [B.S. Kerner, *The Physics of Traffic*, (Springer, Berlin, Heidelberg, New York, 2004); B.S. Kerner, *Introduction to modern traffic flow theory and control*, (Springer, Berlin, New York, 2009); B.S. Kerner, *Breakdown in traffic networks*, (Springer, Berlin, New York, 2017)] as well as in the author's reviews in Springer Encyclopedia [B.S. Kerner (ed.), *Complex Dynamics of Traffic Management*, Encyclopedia of Complexity and Systems Science Series (Springer, New York, NY, 2019)].

The main objective of this book is to explain real traffic phenomena for a broad audience of *nonspecialists* that have never learned about traffic phenomena before. To reach this goal, we have dropped all details that are not needed for understanding real traffic; we hope that through the explanation of any terms and definitions in a way that is accessible to nonspecialists, no additional literature sources are required for the understanding of the book. In its turn, the understanding real traffic allows us to explain the reason for the failure of applications of standard theories of vehicular traffic. Additionally, we discuss briefly general results of Kuhn's historical study of scientific revolutions, which has been introduced by Kuhn in his famous book "The structure of scientific revolutions", to a historical analysis of the paradigm shift in traffic and transportation science, as experienced by the author since 1992. It becomes also understandable why it is very difficult for the most members of the traffic and transportation research community to realize there are empirical traffic phenomena that call into question basically all fundamentals of the standard traffic flow theories and models as well as to accept the three-phase traffic theory's solving the problem.

Stuttgart, Germany Boris S. Kerner
May 2021

Acknowledgements

I thank Michael Schreckenberg, Dietrich Wolf, Florian Mazur, Florian Knorr, Gerhard Hermanns, Dominik Wegerle, and Vincent Wiering of the University of Duisburg-Essen as well as my former colleagues at the Daimler Company, in particular, Hubert Rehborn, Peter Häussermann, Harald Brunini, Ralf-Guido Herrtwich, Ralf Lamberti, Peter Konhäuser, Martin Schilke, Matthias Schulze, Peter Ebel, Viktor Friesen, Mario Aleksić, Jochen Palmer, Micha Koller, Peter Hemmerle, Gerhard Nöcker, Winfried Kronjäger, Sven-Eric Molzahn, Andreas Hiller, and Yidirim Dülgar for support and discussions.

Particular thanks are to Sergey Klenov, Craig Davis, Hubert Rehborn, Jonas Paczia, and Oliver Stein who have read the book and made many useful comments that allowed me to present this book in a considerably more understandable way. I thank Andreas Schadschneider for a discussion of the application of Kuhn's theory of scientific revolutions for the development of transportation science. I am grateful to Sergey Klenov and Stefan Kaufmann for the help with the preparation of illustrations for the book.

I thank our partners for their support in the project "LUKAS—Lokales Umfeldmodell für das Kooperative, Automatisierte Fahren in komplexen Verkehrssituationen" funded by the German Federal Ministry of Economic Affairs and Energy.

I thank Micha Koller, Sven-Eric Molzahn, Hubert Rehborn, Markus Auer, and Yidirim Dülgar for providing anonymized GPS probe data measured on many highways used in the book for presentation of empirical spatiotemporal traffic features. I also thank the authorities of the State Hessen (Germany) for providing real field traffic data measured on German freeways.

Finally, I thank my wife, Tatiana Kerner, for her help and understanding.

Contents

1 Introduction 1
  1.1 Standard Theories of Vehicular Traffic 1
  1.2 Empirical Induced Traffic Breakdown—Empirical
       Anomaly for Standard Traffic Theories 4
  1.3 Objective and Methodology 6
  1.4 Structure 8
  References 9

2 Basic Empirical Spatiotemporal Phenomena in Real Traffic 15
  2.1 Three-Phase Traffic Theory—Framework
       for Understanding Real Traffic 15
       2.1.1 Three Traffic Phases in Empirical Traffic Data 15
       2.1.2 Fronts Between Traffic Phases 18
       2.1.3 Definitions of Synchronized Flow and Wide
              Moving Jam Phases in Congested Traffic 20
       2.1.4 Explanations of Term "Synchronized Flow" 21
  2.2 Empirical Spontaneous Traffic Breakdown at Bottleneck 21
  2.3 Empirical Induced Traffic Breakdown at Bottleneck 22
  2.4 Empirical Emergence of Moving Jams in Synchronized
       Flow 23
  2.5 Common Features of Empirical Traffic Breakdown
       at Bottlenecks 24
  2.6 What Is Necessary for Understanding Real Traffic? 24
  2.7 Why Is the Distinguishing Between the Three Phases
       Needed for Understanding Real Traffic? 25
  References 26

**3 How Can Empirical Spatiotemporal Traffic Dynamics Be
   Reconstructed Through Traffic Measurements?** 27
  3.1 Traffic Dynamics Reconstructed Through Probe Vehicle
       Data 27
       3.1.1 Explanation of Vehicle Trajectory 27
       3.1.2 Can a Driver Resolve Traffic Breakdown? 30

		3.1.3	How Short Should the Average Time Interval Between Probe Vehicles Be for the Resolution of Traffic Breakdown? 32
	3.2	Traffic Dynamics Reconstructed Through Road Detector Data 34
		3.2.1	Road Detector Measurements 34
		3.2.2	Empirical Example of Spatiotemporal Traffic Dynamics 35
	References 36

4 Why Does Traffic Breakdown Occur Mostly at a Bottleneck? 39
    4.1   Local Speed Decrease in Free Flow at Road Bottlenecks 39
           4.1.1   On-Ramp Bottleneck 40
           4.1.2   Off-Ramp Bottleneck 41
    4.2   Empirical Example of Local Speed Decrease in Free Flow at Road Bottleneck 42
    4.3   Local Speed Decrease in Free Flow at Moving Bottleneck 44
    References 46

5 Empirical Spontaneous Traffic Breakdown—Fundamental Problem for Understanding Real Traffic 47
    5.1   Perception of Highway Capacity in Standard Traffic and Transportation Science 48
    5.2   Empirical Fundamental Diagram of Traffic Flow 51
    5.3   Empirical Hysteresis Effect 55
    5.4   Microscopic Spatiotemporal Features of Empirical Spontaneous Traffic Breakdown at Bottleneck 57
    5.5   Ignoring of Phenomenon "Empirical Induced Traffic Breakdown"—Emergence of a Diverse Variety of Traffic Theories 62
    5.6   Ignoring of Phenomenon "Empirical Induced Traffic Breakdown"—Consequences for Transportation Science 64
    References 65

6 Empirical Induced Traffic Breakdown—Nucleation Nature of Traffic Breakdown 69
    6.1   Features of Empirical Induced Traffic Breakdown at Bottleneck 69
           6.1.1   Microscopic Characteristics of Empirical Induced Traffic Breakdown 69
           6.1.2   Macroscopic Characteristics of Empirical Induced Traffic Breakdown 73
           6.1.3   Common Empirical Features of Synchronized Flow Resulting from Spontaneous and Induced Traffic Breakdowns 73

	6.2	Explanation of Nucleation Nature of Traffic Breakdown at Bottleneck	76
		6.2.1 Nucleation of Traffic Breakdown and Metastability of Free Flow with Respect to F→S Transition at Bottleneck	76
		6.2.2 Effect of Empirical Nucleation Nature of Traffic Breakdown at Bottleneck on Definition of Synchronized Flow	79
	6.3	Empirical Induced Traffic Breakdown at Bottleneck: A Summary	79
	6.4	Empirical Proof of Nucleation Nature of Traffic Breakdown Using Opposite Assumption	80
	References		83
7	**Empirical Induced Traffic Breakdown—Understanding Stochastic Highway Capacity**		**85**
	7.1	Empirical Induced Traffic Breakdown as the Usual Reason for Traffic Congestion on Long Highway Sections	85
	7.2	Range of Highway Capacities at Any Time Instant	89
		7.2.1 Minimum and Maximum Highway Capacities	89
		7.2.2 Stochastic Highway Capacity in Three-Phase Traffic Theory	92
	7.3	Empirical Induced Traffic Breakdown at Bottleneck as Empirical Proof for Range of Highway Capacities	94
	7.4	Empirical Induced Traffic Breakdown as One of Consequences of Spill-Over Effect	96
	7.5	Perception of Highway Capacity Resulting from Empirical Induced Traffic Breakdown at Bottleneck	98
	References		99
8	**Empirical Nucleation Nature of Traffic Breakdown—Emergence of Three-Phase Traffic Theory**		**101**
	8.1	Discontinuous Character of Over-Acceleration	102
		8.1.1 Driver Speed Adaptation and Over-Acceleration	102
		8.1.2 Time Delay in Over-Acceleration	103
		8.1.3 Discontinuity of Mean Time Delay in Over-Acceleration	104
		8.1.4 Driver Behaviors Explaining the Range of Highway Capacities at Bottleneck	105
		8.1.5 Explanation of the Choice of the Term "Over-Acceleration"	107
	8.2	Nucleus Occurrence for Spontaneous Traffic Breakdown in Free Flow at Bottleneck	108

	8.2.1	Competition Between Speed Adaptation and Over-Acceleration Within Local Speed Decrease at Bottleneck	108
	8.2.2	Critical Speed Within Local Speed Decrease at Bottleneck	110
8.3		Driver Behaviors Resulting in Nucleation Nature of Traffic Breakdown (F→S Transition) at Bottleneck: A Summary	115
References			116

9 Understanding Empirical Nuclei for Traffic Breakdown (F→S Transition) at Bottleneck 119

9.1	Nucleus for Empirical Spontaneous Traffic Breakdown	119
	9.1.1 Waves in Heterogeneous Free Flow: Qualitative Consideration	120
	9.1.2 Empirical Speed Waves in Heterogeneous Free Flow: Local Speed Decreases at Sequence of Moving Bottlenecks	122
	9.1.3 A Mechanism of Nucleus Occurrence in Heterogeneous Free Flow at Road Bottleneck	124
	9.1.4 Random Occurrence of Nucleus for Empirical Spontaneous Traffic Breakdown	125
9.2	Empirical Transitions from Free Flow to Synchronized Flow and Backwards Before Traffic Breakdown (F→S→F Transitions)	128
9.3	Is There a Difference Between Empirical Spontaneous and Induced Traffic Breakdowns?	131
9.4	Empirical Proof of Time Delay in Over-Acceleration Using Opposite Assumption	132
References		133

10 Origin of Emergence of Empirical Moving Traffic Jams: F→S→J Transitions 135

10.1	Empirical Moving Jam Emergence in Synchronized Flow (S→J Transition)	135
10.2	Qualitative Explanation of Moving Jam Emergence in Synchronized Flow (S→J Instability)	136
	10.2.1 Driver Reaction Time and Classical Traffic Flow Instability	136
	10.2.2 Critical Speed for S→J Instability	139
10.3	Crucial Difference Between Driver Reaction Time and Time Delay in Over-Acceleration—A Difficulty for Understanding of Three-Phase Traffic Theory	141
References		142

11 Basic Types of Empirical Spatiotemporal Congested Traffic Patterns at Bottlenecks 145
  11.1 Synchronized Flow Patterns (SPs) 145
    11.1.1 Emergence of Moving SP (MSP) at Road Bottlenecks 146
    11.1.2 Basic Types of SPs at Road Bottlenecks 146
    11.1.3 Diverse Variety of SPs at Road Bottlenecks 149
    11.1.4 Boomerang Effect 150
    11.1.5 MSP Propagating in Direction of Traffic Flow 150
    11.1.6 Basic Types of SPs at Moving Bottleneck 152
    11.1.7 Diverse Variety of SPs at Moving Bottleneck 152
  11.2 General Congested Traffic Patterns (GPs) 152
    11.2.1 Basic Types of GPs at Road Bottlenecks 153
    11.2.2 Diverse Variety of GPs at Road Bottlenecks 153
    11.2.3 Basic Types of GPs at Moving Bottlenecks 156
  11.3 Empirical Microscopic Structure of Wide Moving Jam and Mega-Jam Phenomenon 157
  References 160

12 Discussion and Outlook 163
  12.1 Kuhn's Structure of Scientific Revolutions in Application to Transportation Science 163
    12.1.1 Normal Science: Cumulative Process in Standard Traffic and Transportation Science 164
    12.1.2 Crisis: Failure of Engineering Applications of Standard Traffic Theories 165
    12.1.3 Anomaly: Empirical Induced Traffic Breakdown at Bottleneck 165
    12.1.4 Response to Crisis: Emergence of Three-Phase Traffic Theory 166
    12.1.5 Incommensurability of Standard Traffic Theories with Three-Phase Traffic Theory 167
    12.1.6 Paradigm Shift in Traffic and Transportation Science 167
    12.1.7 Response of Traffic and Transportation Research Community 168
  12.2 Can Autonomous Driving Improve Traffic? 170
    12.2.1 Mixed Traffic Flow 170
    12.2.2 Can Vehicular Traffic Consisting of 100% Autonomous Vehicles Be Real Option in Near Future? 172
  References 174

Appendix A: Characteristics of Synchronized Flow in Three-Phase Traffic Theory 177

Appendix B: Empirical Features of Wide Moving Jams 205

**Appendix C: Empirical Induced Traffic Breakdown—Failure
of Standard Traffic Theories** 215

Glossary 237

Index 239

About the Author

Boris S. Kerner was born in Moscow in 1947 and graduated from Moscow Technical University MIREA in 1972. He received the Ph.D. and Sc.D. (Doctor of Sciences) degrees from the Academy of Sciences of the Soviet Union in 1979 and 1986, respectively. Between 1972 and 1992, his major research interests included the physics of semiconductors, plasma and solid-state physics as well as the development of a theory of autosolitons—solitary intrinsic states, which form in a broad class of physical, chemical, and biological dissipative systems.

Since 1992, he worked on understanding vehicular traffic at Daimler Company in Stuttgart, Germany. He is the pioneer of the *three-phase traffic theory*, which he introduced and developed in 1996–2002. Between 2000 and 2013, he was Head of *Traffic*, a field of research at Daimler. In 2011, he was appointed Professor at the University of Duisburg-Essen in Germany. Since his retirement from Daimler in 2013, he has been working at the University of Duisburg-Essen.

He authored more than 260 scientific works and patents as well as five books that are devoted to a variety of physical systems, complex dynamics of traffic flow in traffic and transportation networks, applications of diverse intelligent transportation systems for traffic prognosis, traffic control, dynamic traffic assignment as well as to the study of autonomous and connected vehicles in mixed traffic flow.

Acronyms and Symbols

F	Free flow traffic phase (free flow for short)
S	Synchronized flow traffic phase of congested traffic (synchronized flow for short)
J	Wide moving jam traffic phase of congested traffic (wide moving jam for short)
F $\rightarrow$ S transition	Phase transition from free flow (F) to synchronized flow (S) (traffic breakdown)
S $\rightarrow$ J transition	Phase transition from synchronized flow (S) to a wide moving jam (J)
F $\rightarrow$ S $\rightarrow$ J transitions	Sequence of an F$\rightarrow$ S transition with the following S$\rightarrow$ J transition
SP	Synchronized flow traffic pattern
LSP	Localized SP
WSP	Widening SP
MSP	Moving SP
GP	General congested traffic pattern
MB	Moving bottleneck
MGP	Moving GP
ITS	Intelligent transportation systems
x	Road location
t	Time
v	Vehicle speed (km/h) or (m/s)
q	Flow rate (vehicles/h)
ρ	Vehicle density (vehicles/km)
$C_{\max}$	Maximum highway capacity of free flow at a bottleneck in the three-phase traffic theory
$C_{\min}$	Minimum highway capacity of free flow at a bottleneck in the three-phase traffic theory

Chapter 1
Introduction

1.1 Standard Theories of Vehicular Traffic

There are vehicular traffic phenomena that are known for almost any driver:

(i) Vehicular traffic can be either *free traffic flow* (free flow for short) or *congested traffic*. In congested traffic, the average vehicle speed is less than the minimum average speed that is possible in free flow.[1]
(ii) In congested traffic, a *moving traffic jam(s)* (moving jam for short) is often observed.

A transition from free flow to congested traffic, i.e., the onset of congestion in free flow is called *traffic breakdown*, or *breakdown phenomenon*, or else *speed breakdown*. A sequence of moving jams in congested traffic is also called *stop-and-go traffic* or *traffic oscillations* of congested traffic. Already in the first empirical studies of traffic breakdown, congested traffic resulting from traffic breakdown, and the emergence of moving jams in congested traffic have been in the focus of traffic and transportation science.[2]

In particular, moving jams in congested traffic have empirically studied in many works.[3] The term *moving jam* can be defined as follows: It is a moving localized structure of a large vehicle density spatially limited by upstream and downstream jam fronts. At the upstream jam front, vehicles decelerate to a lower speed within the jam; at the downstream jam front, vehicles accelerate escaping from the jam. The moving jam propagates upstream in traffic flow as a whole localized structure. Within the jam (between the upstream and downstream jam fronts) vehicle density is large and speed is very low (sometimes as low as zero). The term *vehicle density*

---

[1] This empirical fact (Sect. 5.2) is often used for the definition of congested traffic (see, e.g., [130]).
[2] See, e.g., references in reviews and books [2, 11, 15, 16, 22, 27, 28, 37–40, 43, 52, 59, 62, 63, 69–73, 76, 82, 83, 97, 99, 100, 103, 107, 123, 123, 139, 141, 142, 144, 145, 147, 151, 153, 156–158, 160, 164, 170, 174, 176, 190–195, 198, 200].
[3] In particular, the empirical moving jams (stop-and-go-traffic) have been studied in classical works by Edie et al. [55–58], Treiterer et al. [201–204], and Koshi et al. [130].

© The Author(s), under exclusive license to Springer Nature Switzerland AG 2021
B. S. Kerner, *Understanding Real Traffic*,
https://doi.org/10.1007/978-3-030-79602-0_1

is a number of vehicles per a road length, for example, per kilometer; usually, the density is measured in units (vehicles/km).

First approaches to traffic and transportation science appeared in 1920s–1930s.[4] Fundamentals of standard vehicular traffic and transportation science that led to the most of the standard approaches for traffic simulation and management were developed in 1950s–1960s.[5] Based on these classical ideas to traffic flow theory and modeling during last 70 years several generations of traffic researchers have developed a huge number of traffic simulation models and tools as well as methods for traffic management that include a diverse variety of microscopic and macroscopic traffic flow models.[6] In this book, we use the term "standard" for traffic and transportation science (in particular, traffic theories and models) whose fundamentals were developed before the three-phase traffic theory was introduced. The standard traffic and transportation theories and models are related to the state-of-the-art in the traffic and transportation research community.[7]

The necessity of traffic management is associated with the occurrence of traffic congestion in a traffic network at a large enough traffic demand. The occurrence of traffic congestion results in a considerable increase in travel time and fuel consumption as well as in a decrease in traffic safety and comfort. Therefore, through the management of vehicular traffic, the negative effects of traffic congestion should be

---

[4] One of the most classical works is widely considered the work made in 1935 by Greenshields et al. [81] about a study of traffic breakdown and highway capacity.

[5] In particular, the theoretical fundamentals of standard traffic and transportation theories have been developed in the classical works by Lighthill and Whitham [140], Richards [189], Prigogine [173], Reuschel [186–188], Pipes [171], Kometani and Sasaki [126–129], Herman, Gazis, Rothery, Montroll, Chandler, and Potts [37, 74, 75, 98], Edie et al. [54–58], Moskowitz [149], Newell [156, 157], Webster [206], and Wardrop [205].

[6] In a microscopic traffic flow model, a car-following behavior of each of the vehicles in traffic flow is simulated; respectively, in the model there are mathematical rules for the motion of each of the vehicles. Microscopic traffic flow models are often called car-following models. The term *car-following* that is very often used in the literature means the dynamic behavior of a vehicle that follows the preceding vehicle. In a macroscopic traffic flow model, a collective dynamic behavior of traffic flow is simulated; respectively, only macroscopic traffic flow characteristics can be found from simulations of the model.

[7] The most prominent standard traffic flow models of the standard traffic and transportation science are, for example, as follows: queuing theory and queuing models of vehicular traffic (see, e.g., references in [76, 158]), traffic stream models as well as hydrodynamic and kinematic traffic models (see, e.g., references in [76]), the macroscopic Lighthill–Whitham–Richards (LWR) model [140, 189], the kinetic model of Prigogine [173], Reuschel's model [186–188], Pipes's model [171], Kometani–Sasaki model [126–129], the General Motors microscopic (car-following) model of Herman, Gazis, Rothery, Montroll, Chandler, and Potts [37, 74, 75, 98], Newell's microscopic model [159], Gipps's model [79, 80], Nagel–Schreckenberg cellular automaton (CA) model [152] and subsequent CA model developments (see, e.g., [12, 13]), Wiedemann's model [208], Fukui–Ishibashi model [65–67], Bando et al. microscopic model [4–6], Payne's macroscopic model [161–163], Daganzo's cell transmission model [41, 42], Nagatani's lattice model [150], Aw–Rascle macroscopic model [3], Treiber's intelligent driver model [199], Krauß model [132, 133]. See, e.g., other references in reviews and books [2, 11, 15, 16, 22, 27, 28, 37–40, 43, 52, 59, 62, 63, 69–73, 76, 82, 83, 97, 99, 100, 103, 107, 123, 123, 139, 141, 142, 144, 145, 147, 151, 153, 156–158, 160, 164, 170, 174, 176, 190–195, 198, 200, 207].

1.1 Standard Theories of Vehicular Traffic

decreased. Different systems for traffic management and control are called intelligent transportation systems (ITS). There is a diverse variety of ITS applications.[8]

As mentioned above, traffic congestion results from traffic breakdown. Therefore, traffic breakdown limits highway capacity of free flow. Traffic breakdown occurs usually at a *bottleneck* in the traffic network. This is because the bottleneck introduces a local speed decrease in free flow that exists permanently at the bottleneck (the term *local speed decrease in free flow at the bottleneck* will be explained in Chap. 4). Examples of the network bottlenecks are on- and off-ramps on highways, traffic signals, road gradients, a decrease in the number of road lanes, etc.

> Traffic breakdown is a transition from free flow to congested traffic in a traffic network. Traffic breakdown occurs usually at a network bottleneck. Traffic breakdown limits highway capacity.

These conclusions about the importance of bottlenecks for traffic breakdown have been made already at the beginning of traffic research. Standard traffic theories include a diverse variety of bottleneck models for the analysis of traffic breakdown.[9]

The standard traffic flow theories and models had a great impact on the understanding of many empirical traffic phenomena. However, users of traffic and transportation networks would expect that traffic breakdown can be prevented in a traffic network through the use of traffic management.

> Any traffic and transportation theory applied for the development of reliable methods of traffic management as well as for the evaluation of autonomous driving in mixed traffic flow consisting of human driving and autonomous driving vehicles should be consistent with the empirical features of traffic breakdown at network bottlenecks.

---

[8] Examples of ITS applications are as follows. Vehicular traffic management can be realized through dynamic traffic assignment and traffic control in a traffic network. Methods of traffic control include on-ramp metering, variable speed limit control, traffic signal control at signalized road intersections, and diverse approaches to cooperative driving systems. Cooperative driving systems include, for example, vehicle assistance systems and automated driving (called also self-driving or autonomous driving) vehicles. It is assumed that future vehicular traffic will be a mixed traffic flow consisting of human driving and autonomous driving vehicles (see Sect. 12.2).

[9] See, e.g., reviews and books [2, 11, 16, 22, 27, 28, 37, 39, 40, 43, 52, 59, 62, 69–73, 76, 82, 83, 97, 99, 100, 103, 107, 123, 123, 139, 141, 142, 144, 145, 147, 151, 153, 156–158, 160, 170, 174, 190–193, 195, 198, 200].

1.2 Empirical Induced Traffic Breakdown—Empirical Anomaly for Standard Traffic Theories

The objective of a traffic theory is the explanation of empirical traffic phenomena observed in real traffic. The term *empirical* for traffic phenomena means that the phenomena are found from an analysis of empirical traffic data. In its turn, the term *empirical traffic data* is a synonym of the term *real field traffic data* measured in real traffic. The empirical traffic data used in this book are solely obtained through measurements in real-world traffic based on road detectors installed along a road, video cameras, global navigation satellite system in vehicles that is often called also Globe Position System (GPS) as well as other methods for the measurement of traffic variables in real traffic.

Traffic occurs in space (on roads of a traffic network) and in time. Therefore, to understand empirical traffic phenomena, spatiotemporal measurements of traffic should be made. The term *spatiotemporal* means that measurements of traffic variables both in space and in time should be made simultaneously (see explanations of the reconstruction of empirical traffic spatiotemporal dynamics through traffic measurements in Chap. 3).

> In this book, the understanding real traffic follows *solely* from the analysis of spatiotemporal empirical traffic data.

Many theoretical conclusions of standard traffic theories resulted from a huge variety of studies of spatiotemporal empirical traffic data.[10] In particular, the empirical traffic breakdown is up to now one of the main subjects of empirical traffic data analysis.[11]

However, from the empirical study of the spatiotemporal behavior of traffic on long enough highway sections (sometimes longer than 20 km) made at the end of 1990s the author recognized that the decisive empirical feature of traffic breakdown at highway bottlenecks was not understood from the earlier empirical studies of vehicular traffic.[12]

---

[10] See, e.g., references in reviews and books [2, 11, 15, 16, 22, 27, 28, 37–40, 43, 52, 59, 62, 63, 69–73, 76, 82, 83, 97, 99, 100, 103, 107, 123, 123, 139, 141, 142, 144, 145, 147, 151, 153, 156–158, 160, 164, 170, 174, 176, 190–195, 198, 200].

[11] Beginning from the classical work by Greenshields et al. [81], there have been a huge number of studies of empirical traffic breakdown. These works include many empirical studies made, in particular, by May et al. [53, 146–148], by Hall et al. [84–94], by Persaud et al. [165–169], by Elefteriadou et al. [60, 61], by Brilon et al. [23–26], by Banks et al. [7–10], by Cassidy et al. [17, 29–32, 34, 35] and many other traffic research groups (see, e.g., [18–20, 36, 44] and references there).

[12] See papers [108–112, 120–122]. A history of this study of empirical traffic data as well as other references can be found in the book by Rehborn et al. [182] and reviews [95, 177–181, 183].

1.2 Empirical Induced Traffic Breakdown—Empirical Anomaly … 5

This fundamental empirical feature of traffic breakdown is *the empirical nucleation nature of traffic breakdown at a highway bottleneck*. The author found that empirical traffic breakdown at a bottleneck is a transition from free flow (F) to so-called synchronized flow (S)[13] (F→S transition), which displays a nucleation-type behavior: While a small enough local speed decrease in free flow does not cause traffic breakdown, a large enough local speed decrease in free flow grows further and leads to traffic breakdown at the bottleneck.

> The ultimate empirical proof of the empirical result that real traffic breakdown is an F→S transition at a bottleneck that exhibits the empirical nucleation nature is provided by observations of the *empirical induced traffic breakdown (F→S transition)* at the bottleneck.

The empirical induced traffic breakdown (F→S transition) at the bottleneck is as follows: A congested traffic pattern that propagates on a road can induce traffic breakdown at the bottleneck at which free flow has been before. The terms *empirical induced traffic breakdown*, *congested traffic pattern*, and *empirical nucleation nature of F→S transition* will be in details explained in Chaps. 2, 6, and 7. To explain the empirical nucleation nature of traffic breakdown, the author introduced a new traffic flow theory called "three-phase traffic theory".[14] The explanation of the term *three-phase traffic theory* will be done in Sect. 2.1.

The conclusion about the empirical nucleation nature of traffic breakdown (F→S transition) proven through the evidence of the empirical induced traffic breakdown has led to the crucial consequences for traffic and transportation science[15]:

- The empirical nucleation nature of traffic breakdown (F→S transition) at a bottleneck cannot be explained by any standard traffic theories and models.
- The empirical nucleation nature of traffic breakdown at the bottleneck changes fundamental understanding real traffic.
- For these reasons, the empirical nucleation nature of traffic breakdown (F→S transition) at the bottleneck can be considered *an empirical fundamental of transportation science*.

> In this book, we will show that the fundamental requirement for a traffic flow theory is that the theory should explain the empirical nucleation nature of traffic breakdown (F→S transition) at a highway bottleneck.

---

[13] The definition of the term *synchronized flow* will be done in Sect. 2.1.3.
[14] The three-phase traffic theory was introduced in 1998–1999 [108–111].
[15] See the books [112, 113].

While working since 1992 in many engineering projects in the field "traffic", the author recognized that no noticeable progress was done in the field of applications of standard traffic and transportation theories for the reduction of traffic congestion in real traffic and transportation networks: The standard traffic and transportation theories as well as standard traffic control methods, which have had a great impact on the understanding of many empirical traffic phenomena, have failed by their applications in the real world. Even several decades of a very intensive effort to improve and validate standard network optimization and control models based on the standard traffic and transportation theories have had no success.

- The standard traffic and transportation theories failed when applied in the real world. These standard traffic and transportation theories[16] are currently the methodologies of teaching programs in most universities and the subject of publications in most transportation research journals and scientific conferences.

The failure of the application of standard traffic and transportation theories for the real world can be explained as follows: The standard traffic and transportation theories cannot explain the empirical nucleation nature of traffic breakdown (F→S transition) at a bottleneck.

The main reason for the three-phase traffic theory is the explanation of the empirical nucleation nature of traffic breakdown (F→S transition) at a bottleneck.

1.3 Objective and Methodology

The three-phase traffic theory and ITS applications based on the theory that are consistent with the empirical nucleation nature of traffic breakdown (F→S transition) at a bottleneck have already been reviewed by the author in books and reviews.[17] These books and reviews have been written for specialists and students learning and/or working in traffic and transportation science.

Contrary to these books and reviews, the main objective of this book is to explain real traffic phenomena for a broad audience of *nonspecialists* that have never learned

---

[16] See, e.g., conference proceedings, reviews, and books [1, 2, 11, 15, 16, 22, 27, 28, 33, 37–40, 43, 52, 59, 62, 63, 69–73, 76–78, 82, 83, 97, 99–101, 103, 107, 123, 123, 135, 136, 139, 141–145, 147, 151, 153, 156–158, 160, 164, 170, 174, 176, 190–195, 198, 200].

[17] See the books [112–114] as well as reviews by the author in Springer Encyclopedia [115].

1.3 Objective and Methodology

about vehicular traffic phenomena before. To understand real traffic, it is sufficient to limit our consideration to real traffic phenomena observed in highway and freeway traffic.[18] It must be mentioned that real vehicular traffic exhibits very complex spatiotemporal traffic phenomena. Already the first empirical traffic data shown at the beginning of Sect. 2.1 are probably extremely difficult to understand. It must be emphasized that empirical traffic data may not be "simplified": The whole complexity of empirical data is presented in the book without any simplification of the real traffic data. Because the real traffic data are not simplified in the book, the question can arise: What is then the methodology of the explanation of this real complex traffic phenomena used in this book?

This methodology is as follows. Rather than simplification of the real traffic data, we make a strong simplification of theoretical explanations of the real (empirical) traffic data. As mentioned above, the book is devoted to nonspecialists that have never learned about traffic phenomena before. For this reason, we do *not* consider mathematical traffic models as well as their simulation results at all.[19] Instead of a consideration of mathematical traffic models and their simulation results, we restrict the analysis by *qualitative explanations* of the empirical traffic phenomena. Therefore, for specialists in the field some of these qualitative explanations might be considered "too simple" ones. However, we emphasize that all statements made in this book with the use of these qualitative explanations have been rigorously proven in simulations of mathematical traffic flow models in the framework of the three-phase traffic theory.[20]

We can assume that at least some of the readers can probably be interested in looking at mathematical traffic flow models in the framework of the three-phase traffic theory that can show the empirical nucleation nature of traffic breakdown (F→S transition) at a highway bottleneck. Therefore, there are a lot of footnotes through the book in which we give explicit references to the associated mathematical three-phase traffic models as well as results of their analysis and simulations. In some of the footnotes, we give also references to literature sources used in this book, in

---

[18] A comparison of traffic phenomena in highway traffic with peculiarities of traffic phenomena at signalized intersections in city traffic can be found in Chap. 9 of the book [114].

[19] It should be noted that the first traffic flow model in the framework of the three-phase traffic theory, which can show the empirical nucleation nature of traffic breakdown (F→S transition) was the Kerner–Klenov car-following model introduced in 2002 [116]. Later, a number of different traffic flow models incorporating some of the hypotheses of the three-phase traffic theory have been developed and many results with the use of these models have been found (see, e.g., [21, 45–51, 64, 68, 95, 96, 102, 104–106, 116–119, 124, 125, 131, 137, 138, 155, 172, 175, 177–185, 196, 197, 209–214]). In spite of the above criticism on standard traffic flow models, many ideas of the standard traffic flow theories are very important and they have been used in the three-phase traffic theory. In particular, in mathematical three-phase traffic flow models [117–119] pioneering ideas for the mathematical description of driver behavioral assumptions of the works by Pipes [171], Herman, Gazis, Rothery, Montroll, Chandler, and Potts [37, 74, 75, 98], Newell [156, 159], Gipps [79], Bando et al. [4–6], Nagel and Schreckenberg [152], Krauß et al. [132, 133], Nagel et al. [154] as well as of many other works (see references in Chap. 1 of [114]) have been used.

[20] Mathematical traffic flow models in the framework of the three-phase traffic theory as well as some of their applications can be found in [117–119].

particular, to the literature sources in which the criticism of standard traffic theories and models made in the book has been grounded.

1.4 Structure

The book is organized as follows. In Chap. 2, we discuss basic empirical spatiotemporal traffic phenomena observed in real traffic. Then we consider how these empirical phenomena can be reconstructed through traffic measurements (Chap. 3), explain the empirical fact that traffic breakdown occurs mostly at a bottleneck (Chap. 4). In Chap. 5, we formulate a fundamental problem for understanding real traffic with the use of standard methodologies of traffic and transportation science. In particular, in Chap. 5, we explain why standard methodologies for studies of empirical traffic data or/and standard traffic theories cannot be used for understanding real traffic.

The empirical nucleation nature of traffic breakdown (F→S transition) at a highway bottleneck is proven in Chap. 6. In Chap. 7, we consider an important consequence of the empirical nucleation nature of traffic breakdown—the existence of a range of stochastic highway capacities at any time instant. In Chap. 8, we discuss the driver behaviors that should be responsible for the empirical nucleation nature of traffic breakdown at the bottleneck. Results of empirical studies of the occurrence of a nucleus for traffic breakdown (F→S transition) at the bottleneck are considered in Chap. 9. In Chap. 10, we discuss the emergence of moving jams in synchronized flow. Spatiotemporal traffic congested patterns that result from traffic breakdown are considered in Chap. 11. In Chap. 12, we consider the application of the famous theory by Kuhn about scientific revolutions[21] for the current development of traffic and transportation science; in this discussion chapter, we consider also a question whether autonomous driving can improve future traffic.

For readers, for which simplified explanations of the three-phase traffic theory made in Chaps. 2, 6–11 might be considered "too simplified", we have written Appendixes A and B, in which some of qualitative results of the three-phase traffic theory have been explained in more details.

In Appendix C, we explain that and why the standard traffic and transportation theories cannot explain the empirical nucleation nature of traffic breakdown and, therefore, the standard theories cannot be used for understanding real traffic. In this Appendix C, we show also that the standard definitions of highway capacity and stochastic highway capacity used in the standard traffic research[22] and, in particular, in "Highway Capacity Manual",[23] which is currently the teaching basis for the most

---

[21] Kuhn's theory of the structure of scientific revolution can be found in the book [134].

[22] See, e.g., conference proceedings, reviews, and books [1, 2, 11, 14–16, 22, 27, 28, 33, 37–40, 43, 52, 59, 62, 63, 69–73, 76–78, 82, 83, 97, 99–101, 103, 107, 123, 123, 135, 136, 139, 141–145, 147, 151, 153, 156–158, 160, 164, 170, 174, 176, 190–195, 198, 200].

[23] See Refs. [99, 100].

traffic engineers, contradict basically the empirical phenomenon *empirical induced traffic breakdown* at a highway bottleneck.

Following the objective of the book written for nonspecialists, Appendixes A–C are also written without mathematical models and results of traffic simulations.

References

1. R.E. Allsop, M.G.H. Bell, B.G. Heydecker (eds.), *Transportation and Traffic Theory 2007* (Elsevier Science Ltd, Amsterdam, 2007)
2. W.D. Ashton, *The Theory of Traffic Flow* (Methuen & Co. London, Wiley, New York, 1966)
3. A. Aw, M. Rascle, SIAM J. Appl. Math. **60**, 916–938 (2000)
4. M. Bando, K. Hasebe, A. Nakayama, A. Shibata, Y. Sugiyama, Jpn. J. Appl. Math. **11**, 203–223 (1994)
5. M. Bando, K. Hasebe, A. Nakayama, A. Shibata, Y. Sugiyama, Phys. Rev. E **51**, 1035–1042 (1995)
6. M. Bando, K. Hasebe, A. Nakayama, A. Shibata, Y. Sugiyama, J. Phys. I France **5**, 1389–1399 (1995)
7. J.H. Banks, Transp. Res. Rec. **1510**, 1–10 (1995)
8. J.H. Banks, Transp. Res. Rec. **1678**, 128–134 (1999)
9. J.H. Banks, Transp. Res. Rec. **1802**, 225–232 (2002)
10. J.H. Banks, Transp. Res. Rec. **2099**, 14–21 (2009)
11. J. Barceló (ed.), *Fundamentals of Traffic Simulation* (Springer, Berlin, 2010)
12. R. Barlović, T. Huisinga, A. Schadschneider, M. Schreckenberg, Phys. Rev. E **66**, 046113 (2002)
13. R. Barlović, L. Santen, A. Schadschneider, M. Schreckenberg, Eur. Phys. J. B **5**, 793–800 (1998)
14. M.G.H. Bell, Ch. Cassir, *Reliability of Transport Networks* (Research Studies Press, Baldock, Hertfordshire, England, 2000)
15. M.G.H. Bell, Y. Iida, *Transportation Network Analysis* (Wiley, Incorporated, Hoboken, NJ 07030-6000 USA, 1997)
16. N. Bellomo, V. Coscia, M. Delitala, Math. Mod. Meth. App. Sc. **12**, 1801–1843 (2002)
17. R.L. Bertini, M.J. Cassidy, Transp. Res. A **36**, 683–697 (2002)
18. R.L. Bertini, S. Boice, K. Bogenberger, Transp. Res. Rec. **1978**, 149–159 (2006)
19. R.L. Bertini, S. Hansen, K. Bogenberger, Transp. Res. Rec. **1934**, 97–107 (2005)
20. R.L. Bertini, M.T. Leal, J. Transp. Eng. **131**, 397–407 (2005)
21. R. Borsche, M. Kimathi, A. Klar, Comput. Math. Appl. **64**, 2939–2953 (2012)
22. M. Brackstone, M. McDonald, Transp. Res. F **2**, 181 (1998)
23. W. Brilon, J. Geistefeldt, M. Regler, in *Traffic and Transportation Theory*, ed. by H.S. Mahmassani (Elsevier Science, Amsterdam, 2005), pp. 125–144
24. W. Brilon, J. Geistefeld, H. Zurlinden, Transp. Res. Record **2027**, 1–8 (2007)
25. W. Brilon, M. Regler, J. Geistefeld, Straßenverkehrstechnik. Heft **3**, 136 (2005)
26. W. Brilon, H. Zurlinden, Straßenverkehrstechnik. Heft **4**, 164 (2004)
27. E. Brockfeld, R.D. Kühne, A. Skabardonis, P. Wagner, Trans. Res. Rec. **1852**, 124–129 (2003)
28. A.P. Buslaev, A.G. Tatashev, M.V. Yashina, *Mathematical Physics of Traffic* (Technical Center Publisher, Moscow, 2013) (in Russian)
29. M.J. Cassidy, in *2001 IEEE Inter. Transp. Sys. Proc.* (IEEE, Oakland, USA, 2001), pp. 513–535
30. M.J. Cassidy, S. Ahn, Transp. Res. Rec. **1934**, 140–147 (2005)
31. M.J. Cassidy, R.L. Bertini, Transp. Res. B **33**, 25–42 (1999)
32. M.J. Cassidy, B. Coifman, Transp. Res. Rec. **1591**, 1–6 (1997)

33. M.J. Cassidy, A. Skabardonis (eds.), *Papers selected for the 19th International Symposium on Transportation and Traffic Theory*, Procedia – Social and Behavioral Sciences, vol. 17 (2011), pp. 1–716
34. M.J. Cassidy, A. Skabardonis, A.D. May, Transp. Res. Rec. **1225**, 61–72 (1989)
35. M.J. Cassidy, J.R. Windover, Transp. Res. Rec. **1484**, 73–79 (1995)
36. A. Ceder, A.D. May, Transp. Res. Rec. **567**, 1–15 (1976)
37. R.E. Chandler, R. Herman, E.W. Montroll, Oper. Res. **6**, 165–184 (1958)
38. Y.-C. Chiu, J. Bottom, M. Mahut, A. Paz, R. Balakrishna, T. Waller, J. Hicks, Dynamic traffic assignment, a primer. Trans. Res. Circular E-C153 (2011), http://onlinepubs.trb.org/onlinepubs/circulars/ec153.pdf
39. D. Chowdhury, L. Santen, A. Schadschneider, Phys. Rep. **329**, 199 (2000)
40. M. Cremer, *Der Verkehrsfluss Auf Schnellstrassen* (Springer, Berlin, 1979)
41. C.F. Daganzo, Transp. Res. B **28**, 269–287 (1994)
42. C.F. Daganzo, Transp. Res. B **29**, 79–93 (1995)
43. C.F. Daganzo, *Fundamentals of Transportation and Traffic Operations* (Elsevier Science Inc., New York, 1997)
44. C.F. Daganzo, M.J. Cassidy, R.L. Bertini, Transp. Res. A **33**, 365–379 (1999)
45. L.C. Davis, Phys. Rev. E **69**, 066110 (2004)
46. L.C. Davis, Physica A **405**, 128–139 (2014)
47. L.C. Davis, Phys. Rev. E **69**, 016108 (2004)
48. L.C. Davis, Physica A **368**, 541–550 (2006)
49. L.C. Davis, Physica A **361**, 606–618 (2006)
50. L.C. Davis, Physica A **379**, 274–290 (2007)
51. L.C. Davis, Physica A **387**, 6395–6410 (2008)
52. D.R. Drew, *Traffic Flow Theory and Control* (McGraw Hill, New York, NY, 1968)
53. S.M. Easa, A.D. May, Transp. Res. Rec. **772**, 24–37 (1980)
54. L.C. Edie, Oper. Res. **2**, 107–138 (1954)
55. L.C. Edie, Oper. Res. **9**, 66–77 (1961)
56. L.C. Edie, R.S. Foote, in *Highway Research Board Proceedings*, vol. 37 (HRB, National Research Council, Washington, D.C., 1958), pp. 334–344
57. L.C. Edie, R.S. Foote, in *Highway Research Board Proceedings*, vol. 39 (HRB, National Research Council, Washington, D.C., 1960), pp. 492–505
58. L.C. Edie, R. Herman, T.N. Lam, Transp. Sci. **14**, 55–76 (1980)
59. L. Elefteriadou, *An Introduction to Traffic Flow Theory* (Springer, Berlin, 2014)
60. L. Elefteriadou, A. Kondyli, W. Brilon, F.L. Hall, B. Persaud, S. Washburn, J. Transp. Eng. **140**, 04014003 (2014)
61. L. Elefteriadou, R.P. Roess, W.R. McShane, Transp. Res. Rec. **1484**, 80–89 (1995)
62. A. Ferrara, S. Sacone, S. Siri, *Freeway Traffic Modelling and Control* (Springer, Berlin, 2018)
63. T.L. Friesz, D. Bernstein, *Foundations of Network Optimization and Games, Complex Networks and Dynamic Systems 3* (Springer, New York, Berlin, 2016)
64. D.-J. Fu, Q.-L. Li, R. Jiang, B.-H. Wang, Physica A **559**, 125075 (2020)
65. M. Fukui, Y. Ishibashi, J. Phys. Soc. Jpn. **62**, 3841–3844 (1993)
66. M. Fukui, Y. Ishibashi, J. Phys. Soc. Jpn. **63**, 2882 (1994)
67. M. Fukui, Y. Ishibashi, J. Phys. Soc. Jpn. **65**, 1868–1870 (1996)
68. K. Gao, R. Jiang, S.-X. Hu, B.-H. Wang, Q.-S. Wu, Phys. Rev. E **76**, 026105 (2007)
69. N.H. Gartner, C.J. Messer, A.K. Rathi (eds.), *Special Report 165: Revised Monograph on Traffic Flow Theory* (Trans. Res. Board, Washington, DC, 1997)
70. N.H. Gartner, C.J. Messer, A.K. Rathi (eds.), *Traffic Flow Theory: A State-of-the-Art Report* (Transportation Research Board, Washington DC, 2001)
71. A.V. Gasnikov, S.L. Klenov, E.A. Nurminski, Y.A. Kholodov, N.B. Shamray, *Introduction to Mathematical Simulations of Traffic Flow* (MCNMO, Moscow, 2013) (in Russian)
72. D.C. Gazis, *Traffic Science* (Wiley, New York, 1974)
73. D.C. Gazis, *Traffic Theory* (Springer, Berlin, 2002)
74. D.C. Gazis, R. Herman, R.B. Potts, Oper. Res. **7**, 499–505 (1959)

75. D.C. Gazis, R. Herman, R.W. Rothery, Oper. Res. **9**, 545–567 (1961)
76. D.L. Gerlough, M.J. Huber, *Traffic Flow Theory Special Report 165* (Transp. Res. Board, Washington D.C., 1975)
77. N. Geroliminis, L. Leclercq (eds.), Transp. Res. C **113**, 1–388 (2020)
78. N. Geroliminis, L. Leclercq (eds.), Transp. Res. B **132**, 1–378 (2020)
79. P.G. Gipps, Trans. Res. B **15**, 105–111 (1981)
80. P.G. Gipps, Trans. Res. B **20**, 403–414 (1986)
81. B.D. Greenshields, J.R. Bibbins, W.S. Channing, H.H. Miller, in *Highway Research Board Proceedings*, vol. 14 (1935), pp. 448–477
82. M. Guerrieri, R. Mauro, *A Concise Introduction to Traffic Engineering* (Springer, Berlin, 2021)
83. F.A. Haight, *Mathematical Theories of Traffic Flow* (Academic Press, New York, 1963)
84. F.L. Hall, Transp. Res. A **21**, 191–201 (1987)
85. F.L. Hall, in [70], pp. 2-1–2-36
86. F.L. Hall, K. Agyemang-Duah, Transp. Res. Rec. **1320**, 91–98 (1991)
87. F.L. Hall, B.L. Allen, M.A. Gunter, Transp. Res. A **20**, 197–210 (1986)
88. F.L. Hall, D. Barrow, Transp. Res. Rec. **1194**, 55–65 (1988)
89. F.L. Hall, W. Brilon, Transp. Res. Rec. **1457**, 35–42 (1994)
90. F.L. Hall, M.A. Gunter, Transp. Res. Rec. **1091**, 1–9 (1986)
91. F.L. Hall, L.M. Hall, Transp. Res. Rec. **1287**, 108–118 (1990)
92. F.L. Hall, V.F. Hurdle, J.H. Banks, Transp. Res. Rec. **1365**, 12–18 (1992)
93. F.L. Hall, B.N. Persaud, Transp. Res. Rec. **1232**, 9–16 (1989)
94. F.L. Hall, A. Pushkar, Y. Shi, Transp. Res. Rec. **1398**, 24–30 (1993)
95. K. Hausken, H. Rehborn, in *Game Theoretic Analysis of Congestion, Safety and Security*, ed. by K. Hausken, J. Zhuang (Springer, Berlin, 2015), pp. 113–141
96. S. He, W. Guan, L. Song, Physica A **389**, 825–836 (2010)
97. D. Helbing, Rev. Mod. Phys. **73**, 1067–1141 (2001)
98. R. Herman, E.W. Montroll, R.B. Potts, R.W. Rothery, Oper. Res. **7**, 86–106 (1959)
99. *Highway Capacity Manual 2000* (National Research Council, Transportation Research Board, Washington, DC, 2000)
100. *Highway Capacity Manual 2016*, 6th edn. (National Research Council, Transportation Research Board, Washington, DC, 2016)
101. S.P. Hoogendoorn, V.L. Knoop, H. van Lint (eds.), *20th International Symposium on Transportation and Traffic Theory (ISTTT 2013)*, Procedia – Social and Behavioral Sciences, vol. 80 (2013), pp. 1–996
102. S. Hoogendoorn, H. van Lint, V.L. Knoop, Trans. Res. Rec. **2088**, 102–108 (2008)
103. A. Horni, K. Nagel, K.W. Axhausen (eds.), *The Multi-Agent Transport Simulation MATSim* (Ubiquity, London, 2016), http://matsim.org/the-book. https://doi.org/10.5334/baw
104. X.-J. Hu, X.-T. Hao, H. Wang, Z. Su, F. Zhang, Physica A **545**, 123725 (2020)
105. X.-J. Hu, F. Zhang, J. Lu, M.-Y. Liu, Y.-F. Ma, Q. Wan, Physica A **527**, 121176 (2019)
106. X.-j. Hu, H. Liu, X. Hao, Z. Su Z. Yang, Physica A **563**, 125495 (2021)
107. P. Kachroo, K.M.A. Özbay, *Feedback Control Theory for Dynamic Traffic Assignment* (Springer, Berlin, 2018)
108. B.S. Kerner, Phys. Rev. Lett. **81**, 3797–3800 (1998)
109. B.S. Kerner, in *Proceedings of the 3rd Symposium on Highway Capacity and Level of Service*, ed. by R. Rysgaard (Road Directorate, Copenhagen, Ministry of Transport – Denmark 1998), pp. 621–642
110. B.S. Kerner, Transp. Res. Rec. **1678**, 160–167 (1999)
111. B.S. Kerner, in *Transportation and Traffic Theory*, ed. by A. Ceder (Elsevier Science, Amsterdam, 1999), pp. 147–171
112. B.S. Kerner, *The Physics of Traffic* (Springer, Berlin, Heidelberg, New York, 2004)
113. B.S. Kerner, *Introduction to Modern Traffic Flow Theory and Control* (Springer, Berlin, New York, 2009)
114. B.S. Kerner, *Breakdown in Traffic Networks* (Springer, Berlin, New York, 2017)

115. B.S. Kerner (ed.), *Complex Dynamics of Traffic Management, Encyclopedia of Complexity and Systems Science Series* (Springer, New York, NY, 2019)
116. B.S. Kerner, S.L. Klenov, J. Phys. A: Math. Gen. **35**, L31–L43 (2002)
117. B.S. Kerner, S.L. Klenov, Phys. Rev. E **68**, 036130 (2003)
118. B.S. Kerner, S.L. Klenov, J. Phys. A: Math. Gen. **39**, 1775–1809 (2006)
119. B.S. Kerner, S.L. Klenov, D.E. Wolf, J. Phys. A: Math. Gen. **35**, 9971–10013 (2002)
120. B.S. Kerner, H. Rehborn, Phys. Rev. E **53**, R1297–R1300 (1996)
121. B.S. Kerner, H. Rehborn, Phys. Rev. E **53**, R4275–R4278 (1996)
122. B.S. Kerner, H. Rehborn, Phys. Rev. Lett. **79**, 4030–4033 (1997)
123. F. Kessels, *Traffic Flow Modelling* (Springer, Berlin, 2019)
124. S.L. Klenov, in *Proceedings of Moscow Institute of Physics and Technology (State University)*, vol. 2, ed. by V.V. Kozlov (Moscow Institute of Physics and Technology, Moscow 2010), pp. 75–90 (in Russian)
125. S. Kokubo, J. Tanimoto, A. Hagishima, Physica A **390**, 561–568 (2011)
126. E. Kometani, T. Sasaki, J. Oper. Res. Soc. Jap. **2**, 11–26 (1958)
127. E. Kometani, T. Sasaki, Oper. Res. **7**, 704–720 (1959)
128. E. Kometani, T. Sasaki, Oper. Res. Soc. Jap. **3**, 176–190 (1961)
129. E. Kometani, T. Sasaki, in *Theory of Traffic Flow*, ed. by R. Herman (Elsevier, Amsterdam, 1961), pp. 105–119
130. M. Koshi, M. Iwasaki, I. Ohkura, in *Proceedings of the 8th International Symposium on Transportation and Traffic Theory*, ed. by V.F. Hurdle (University of Toronto Press, Toronto, Ontario, 1983), pp. 403
131. F. Knorr, M. Schreckenberg, J. Stat. Mech. **P07002** (2013)
132. S. Krauß, PhD thesis, (DRL-Forschungsbericht 98-08, 1998), http://www.zaik.de/~paper
133. S. Krauß, P. Wagner, C. Gawron, Phys. Rev. E **54**, 3707–3712 (1996)
134. T.S. Kuhn, *The Structure of Scientific Revolutions*, 4th edn. (The University of Chicago Press, Chicago, London, 2012)
135. M. Kuwahara, H. Kita, Y. Asakura (eds.), *21st International Symposium on Transportation and Traffic Theory*, Transp. Res. Procedia **7**, 1–704 (2015)
136. W.H.K. Lam, S.C. Wong, H.K. Lo (eds.), *Transportation and Traffic Theory 2009* (Springer, Dordrecht, Heidelberg, London, New York, 2009)
137. H.K. Lee, R. Barlović, M. Schreckenberg, D. Kim, Phys. Rev. Lett. **92**, 238702 (2004)
138. H.-K. Lee, B.-J. Kim, Physica A **390**, 4555–4561 (2011)
139. W. Leutzbach, *Introduction to the Theory of Traffic Flow* (Springer, Berlin, 1988)
140. M.J. Lighthill, G.B. Whitham, Proc. Roy. Soc. A **229**, 281–345 (1955)
141. S. Maerivoet, B. De Moor, Phys. Rep. **419**, 1–64 (2005)
142. H.S. Mahmassani, Netw. Spat. Econ. **1**, 267–292 (2001)
143. H.S. Mahmassani, N. Yu, K. Smilowitz (eds.), Trans. Res. C **95**, 1–354 (2018)
144. R. Mahnke, J. Kaupužs, I.A. Lubashevsky, *Physics of Stochastic Processes: How Randomness Acts in Time* (Wiley-VCH, Weinheim, 2009)
145. F.L. Mannering, W.P. Kilareski, *Principles of Highway Engineering and Traffic Analysis*, 2nd edn. (Wiley, New York, 1998)
146. A.D. May, Highway Res. Rec. **59**, 9–38 (1964)
147. A.D. May, *Traffic Flow Fundamentals* (Prentice-Hall Inc., New Jersey, 1990)
148. A.D. May, P. Athol, W. Parker, J.B. Rudden, Highway Res. Rec. **21**, 48–70 (1963)
149. K. Moskowitz, Proc. Highw. Res. Board **33**, 385–395 (1954)
150. T. Nagatani, Phys. Rev. E **59**, 4857–4864 (1999)
151. T. Nagatani, Rep. Prog. Phys. **65**, 1331–1386 (2002)
152. K. Nagel, M. Schreckenberg, J. Phys. (France) I **2**, 2221–2229 (1992)
153. K. Nagel, P. Wagner, R. Woesler, Oper. Res. **51**, 681–716 (2003)
154. K. Nagel, D.E. Wolf, P. Wagner, P. Simon, Phys. Rev. E **58**, 1425–1437 (1998)
155. J.P.L. Neto, M.L. Lyra, C.R. da Silva, Physica A **390**, 3558–3565 (2011)
156. G.F. Newell, Oper. Res. **9**, 209–229 (1961)

157. G.F. Newell, in *Proceedings of the Second International Symposium on Traffic Road Traffic Flow* (OECD, London, 1963), pp. 73–83
158. G.F. Newell, *Applications of Queuing Theory* (Chapman Hall, London, 1982)
159. G.F. Newell, Trans. Res. B **36**, 195–205 (2002)
160. M. Papageorgiou, I. Papamichail, Transp. Res. Rec. **2047**, 28–36 (2008)
161. H.J. Payne, in *Mathematical Models of Public Systems*, vol. 1, ed. by G.A. Bekey (Simulation Council, La Jolla, 1971), pp. 51–60
162. H.J. Payne, in *Research Directions in Computer Control of Urban Traffic Systems*, ed. by W.S. Levine (New York, American Society of Civil Engineers, 1979), pp. 251–265
163. H.J. Payne, Transp. Res. Rec. **772**, 68 (1979)
164. S. Peeta, A.K. Ziliaskopoulos, Netw. Spat. Econ. **1**, 233–265 (2001)
165. B.N. Persaud, F.L. Hall, Transp. Res. A **23**, 103–113 (1989)
166. B.N. Persaud, F.L. Hall, L.M. Hall, Transp. Res. Rec. **1287**, 167–175 (1990)
167. B.N. Persaud, V.F. Hurdle, Transp. Res. Rec. **1194**, 191–198 (1988)
168. B.N. Persaud, S. Yagar, R. Brownlee, Transp. Res. Rec. **1634**, 64–69 (1998)
169. B.N. Persaud, S. Yagar, D. Tsui, H. Look, Transp. Res. Rec. **1748**, 110–115 (2001)
170. B. Piccoli, A. Tosin, in *Encyclopedia of Complexity and System Science*, ed. by R.A. Meyers (Springer, Berlin, 2009), pp. 9727–9749
171. L.A. Pipes, J. Appl. Phys. **24**, 274–287 (1953)
172. A. Pottmeier, C. Thiemann, A. Schadschneider, M. Schreckenberg, in *Traffic and Granular Flow'05*, ed. by A. Schadschneider, T. Pöschel, R. Kühne, M. Schreckenberg, D.E. Wolf (Springer, Berlin, 2007), pp. 503–508
173. I. Prigogine, in *Theory of Traffic Flow*, ed. by R. Herman (Elsevier, Amsterdam 1961), pp. 158–164
174. I. Prigogine, R. Herman, *Kinetic Theory of Vehicular Traffic* (American Elsevier, New York, 1971)
175. Y.-S. Qian, X. Feng, J.-W. Zeng, Physica A **479**, 509–526 (2017)
176. B. Ran, D. Boyce, *Modeling Dynamic Transportation Networks* (Springer, Berlin, 1996)
177. H. Rehborn, S.L. Klenov, in *Encyclopedia of Complexity and System Science*, ed. by R.A. Meyers (Springer, Berlin, 2009), pp. 9500–9536
178. H. Rehborn, S.L. Klenov, M. Koller, in *Complex Dynamics of Traffic Management*, ed. by B.S. Kerner, Encyclopedia of Complexity and Systems Science Series (Springer, New York, NY, 2019), pp. 501–557
179. H. Rehborn, S.L. Klenov, J. Palmer, Physica A **390**, 4466–4485 (2011)
180. H. Rehborn, S.L. Klenov, J. Palmer, I.E.E.E. Intell, Veh. Sym. (IV) **19–24** (2011)
181. H. Rehborn, M. Koller, J. Adv. Transp. **48**, 1107–1120 (2014)
182. H. Rehborn, M. Koller, S. Kaufmann, *Data-Driven Traffic Engineering: Understanding of Traffic and Applications Based on Three-Phase Traffic Theory* (Elsevier, Amsterdam, 2021)
183. H. Rehborn, J. Palmer, Intell. Veh. Sym. IEEE **186–191** (2008)
184. F. Rempe, P. Franeck, U. Fastenrath, K. Bogenberger, in *Proceedings of the IEEE 19th International Conference on ITS* (2016), pp. 1838–1843
185. F. Rempe, P. Franeck, U. Fastenrath, K. Bogenberger, Transp. Res. C **85**, 644–663 (2017)
186. A. Reuschel, Österreichisches Ingenieur-Archiv **4**(3/4), 193–215 (1950)
187. A. Reuschel, Z. Österr, Ing.-Archit.-Ver. **95**, 59–62 (1950)
188. A. Reuschel, Z. Österr, Ing.-Archit.-Ver. **95**, 73–77 (1950)
189. P.I. Richards, Oper. Res. **4**, 42–51 (1956)
190. R.P. Roess, E.S. Prassas, *The Highway Capacity Manual: A Conceptual and Research History* (Springer, Berlin, 2014)
191. M. Saifuzzaman, Z. Zheng, Transp. Res. C **48**, 379–403 (2014)
192. T. Seo, A.M. Bayen, T. Kusakabe, Y. Asakura, Annu. Rev. Cont. **43**, 128–151 (2017)
193. A. Schadschneider, D. Chowdhury, K. Nishinari, *Stochastic Transport in Complex Systems* (Elsevier Science Inc., New York, 2011)
194. Y. Sheffi, *Urban Transportation Networks: Equilibrium Analysis with Mathematical Programming Methods* (Prentice-Hall, New Jersey, 1984)

195. V.I. Shvetsov, Autom. Remote Control **64**, 1651–1689 (2003)
196. F. Siebel, W. Mauser, Phys. Rev. E **73**, 066108 (2006)
197. J.-F. Tian, C.-Q. Zhu, R. Jiang, in *Complex Dynamics of Traffic Management*, ed. by B.S. Kerner, Encyclopedia of Complexity and Systems Science Series (Springer, New York, NY, 2019), pp. 313–342
198. TRANSIMS (U.S. Department of Transportation, Federal Highway Administration, Washington, DC, 2017), https://www.fhwa.dot.gov/planning/tmip/resources/transims/
199. M. Treiber, A. Hennecke, D. Helbing, Phys. Rev. E **62**, 1805–1824 (2000)
200. M. Treiber, A. Kesting, *Traffic Flow Dynamics* (Springer, Berlin, 2013)
201. J. Treiterer, Transp. Res. **1**, 231–251 (1967)
202. J. Treiterer, *Investigation of Traffic Dynamics by Aerial Photogrammetry Techniques*, Ohio State University Technical Report PB 246 094 (Columbus, Ohio, 1975)
203. J. Treiterer, J.A. Myers, in *Proceedings of the 6th International Symposium on Transportation and Traffic Theory*, ed. by D.J. Buckley (A.H. & AW Reed, London, 1974), pp. 13–38
204. J. Treiterer, J.I. Taylor, Highway Res. Rec. **142**, 1–12 (1966)
205. J.G. Wardrop, in *Proc. of Inst. of Civil Eng. II.* **1**, 325–378 (1952)
206. F.V. Webster, *Road Research Technical Paper No. 39* (Road Research Laboratory, London, UK, 1958)
207. G.B. Whitham, *Linear and Nonlinear Waves* (Wiley, New York, 1974)
208. R. Wiedemann, *Simulation Des Verkehrsflusses* (University of Karlsruhe, Karlsruhe, 1974)
209. J.J. Wu, H.J. Sun, Z.Y. Gao, Phys. Rev. E **78**, 036103 (2008)
210. H. Yang, J. Lu, X.-J. Hu, J. Jiang, Physica A **392**, 4009–4018 (2013)
211. H. Yang, X. Zhai, C. Zheng, Physica A **509**, 567–577 (2018)
212. J.-W. Zeng, Y.-S. Qian, Z. Lv, F. Yin, L. Zhu, Y. Zhang, D. Xu, Physica A **574**, 125918 (2021)
213. J.-W. Zeng, Y.-S. Qian, S.-B. Yu, X.-T. Wei, Physica A **530**, 121567 (2019)
214. H.-T. Zhao, L. Lin, C.-P. Xu, Z.-X. Li, X. Zhao, Physica A **553**, 124213 (2020)

Chapter 2
Basic Empirical Spatiotemporal Phenomena in Real Traffic

In this chapter, we try to answer the following question[1]:

- What are spatiotemporal traffic phenomena observed in real traffic?

Another objective of this chapter is to make basic definitions needed for understanding this book. With the use of these definitions, in the next chapters of the book we make a subsequent consideration of empirical traffic data to achieve the main objective of the book—understanding real traffic.

2.1 Three-Phase Traffic Theory—Framework for Understanding Real Traffic

2.1.1 Three Traffic Phases in Empirical Traffic Data

Many of the common important empirical spatiotemporal features of real traffic measured during 1996–2020 on different highways in different countries can already be studied from an analysis of a typical empirical example measured simultaneously in space and time on a long enough highway section with several bottlenecks presented in Fig. 2.1a, b. On the three-lane highway section shown in Fig. 2.1a, there are three bottlenecks. Each of the curves shown in Fig. 2.1b is a trajectory of a probe vehicle along the road location (vertical axis) over time (horizontal axis). In other words, the empirical data in Fig. 2.1b have been derived from probe vehicle data (called also floating car data (FCD)) in a traffic data center. The probe vehicle data include locations of probe vehicles moving in real traffic at subsequent time instants. The time dependence of the road location of a vehicle is called a vehicle trajectory in

---

[1] Real spatiotemporal traffic phenomena are presented in this chapter based on results of papers [1–13]; for more details, see books and reviews [14–20, 23].

© The Author(s), under exclusive license to Springer Nature Switzerland AG 2021
B. S. Kerner, *Understanding Real Traffic*,
https://doi.org/10.1007/978-3-030-79602-0_2

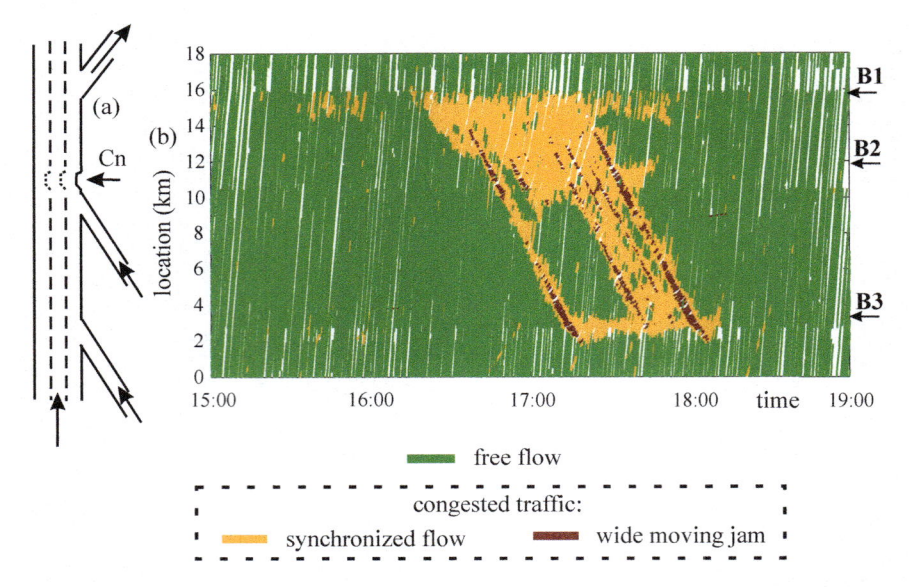

Fig. 2.1 Empirical example of real traffic reconstructed in space and time through the use of probe vehicles moving on November 04, 2016, on a highway section A8-East in Germany. **a** Simplified schema of three-lane section of highway A8-East. **b** Empirical trajectories of probe vehicles in space and time. There are three highway bottlenecks on the highway section: bottleneck B1 (off-ramp bottleneck in highway intersection "Wendlingen"; bottleneck location is about at 15.7 km), bottleneck B2 (bottleneck in highway intersection "Esslingen" caused by two road inhomogeneities (i) on-ramp (at location about 11 km) and (ii) subsequent lane-construction on the length about 1 km denoted by "Cn"; bottleneck location is about at 12 km), and bottleneck B3 (on-ramp bottleneck in highway intersection "Stuttgart-Möhringen"; the bottleneck location is about at 3.4 km). The average time interval between probe vehicles whose trajectories are shown in **b** is equal to 14 s; it should be noted that for the empirical example shown in **b** trajectories of probe vehicles have been measured only, i.e., no trajectories of other vehicles and no flow rates are known (see explanations of probe vehicle measurements in Sect. 3.1)

space and time. The terms *probe vehicle data*, *vehicle trajectory*, and *bottleneck* will be in details discussed, respectively, in Sect. 3.1.1 and Chap. 4.

The time dependence of the vehicle speed along a trajectory of a probe vehicle calculated in the traffic data center can be used for the approximate distinguishing of the three traffic phases (Figs. 2.1b and 2.2):

1. The free flow traffic phase (free flow (F) for short).
2. The synchronized flow traffic phase (synchronized flow (S) for short).
3. The wide moving jam traffic phase (wide moving jam (J) for short).[2]

---

[2] The word *wide* in the term *wide moving jam* corresponds to the longitudinal direction (traffic direction). The use of the word *wide* comes back to a *characteristic feature* of the wide moving jam to maintain the mean velocity of the downstream jam front while propagating through any traffic states and highway bottlenecks (see Chap. 10). Many other localized patterns called *wide*

2.1 Three-Phase Traffic Theory—Framework for Understanding Real Traffic

In Figs. 2.1b and 2.2, free flow (green color) is approximately related to the vehicle speed that is larger than 60 km/h; synchronized flow (yellow color) is related to the vehicle speed that is approximately within the speed range 10–60 km/h; the wide moving jam (claret-red color) is approximately related to the vehicle speed that is within the speed range 0–10 km/h. In general, the vehicle speed in free flow is on average larger than the speed in synchronized flow. The vehicle speed in synchronized flow is on average larger than the speed within a wide moving jam.[3]

> The three-phase traffic theory is the framework for the description of empirical traffic states in three traffic phases: (i) free flow (F), (ii) synchronized flow (S), and (iii) wide moving jam (J). The synchronized flow and wide moving jam phases belong to congested traffic.

Another empirical example of spatiotemporal traffic phenomena is shown in Fig. 2.3. In this case, traffic dynamics has been reconstructed through road detector data that will be defined and explained in Sect. 3.2.

It must be emphasized that threshold speeds used for the distinguishing between the three traffic phases in Figs. 2.1b, 2.2, and 2.3b are used for the approximate visualization of the traffic phases, *not* for the traffic phase definition. The traffic phase definition is made in the three-phase traffic theory based on some *characteristic spatiotemporal features* of the traffic phases (Sect. 2.1.3). We should mention that in different sets of empirical data the speed thresholds for the approximate visualization of the traffic phases in the data can considerably depend on speed limitations, weather conditions, the share of long vehicles in traffic flow, road infrastructure, etc. Therefore, for each particular data set the correct distinguishing between the traffic phases can be made based only on the traffic phase definition. Only then the speed thresholds can be estimated in the data set, to make the visualization of the traffic phases in empirical data in accordance with the traffic phase definition.

---

autosolitons exhibit also some characteristic features, which are observed in biological, chemical, and non-equilibrium physical systems [21].

[3] For an approximate distinguishing between the free flow, synchronized flow, and wide moving jam traffic phases in Figs. 2.1b and 2.2, a model for the detection of so-called phase transition points on a vehicle trajectory of Ref. [22] has been applied (the term *phase transition points* on the vehicle trajectory will be explained in Sect. 3.1.2). Parameters of the model, in which some speed and time thresholds for the distinguishing between the three traffic phases have been used, are related to Table 2 of Ref. [22]. The approximate values for the speed ranges related to the free flow, synchronized flow, and wide moving jam traffic phases in Figs. 2.1b and 2.2 mentioned in the main text correspond to these model parameters. It should be emphasized that we will not discuss the model for the detection of phase transition points on vehicle trajectories any more because the detailed knowledge of this model has no importance for understanding real traffic.

18   2 Basic Empirical Spatiotemporal Phenomena in Real Traffic

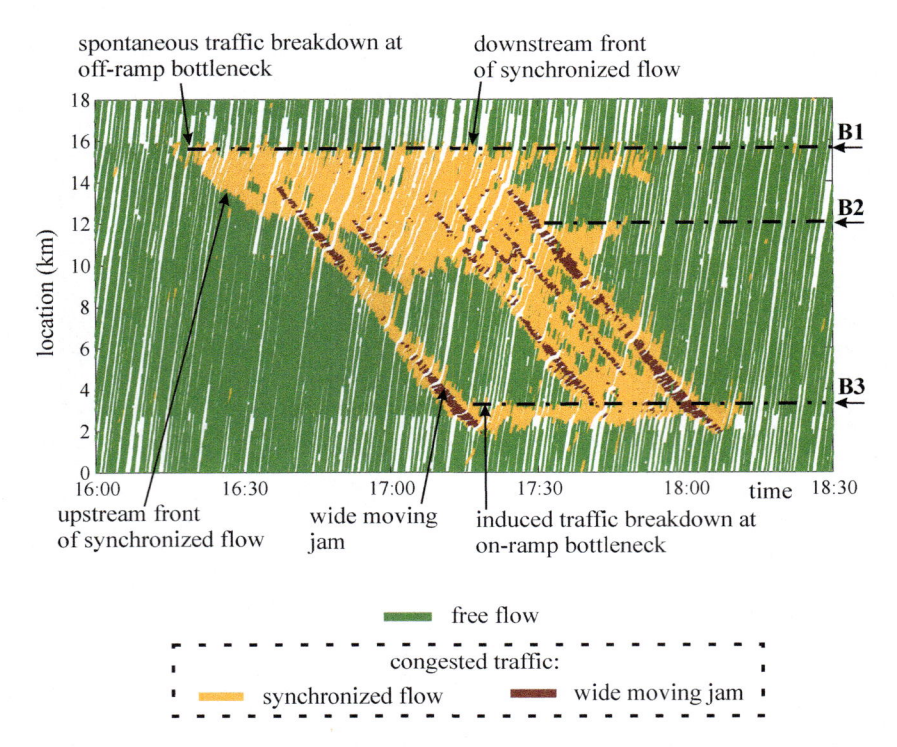

Fig. 2.2 Continuation of Fig. 2.1b. Explanation of the localization of the downstream front of synchronized flow at related bottlenecks B1, B2, and B3 (dotted-dashed horizontal lines). Empirical trajectories of probe vehicles that are the same as those in Fig. 2.1b presented in a shorter time interval

2.1.2 Fronts Between Traffic Phases

For a further study of empirical traffic data, we define the term *front* (boundary) between traffic phases. The front between traffic phases separates two different traffic phases on the road one from another. A front separating two traffic phases can be either motionless or moving. Empirical examples of a motionless front (boundary) between traffic phases are (i) the motionless front between free flow and synchronized flow localized at bottleneck B1 in Fig. 2.2 and (ii) the motionless front between free flow and synchronized flow localized at bottleneck B2 in Fig. 2.3b. This motionless front can also be considered the downstream front of synchronized flow (labeled by "downstream front of synchronized flow" in Figs. 2.2 and 2.3b): Moving through this downstream front of synchronized flow, a vehicle accelerates from a synchronized flow speed upstream of the bottleneck to a higher speed in free flow downstream of the bottleneck. There is also the upstream front of synchronized flow between free flow and synchronized flow; after traffic breakdown (F→S transition) has occurred at a bottleneck, the upstream front of synchronized flow begins to move

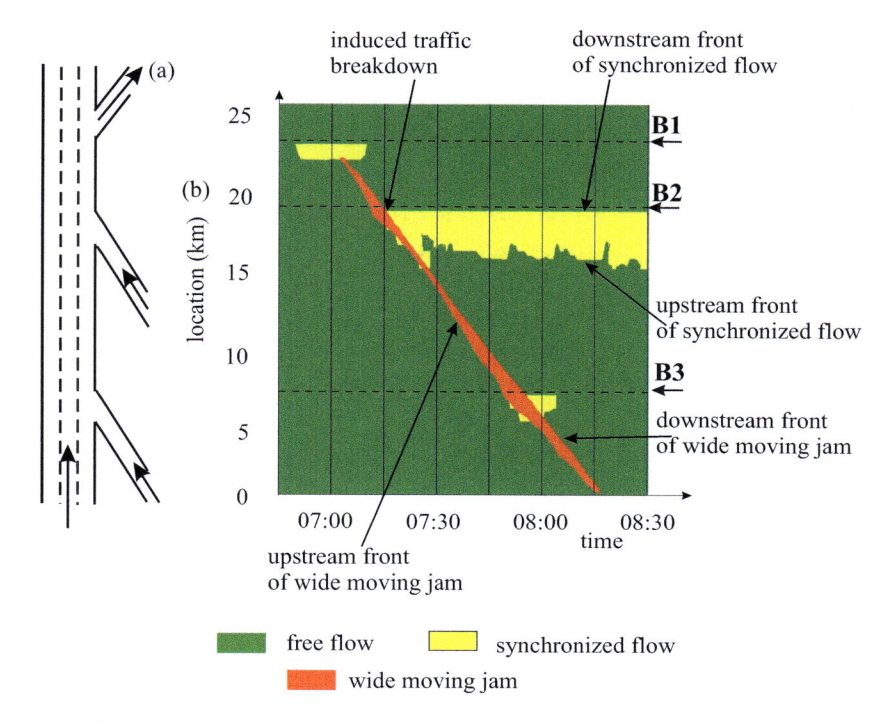

2.1 Three-Phase Traffic Theory—Framework for Understanding Real Traffic 19

Fig. 2.3 Empirical reconstruction of spatiotemporal traffic dynamics made through the use of road detector data (1 min averaged data): **a** Simplified schema of highway section. **b** Empirical speed data presented in space and time through the use of averaging method described in Sect. C.2 of [22]. Arrows B1, B2, and B3 show locations of an off-ramp bottleneck (B1) and two upstream on-ramp bottlenecks (B2 and B3) at highway section; road detectors are installed at distances about 1 km each of other. Empirical traffic data were measured on three-lane freeway A5-South in Germany on June 23, 1998. Adapted from [14]

in the upstream direction from the bottleneck.[4] While approaching the upstream front of synchronized flow, vehicles should decelerate from a higher free flow speed to a lower synchronized flow speed (labeled by "upstream front of synchronized flow" in Figs. 2.2 and 2.3b).

Other examples of moving fronts between traffic phases are downstream and upstream jam fronts. The downstream jam front separates a wide moving jam from another traffic phase (S or F) downstream of the jam: A vehicle that is in a standstill within the wide moving jam escapes from the jam to free flow or synchronized flow downstream of the jam (labeled by "downstream front of wide moving jam" in Fig. 2.3b). The upstream jam front separates the wide moving jam from another

---

[4] In the empirical example shown in Fig. 2.2, the upstream front of synchronized flow moves upstream between bottlenecks B1 and B2 at the velocity about − 18 km/h. In other empirical examples that we have studied, the velocity of the upstream front of synchronized flow is between − 12 and − 22 km/h.

traffic phase (S or F) upstream of the jam: A vehicle approaching the wide moving jam should decelerate within the upstream jam front to a low speed that can be as low as zero within the jam (labeled by "upstream front of wide moving jam" in Fig. 2.3b).

2.1.3 Definitions of Synchronized Flow and Wide Moving Jam Phases in Congested Traffic

Definitions of the synchronized flow and wide moving jam traffic phases in congested traffic are based on qualitative different *spatiotemporal* features of these two traffic phases. The distinguishing between the synchronized flow and wide moving jam traffic phases in congested traffic is as follows:

- A wide moving jam is low-speed congestion that cannot be localized at a highway bottleneck. Indeed, in Fig. 2.2 any of the wide moving jams propagates through the associated bottleneck location.[5] The same conclusion can be made for the wide moving jam shown in Fig. 2.3b: The jam propagates through bottleneck B2.
- The phase definitions in the three-phase traffic theory mean that if in a set of empirical traffic data, we have identified congested traffic states associated with the wide moving jam traffic phase, then all remaining congested states in the empirical data set are related to the synchronized flow traffic phase.
- In contrast to wide moving jams, synchronized flow is congested traffic whose downstream front is usually localized at the bottleneck.[6] Indeed, the downstream front of synchronized flow are localized at related bottlenecks B1, B2, and B3 as shown in Fig. 2.2 by dotted-dashed horizontal lines. The same effect can be seen in Fig. 2.3b, in which the downstream front of synchronized flow is localized at bottleneck B2.

From the phase definitions, we can conclude that any vehicle moves always in one of the three traffic phases except when the vehicle is at a front between the traffic phases: In real traffic, there are only the three traffic phases and fronts between them.

In the book, we use also the following terms associated with the traffic phase definitions made above. The term a *traffic pattern* means a particular distribution of traffic variables. Examples of the traffic variables are the vehicle density and vehicle speed (other traffic variables will be considered and defined below in the book in the order of their application). The term *spatiotemporal traffic pattern* means a traffic pattern that presents a particular development of at least one of the three

---

[5] A precise definition of a wide moving jam is as follows: The wide moving jam is a moving jam that propagates through any states of free flow and synchronized flow as well as through any bottleneck while maintaining the mean velocity of the downstream jam front; this jam characteristic feature is called the jam feature [J]. The jam feature [J] can clear be seen in Fig. 2.3b: While propagating through bottleneck B2, the mean velocity of the downstream front of the wide moving jam does not change. A more detailed explanation of the jam feature [J] will be done in Sect. B.1 of Appendix B.

[6] The effect of the empirical nucleation nature of traffic breakdown (F→S transition) on the definition of the synchronized flow traffic phase will be considered in Sect. 6.2.1.

traffic phases in space and time. The term *spatiotemporal congested traffic pattern* (congested traffic pattern for short) is a spatiotemporal traffic pattern within which there are either both phases S and J or at least one of the phases (S or J) that belong to congested traffic. Figures 2.2 and 2.3b present empirical examples of spatiotemporal congested traffic patterns.

2.1.4 Explanations of Term "Synchronized Flow"

The above definition of the synchronized flow traffic phase does not explain the choice of the word *synchronized* and the word *flow* in the term *synchronized flow*. These words reflect the following empirical features of synchronized flow:

(i) It is a continuous traffic flow with no significant stoppage, as often occurs within a wide moving jam. The word *flow* reflects this feature.
(ii) Although in empirical synchronized flow vehicle speeds across different lanes on a multi-lane road should not be necessarily synchronized, there is a *tendency* toward synchronization of these speeds in this flow. In addition, there is a *tendency* towards synchronization of vehicle speeds in each of the road lanes (bunching of vehicles) in synchronized flow. This is due to a relatively small probability of passing in synchronized flow. The word *synchronized* reflects these speed synchronization effects.

To avoid possible confusions, it should be emphasized these empirical features of synchronized flow have led only to the choice of the words *synchronized* and *flow* in the term *synchronized flow*, not to the definition of the synchronized flow *traffic phase* in congested traffic. Indeed, some other empirical spatiotemporal features of synchronized flow are responsible for the *definition* of the synchronized flow traffic phase in congested traffic made in Sects. 2.1.3 and 6.2.1.

2.2 Empirical Spontaneous Traffic Breakdown at Bottleneck

In Fig. 2.2 at $t \approx 16{:}19$ traffic breakdown occurs (labeled by "spontaneous traffic breakdown" in Fig. 2.2): Free flow (green color) transforms suddenly into synchronized flow (yellow color) (F→S transition) at the location of off-ramp bottleneck B1. It can be seen that after traffic breakdown has occurred the downstream front of synchronized flow separating synchronized flow at the off-ramp bottleneck and free flow downstream of the bottleneck is localized in a vicinity of the off-ramp bottleneck (this localization of the downstream front of synchronized flow is marked by a dotted-dashed horizontal line at the location of off-ramp bottleneck B1 in Fig. 2.2). Indeed, probe vehicles in synchronized flow upstream of the off-ramp bottleneck (bottleneck B1 in Fig. 2.2) can accelerate to free flow downstream of the off-ramp bottleneck.

The empirical feature of the localization of the downstream front of synchronized flow corresponds to the definition of the synchronized flow phase in congested traffic made in Sect. 2.1.3.

Before traffic breakdown has occurred, free flow has been at off-ramp bottleneck B1 as well as both upstream and downstream in a neighborhood of the off-ramp bottleneck (Figs. 2.1b and 2.2). Such empirical traffic breakdown is called *empirical spontaneous* traffic breakdown.

> If before traffic breakdown (F→S transition) has occurred at a highway bottleneck, free flow exists at the bottleneck as well as upstream and downstream in a neighborhood of the bottleneck, then such an empirical traffic breakdown at the bottleneck is called *empirical spontaneous traffic breakdown* at the bottleneck.

2.3 Empirical Induced Traffic Breakdown at Bottleneck

In contrast to synchronized flow (yellow color in Figs. 2.1b and 2.2), the wide moving jam (claret-red color in Figs. 2.1b and 2.2) is a moving jam that can propagate through highway bottlenecks (see propagation of wide moving jams through bottleneck B2 in Fig. 2.2).

While propagating through upstream on-ramp bottleneck B3, the wide moving jam labeled by "wide moving jam" in Fig. 2.2 induces traffic breakdown at upstream on-ramp bottleneck B3. Indeed, after the wide moving jam is already upstream of bottleneck B3, synchronized flow remains at bottleneck B3 during a long time interval (between $t \approx 17{:}18$ and $t \approx 17{:}35$ in Fig. 2.2): The empirical induced traffic breakdown is the induced F→S transition. The downstream front of synchronized flow is localized at bottleneck B3 (this localization of the downstream front of synchronized flow is marked by a dotted-dashed horizontal line at the location of on-ramp bottleneck B3 in Fig. 2.2).

Another empirical example of induced traffic breakdown is shown in Fig. 2.3b: While propagating through on-ramp bottleneck B2, the wide moving jam induces traffic breakdown at bottleneck B2 (labeled by "induced traffic breakdown" in Fig. 2.3b). Due to the empirical induced traffic breakdown, synchronized flow emerges at on-ramp bottleneck B2. The synchronized flow is self-maintained during a long time interval at on-ramp bottleneck B2.

> *Empirical induced traffic breakdown* at a highway bottleneck is traffic breakdown (F→S transition) that is induced by the propagation of a moving spatiotemporal congested traffic pattern to the location of the bottleneck.

A *moving spatiotemporal congested traffic pattern* (moving congested traffic pattern for short) is a spatiotemporal congested traffic pattern that downstream and upstream fronts are moving ones. Within the fronts of the moving congested traffic pattern traffic variables can change considerably.

2.4 Empirical Emergence of Moving Jams in Synchronized Flow

It must be emphasized that the spontaneous emergence of empirical moving jams is *not* observed in real free flow. In contrast, it has been found out that empirical moving jams can emerge spontaneously only in synchronized flow. In many empirical traffic data, after synchronized flow resulting from traffic breakdown has formed at a bottleneck, moving jams appear spontaneously in the emergent synchronized flow. In other words, first traffic breakdown (F→S transition) occurs at the bottleneck. Later moving jams can emerge spontaneously in this synchronized flow. Some of the growing moving jams can transform into wide moving jams. The emergence of a wide moving jam in synchronized flow is called an S→J transition. Empirical S→J transitions (formation of wide moving jams) can be clear seen in Fig. 2.2.

In real traffic, the emergence of a wide moving jam is a sequence of F→S→J transitions: The first F→S transition within this sequence occurring at a bottleneck causes the emergence of synchronized flow at the bottleneck; the second S→J transition within this sequence occurs later within the synchronized flow.

Thus, the emergence of moving jams is realized spontaneously in synchronized flow *only*. In other words, understanding real traffic is only possible when the origin of empirical traffic breakdown (F→S transition) can be revealed. As we will show in the book (see Chap. 5 and Appendix C), the failure of standard traffic flow theories is basically associated with the invalid understanding of the origin of empirical traffic breakdown. For this reason, most of the results of the book are devoted to understanding real traffic breakdown. The exception is Chaps. 10 and 11, in which empirical traffic phenomena associated with the emergence of moving jams in synchronized flow will be explained.

2.5 Common Features of Empirical Traffic Breakdown at Bottlenecks

> In all empirical studies of traffic breakdown at highway bottlenecks, traffic breakdown is a transition from the free flow traffic phase (F) to the synchronized flow traffic phase (S) called as an F→S transition. The term *traffic breakdown* at a bottleneck is a synonym of the term *F→S transition* at the bottleneck.

> Empirical traffic breakdown (F→S transition) at a highway bottleneck can be either *spontaneous* or *induced* traffic breakdown.

> Empirical induced traffic breakdown (empirical induced F→S transition) at a bottleneck is a common empirical phenomenon in vehicular traffic.

2.6 What Is Necessary for Understanding Real Traffic?

From a brief consideration of empirical spatiotemporal traffic data (Figs. 2.1, 2.2, and 2.3) made above, we can give the following answer to the question in the title of this Sect. 2.6: For understanding real vehicular traffic, it is necessary to disclose the empirical nature of real traffic breakdown (F→S transition) at a bottleneck. Indeed, it turns out (and this will be shown in this book) that all other traffic phenomena can be sufficiently understood only when the empirical nature of traffic breakdown (F→S transition) at the bottleneck has been revealed.

> To understand real traffic, the empirical nature of traffic breakdown (F→S transition) at a highway bottleneck should be disclosed.

However, the empirical nature of traffic breakdown (F→S transition) at the bottleneck will be disclosed only in Chap. 6. This is because before it is necessarily to discuss the following subjects:

- How empirical spatiotemporal traffic dynamics can be reconstructed through traffic measurements (Chap. 3).

- Why traffic breakdown occurs mostly at bottlenecks (Chap. 4).
- Why standard methodologies for a study of empirical traffic data or/and standard traffic theories cannot be used for understanding real traffic (Chap. 5).

2.7 Why Is the Distinguishing Between the Three Phases Needed for Understanding Real Traffic?

From the author's experience, one of the most frequent questions to the three-phase traffic theory is as follows:

- Why is the distinguishing between the three traffic phases (the free flow, synchronized flow, and wide moving jam traffic phases) in real vehicular traffic needed to understand real traffic?

To answer this question, we recall the conclusion of Sect. 2.6: To understand real traffic, the empirical nature of traffic breakdown (F→S transition) at a highway bottleneck should be disclosed.

As mentioned in Sect. 1.2, traffic breakdown (F→S transition) at the bottleneck exhibits the empirical *nucleation* nature: While a small enough local speed decrease in free flow does not cause traffic breakdown at the bottleneck, a large enough local speed decrease in free flow grows further and leads to the emergence of synchronized flow at the bottleneck.

> The empirical nucleation nature of the F→S transition at a bottleneck leads to the conclusion made in the three-phase traffic theory that free flow and synchronized flow should be considered two different traffic phases.

Empirical features of synchronized flow are qualitatively different from empirical spatiotemporal features of a wide moving jam (Sect. 2.1):

(i) In contrast to synchronized flow, the wide moving jam does not emerge spontaneously in real free flow: In the empirical traffic data, the wide moving jam can spontaneously emerge in synchronized flow only.
(ii) While the downstream front of synchronized flow can be localized at the bottleneck, the wide moving jam *cannot* be localized at the bottleneck: The wide moving jam always propagates through the bottleneck.

> The synchronized flow and the wide moving jam should be considered two different traffic phases of congested traffic.

For understanding real traffic (see, for example, Fig. 2.2), the distinguishing between the three qualitatively different traffic phases (the free flow, the synchronized flow, and the wide moving jam traffic phases) in real vehicular traffic is needed.

References

1. B.S. Kerner, Phys. Rev. Lett. **81**, 3797–3800 (1998)
2. B.S. Kerner, in *Proceedings of the 3rd Symposium on Highway Capacity and Level of Service*, ed. by R. Rysgaard (Road Directorate, Copenhagen, Ministry of Transport – Denmark 1998), pp. 621–642
3. B.S. Kerner, Transp. Res. Rec. **1678**, 160–167 (1999)
4. B.S. Kerner, in *Transportation and Traffic Theory*, ed. by A. Ceder (Elsevier Science, Amsterdam, 1999), pp. 147–171
5. B.S. Kerner, Phys. World **12**, 25–30 (August 1999)
6. B.S. Kerner, J. Phys. A: Math. Gen. **33**, L221–L228 (2000)
7. B.S. Kerner, in *Traffic and Granular Flow '99: Social, Traffic and Granular Dynamics*, ed. by D. Helbing, H.J. Herrmann, M. Schreckenberg, D.E. Wolf (Springer, Berlin, 2000), pp. 253–284
8. B.S. Kerner, Transp. Res. Rec. **1710**, 136–144 (2000)
9. B.S. Kerner, Netw. Spat. Econ. **1**, 35–76 (2001)
10. B.S. Kerner, Transp. Res. Rec. **1802**, 145–154 (2002)
11. B.S. Kerner, Math. Comput. Model. **35**, 481–508 (2002)
12. B.S. Kerner, in *Traffic and Transportation Theory in the 21st Century*, ed. by M.A.P. Taylor (Elsevier Science, Amsterdam, 2002), pp. 417–439
13. B.S. Kerner, Phys. Rev. E **65**, 046138 (2002)
14. B.S. Kerner, *The Physics of Traffic* (Springer, Berlin, 2004)
15. B.S. Kerner, *Introduction to Modern Traffic Flow Theory and Control* (Springer, Heidelberg, 2009)
16. B.S. Kerner, Phys. A **392**, 5261–5282 (2013)
17. B.S. Kerner, Elektrotech. Inf. **132**, 417–433 (2015)
18. B.S. Kerner, Phys. A **450**, 700–747 (2016)
19. B.S. Kerner, *Breakdown in Traffic Networks* (Springer, Berlin, 2017)
20. B.S. Kerner, in *Complex Dynamics of Traffic Management*. Encyclopedia of Complexity and Systems Science Series, ed. by B.S. Kerner (Springer, New York, 2019), pp. 21–77
21. B.S. Kerner, V.V. Osipov, *Autosolitons* (Kluwer Academic Publishers, Dordrecht, 1994)
22. B.S. Kerner, H. Rehborn, R.-P. Schäfer, S.L. Klenov, J. Palmer, S. Lorkowski, N. Witte, Phys. A **392**, 221–251 (2013)
23. H. Rehborn, M. Koller, S. Kaufmann, *Data-Driven Traffic Engineering: Understanding of Traffic and Applications Based on Three-Phase Traffic Theory* (Elsevier, Amsterdam, 2021)

Chapter 3
How Can Empirical Spatiotemporal Traffic Dynamics Be Reconstructed Through Traffic Measurements?

There are different methodologies for the measurement of real traffic data that can be used for the reconstruction of empirical spatiotemporal traffic dynamics. Some of these methodologies used in this book are briefly discussed below.[1]

3.1 Traffic Dynamics Reconstructed Through Probe Vehicle Data

3.1.1 Explanation of Vehicle Trajectory

A vehicle trajectory is a time dependence of the vehicle coordinate (road location) $x(t)$ on a road. A hypothetical trajectory of a vehicle labeled by colored black in Fig. 3.1a is shown in Fig. 3.1b (labeled by "vehicle trajectory"). Each point (x, t) on the vehicle trajectory gives the road location x that the vehicle reaches at time instant t. During a time interval δt the vehicle drives through some road distance δx. Respectively, the average vehicle speed during this time interval δt is

$$v = \frac{\delta x}{\delta t}. \tag{3.1}$$

When the coordinate of the vehicle $x(t)$ (Fig. 3.1b) and the vehicle speed $v(t)$ (3.1) are known at subsequent time instants, the dependence of the speed on the road location $v(x)$ can also be found as shown in Fig. 3.1c.

---

[1] Results of this chapter are based on methodologies for the measurement of real traffic data that can be found in the books [3, 4, 10, 15].

© The Author(s), under exclusive license to Springer Nature Switzerland AG 2021
B. S. Kerner, *Understanding Real Traffic*,
https://doi.org/10.1007/978-3-030-79602-0_3

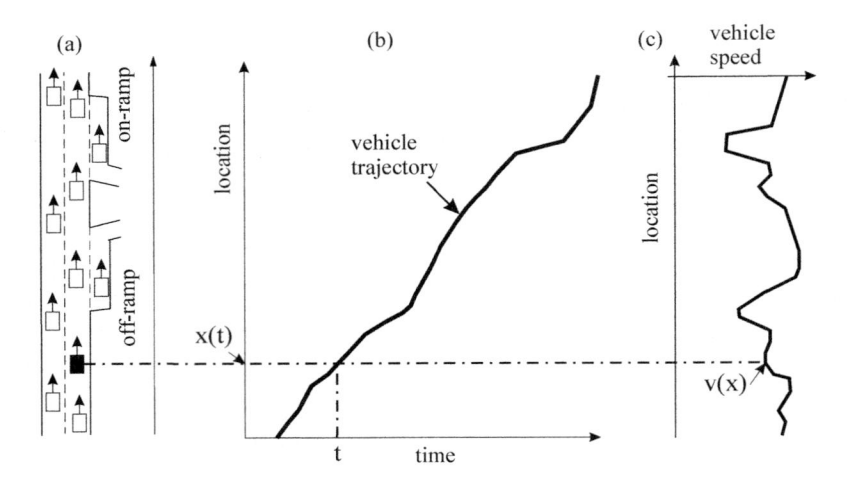

Fig. 3.1 Qualitative explanation of vehicle trajectory: **a** Schema of two-lane highway section with off-ramp and on-ramp. **b** Qualitative trajectory in space and time of one of the vehicles in **a**. **c** Qualitative vehicle speed related to the vehicle trajectory in **b**

However, for an empirical study of traffic flow we need trajectories of many vehicles moving in the same traffic flow. For this reason, during recent years some car-developing companies produce vehicles that are connected with a traffic data center (Fig. 3.2). These connected vehicles can automatically send their coordinates to the data center.[2] These connected vehicles moving in real traffic are called *probe vehicles*.[3] Probe vehicles are randomly distributed in traffic flow.

To reduce communication costs, a probe vehicle sends the coordinates at discrete time instants t_n with $n = 1, 2, \ldots$ to the data center (Fig. 3.3a), where time interval $\Delta t = t_n - t_{n-1}$ is usually between 5 and 10 s. We denote the vehicle coordinate on the road measured by the probe vehicle at time instant t_n by x_n. Then, the vehicle speed $v_n, n = 1, 2, \ldots$ (Fig. 3.3) related to the coordinate x_n can be estimated through

---

[2] Vehicle coordinates are also called GPS locations of vehicles. This is because vehicle coordinates are measured in probe vehicles through the use of Global Positioning System (GPS). GPS is a satellite-based radio navigation system owned by the United States government and operated by the United States Space Force. It is one of the global navigation satellite systems that provides geolocation and time information to a GPS receiver anywhere on the Earth [5]. A more detailed consideration of features and parameters of probe vehicle data can be found, e.g., in [2, 11] as well as in the book [15]. Probe vehicle data are also called floating car data (FCD).

[3] Additionally to probe vehicle data, vehicle trajectories can be found from microscopic measurements of traffic flow made through aerial observations of real traffic or/and video cameras. This method could permit to find trajectories of all vehicles in traffic flow. Therefore, this measurement method can be the most precise one for the empirical study of traffic. However, this method is considerably more expensive one than the use of probe vehicle data or data measured through road detectors (see Sect. 3.2). Therefore, up to now there are only few sets of empirical data measured with this method (see, e.g., [1, 6, 13, 14, 16–18]). Unfortunately, currently there are no data sets available that can be used for a study of traffic breakdown at highway bottlenecks.

3.1 Traffic Dynamics Reconstructed Through Probe Vehicle Data

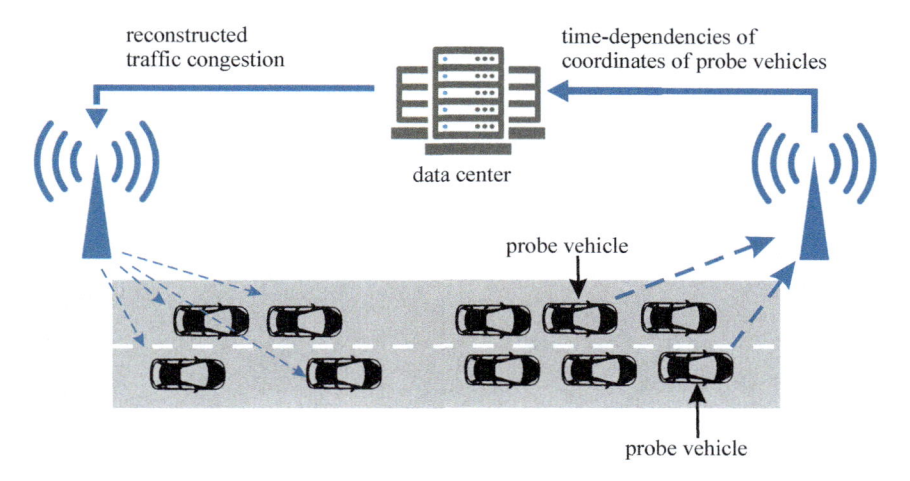

Fig. 3.2 Explanation of traffic data measurements through the use of probe vehicles that are randomly distributed in traffic flow. Probe vehicles send their coordinates to a data center. In the data center, real traffic is reconstructed based on the vehicle coordinates. The information about traffic congestion derived in the data center is sent to all vehicles moving in traffic flow

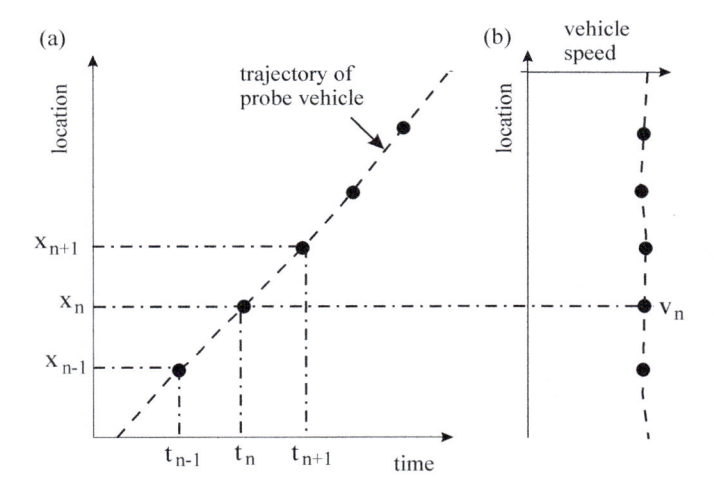

Fig. 3.3 Qualitative explanation of speed calculation on trajectory of probe vehicle: **a** Road coordinates x_n measured by probe vehicle at time instants t_n, $n = 1, 2, \ldots$ (black circles) and related approximate trajectory of probe vehicle (dashed curve). **b** Speeds of probe vehicle v_n related to **a** that are calculated via formula (3.2) (black circles) and continuous speed approximation of probe vehicle (dashed curve)

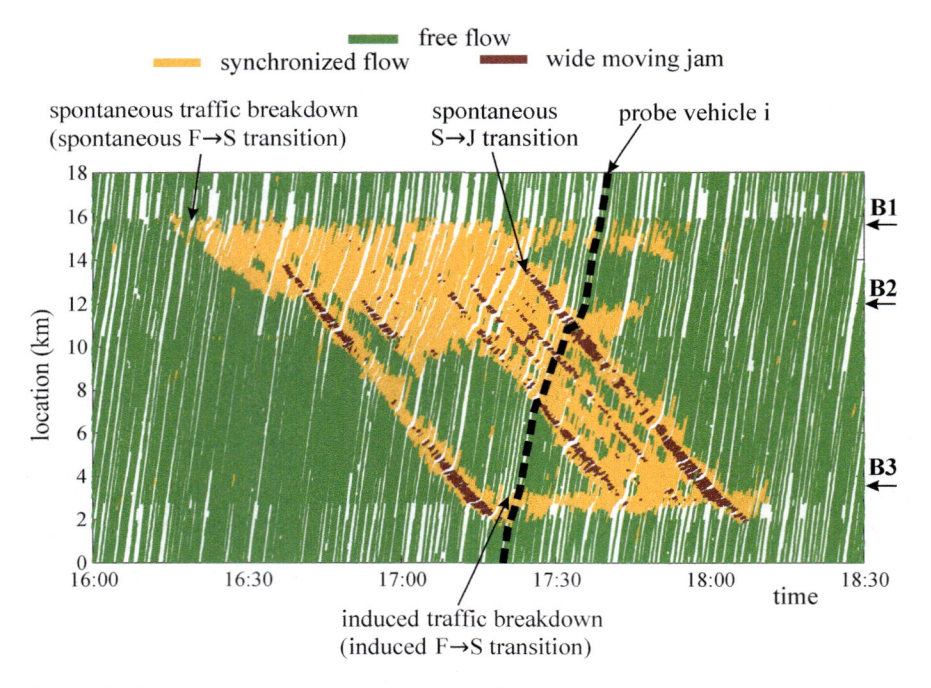

30   3 How Can Empirical Spatiotemporal Traffic Dynamics Be Reconstructed …

Fig. 3.4 Trajectories of probe vehicles taken from Fig. 2.2 in which probe vehicle i is marked by a dashed curve

formula

$$v_n = \frac{x_{n+1} - x_{n-1}}{2\Delta t}. \tag{3.2}$$

Indeed, formula (3.2) is found from formula (3.1) in which in accordance with Fig. 3.3 values $\delta x = x_{n+1} - x_{n-1}$ and $\delta t = 2\Delta t$.

3.1.2 Can a Driver Resolve Traffic Breakdown?

To answer the question whether a driver can resolve traffic breakdown, we consider one of the empirical vehicle trajectories of probe vehicles denoted by "probe vehicle i" in Fig. 3.4. To understand features of empirical vehicle trajectory i, in Fig. 3.5a, we present trajectory i separately from other vehicle trajectories. The speed of probe vehicle i as a road-location function calculated with formula (3.2) is shown in Fig. 3.5b. In this case, we can see that a driver of probe vehicle i must several times decelerate while approaching congested traffic and then accelerate while escaping from congested traffic.

3.1 Traffic Dynamics Reconstructed Through Probe Vehicle Data 31

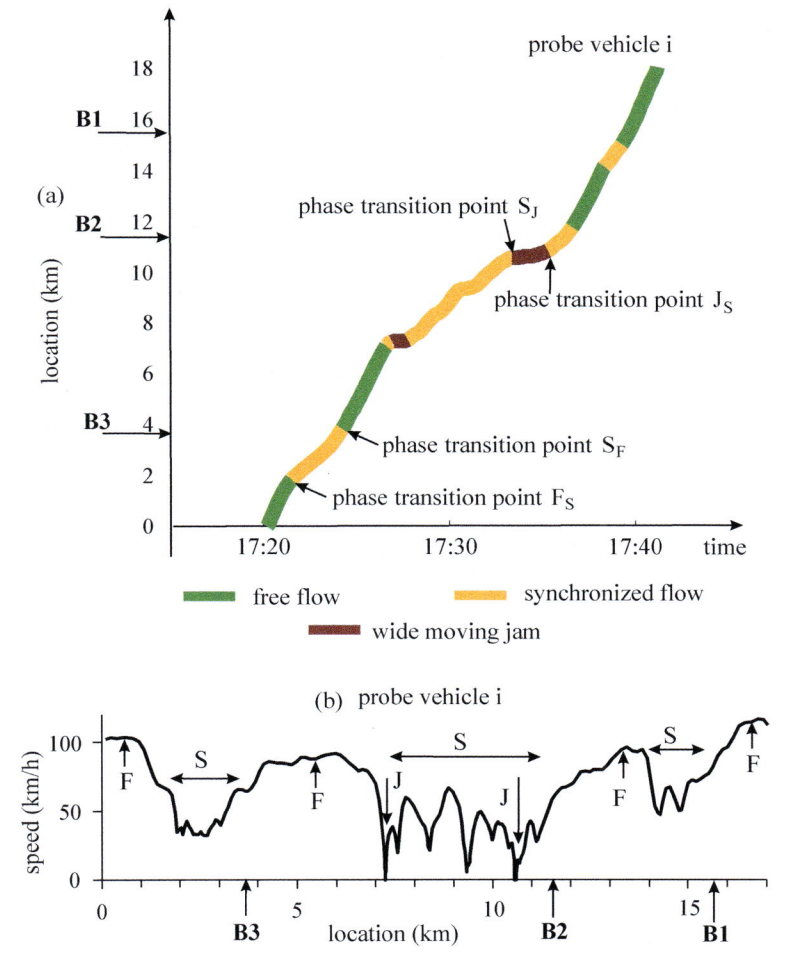

Fig. 3.5 Continuation of consideration of Fig. 3.4. Sequence of traffic phases observed by the driver of probe vehicle i moving on a long highway section: **a** Trajectory of probe vehicle i taken from Fig. 3.4; some of the phase transition points on probe vehicle trajectory i: F_S—from the phase F to the phase S, S_F—from the phase S to the phase F, S_J—from the phase S to the phase J, J_S—from the phase J to the phase S. **b** Road location dependence of the speed of probe vehicle i shown in **a**. B1, B2, and B3 are, respectively, the same bottleneck locations as those shown in Fig. 3.4 and explained in caption to Fig. 2.1

It can be seen from Fig. 3.5 that the driver does observe a diverse sequence of the three traffic phases F, S, and J as well as the transitions between these traffic phases. However, are these transitions the phase transitions that occur in real traffic? The answer to this question is that it is *not*.

Indeed, traffic breakdown (F→S transition) at bottleneck B1 has occurred considerably earlier ($t \approx$ 16:19 in Fig. 3.4) than the time when the driver moves through bottleneck B1 ($t \approx$ 17:40 in Fig. 3.4). Respectively, the empirical induced traffic breakdown (induced F→S transition) at bottleneck B3 has also occurred earlier ($t \approx$ 17:14 in Fig. 3.4) than the time when the driver moves through bottleneck B3 ($t \approx$ 17:24 in Fig. 3.4). In other words, although the driver observes a diverse sequence of the three phases F, S, and J, nevertheless the driver can resolve *neither* spontaneous traffic breakdown (spontaneous F→S transition) at bottleneck B1 *nor* induced traffic breakdown (induced F→S transition) at bottleneck B3 (Fig. 3.4).

> Usually, a single driver cannot resolve traffic breakdown at a bottleneck.

Rather than real road locations at which different phase transitions occur (see road locations of empirical examples of F→S and S→J transitions marked in Fig. 3.4), a driver observes so-called *phase transition points* on the vehicle trajectory.[4] A phase transition point is a transition from one traffic phase to another traffic phase while the driver moves through the associated front (boundary) between these two traffic phases. Some of the phase transition points on vehicle trajectory i have been labeled in Fig. 3.5a. The phase transition points separate the phases F, S, and J observed by the driver in vehicular traffic. To distinguish real phase transitions in traffic flow from phase transition points on vehicle trajectories, in the three-phase traffic theory the phase transition points are denoted differently in comparison with the phase transitions: Whereas empirical examples of real phase transitions in traffic flow are denoted in Fig. 3.4 by F→S and S→J transitions (as they have been designated in Chap. 2), empirical phase transition points for the same transitions on vehicle trajectory i are denoted, respectively, by F_S and S_J, as shown in Fig. 3.5a.

3.1.3 How Short Should the Average Time Interval Between Probe Vehicles Be for the Resolution of Traffic Breakdown?

As found in Sect. 3.1.2, a single driver cannot usually resolve traffic breakdown at a bottleneck. Herewith the following question can be raised: How short should the average time interval between probe vehicles be to resolve traffic breakdown?

An analysis of the quality of traffic reconstruction at different average time intervals between probe vehicles is presented in Fig. 3.6. We can see that when the average time interval between probe vehicles is noticeably longer than 60 s (Fig. 3.6a, b),

---

[4] A detailed consideration of phase transition points on vehicle trajectories has been done in [8]. Some applications of the approach of phase transition points [8] for the recognition of traffic flow dynamics in probe vehicle data can be found in [2, 9, 11, 12, 15, 19].

3.1 Traffic Dynamics Reconstructed Through Probe Vehicle Data 33

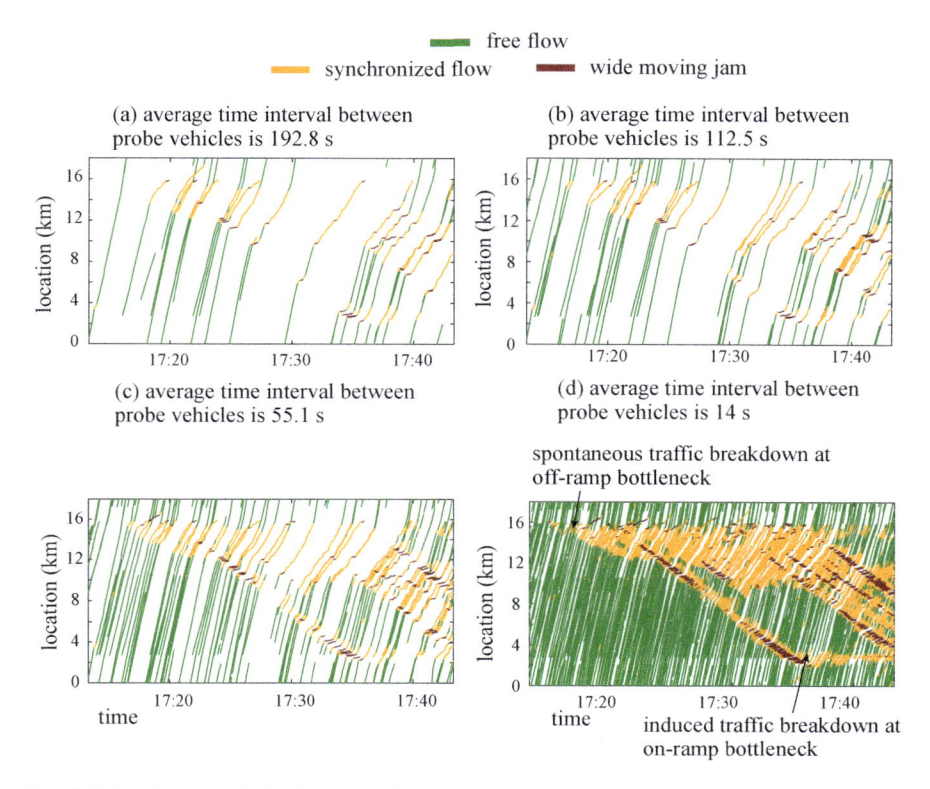

Fig. 3.6 Empirical analysis of average time interval between probe vehicles required for the resolution of traffic breakdowns at highway bottlenecks. **a–c** Trajectories of only some of the probe vehicles chosen in a fragment of Fig. 3.4 with different average time intervals between probe vehicles: 192.8 s (**a**), 112.5 s (**b**), and 55.1 s (**c**). **d** Trajectories of all probe vehicles in which average time interval between probe vehicles is equal to 14 s in the same fragment of Fig. 3.4 as that in **a–c**

it is difficult make a decision about the time instant of spontaneous traffic breakdown. Contrarily, already at the average time interval between probe vehicles about or less than 60 s both empirical spontaneous and induced traffic breakdowns can be resolved at highway bottlenecks (Fig. 3.6c, d). Clearly, the less the average time interval between probe vehicles is, the more precisely the time instant of traffic breakdown can be found.

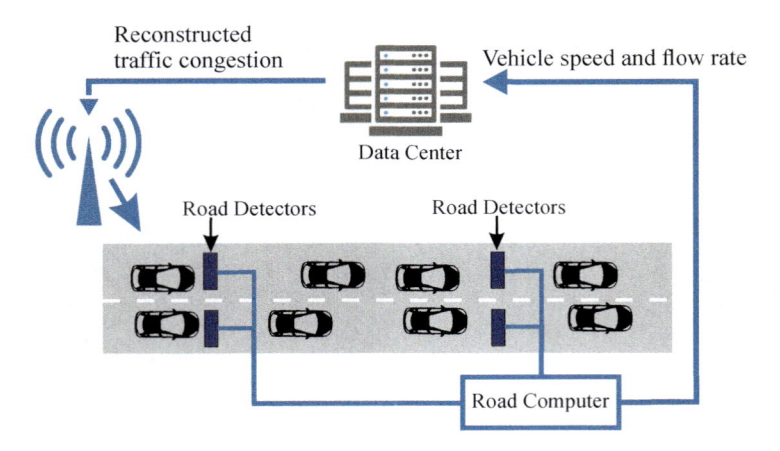

Fig. 3.7 Qualitative explanation of traffic measurements through the use of road detectors

3.2 Traffic Dynamics Reconstructed Through Road Detector Data

3.2.1 Road Detector Measurements

An empirical analysis of traffic breakdown and the resulting spatiotemporal dynamics of vehicular traffic can also performed based on traffic measurements made through the use of detectors installed along roads of a traffic network (Fig. 3.7).[5]

A road detector measures the speed of a vehicle passing the detector.[6] Because the detector can register all vehicles passing the detector, the detector measures also the flow rate. The term *flow rate* means the number of vehicles passing a road location (in our case, the detector location) during a given time unit (hour or minute). Usually, the units of the flow rate are (vehicles/h) or (vehicles/min). We denote the flow rate by q.

It must be emphasized that to resolve empirical features of traffic breakdown (F→S transition) at a highway bottleneck, real traffic at several road detectors should be simultaneously measured. There can be many different bottlenecks on a highway section. Moreover, initially, it is often not clear at which of the highway bottlenecks traffic breakdown has occurred. Therefore, to resolve empirical features of traffic breakdown (F→S transition), usually a long enough road section is required at which many road detectors installed with short enough distances to each other. For example, for the spatiotemporal resolution of traffic dynamics shown in Figs. 2.3 and 3.8

---

[5] Road detector measurements are widely used in traffic and transportation engineering (see, e.g., [3, 4, 10, 15]).

[6] The detector can also measure some other traffic variables like the vehicle length; however, for simplicity of the analysis made in this book we omit a consideration of these measurements.

3.2 Traffic Dynamics Reconstructed Through Road Detector Data 35

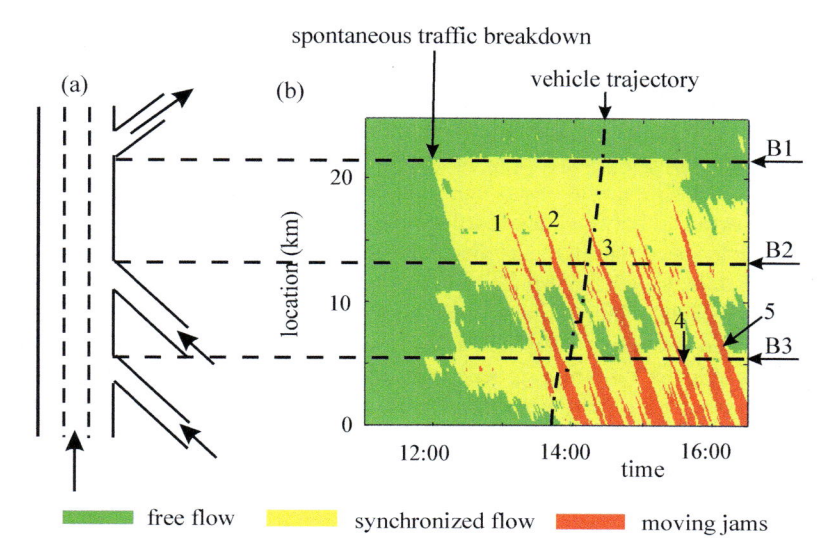

Fig. 3.8 Empirical reconstruction of spatiotemporal traffic dynamics made through the use of road detector data: **a** Sketch of highway section with off-ramp bottleneck B1 and two on-ramp bottlenecks B2 and B3. **b** Empirical speed data presented in space and time with the use of averaging method described in Sect. C.2 of [8]; moving jams marked by 1–5 are wide moving jams. Empirical example of a congested traffic pattern measured on three-lane highway A5-North in Germany on March 23, 2001. Adapted from [7]

highway sections of about 30 km long with more than 30 road detectors installed on each of the highway sections have been needed.[7]

3.2.2 Empirical Example of Spatiotemporal Traffic Dynamics

One of the empirical examples of the reconstruction of spatiotemporal traffic dynamics made through the use of road detector data has already been presented in Fig. 2.3. Here, we consider another empirical example shown in Fig. 3.8. As in Fig. 2.3b, in Fig. 3.8b vehicle speeds between detector locations have been estimated through the use of some averaging method in the framework of the three-phase traffic theory.

In this empirical example, spontaneous traffic breakdown (F→S transition) occurring at off-ramp bottleneck B1 leads to the emergence of a complex spatiotemporal congested pattern. Later, upstream of the bottleneck many moving jams emerge in synchronized flow. Some of the moving jams propagate through upstream on-ramp bottlenecks B2 and B3 while maintaining the mean velocity of the downstream jam

---
[7] Schemes of road detectors installed on highway sections related, respectively, to Figs. 2.3 and 3.8 are shown in Figs. 2.1 and 2.2 of the book [7].

fronts. In accordance with the definition of the wide moving jam phase in congested traffic (Sect. 2.1), the moving jams are wide moving jams.

It should be noted that road detector data exhibit the following disadvantage in comparison with probe vehicle data: While the vehicle speed can be found with a good accuracy along the whole trajectory of a probe vehicle (Fig. 3.5b), traffic flow variables cannot be measured between road detectors; therefore, the vehicle speed should be estimated between detector locations (Fig. 3.8b).

The vehicle speed within the traffic congested pattern can furthermore be used to estimate the trajectory that a vehicle can show moving through the congested pattern. One of the possible vehicle trajectories is shown by dashed-dotted curve labeled by "vehicle trajectory" in Fig. 3.8b: In free flow, the vehicle moves with a free flow speed that average value is 120 km/h (green region in Fig. 3.8b); in synchronized flow, the vehicle moves with a synchronized flow speed that average value is 40 km/h (yellow region in Fig. 3.8b); within the wide moving jam, the vehicle moves on average with a low speed about 5 km/h. However, in comparison with vehicle trajectories reconstructed from probe vehicle data discussed in Sect. 3.1, vehicle trajectories estimated from the road detector data can only be considered as the average result for many vehicles that is too uncertain to be used for a particular vehicle.

Nevertheless, road detector data exhibit some important advantage in comparison with probe vehicle data: While probe vehicle data are related to only a share of vehicles moving in traffic flow, it is not possible to measure the flow rate based solely on probe vehicle data. Contrarily, a road detector measures both the speed of all vehicles passing the detector and the flow rate. We will use this advantage of road detector data below in the book.

References

1. E. Barmpounakis, N. Geroliminis, Transp. Rec. C **111**, 50–71 (2020)
2. Y. Dülgar, S.-E. Molzahn, H. Rehborn, M. Koller, B.S. Kerner, D. Wegerle, M. Schreckenberg, M. Menth, S.L. Klenov, J. Int. Transp. Syst. **24**, 539–555 (2020)
3. L. Elefteriadou, *An Introduction to Traffic Flow Theory* (Springer, Berlin, 2014)
4. N.H. Gartner, C.J. Messer, A.K. Rathi (eds.), *Traffic Flow Theory: a State-of-the-Art Report* (Transportation Research Board, Washington, 2001)
5. https://en.wikipedia.org/wiki/Global_Positioning_System
6. S. Kaufmann, B.S. Kerner, H. Rehborn, M. Koller, S.L. Klenov, Transp. Res. C **86**, 393–406 (2018)
7. B.S. Kerner, *Breakdown in Traffic Networks* (Springer, Berlin, 2017)
8. B.S. Kerner, H. Rehborn, R.-P. Schäfer, S.L. Klenov, J. Palmer, S. Lorkowski, N. Witte, Phys. A **392**, 221–251 (2013)
9. S.L. Klenov, D. Wegerle, B.S. Kerner, M. Schreckenberg, Comput. Res. Model. **13**, 319–363 (2021)
10. A.D. May, *Traffic Flow Fundamentals* (Prentice-Hall Inc., New Jersey, 1990)
11. S.-E. Molzahn, B.S. Kerner, H. Rehborn, S.L. Klenov, M. Koller, IET Int. Transp. Syst. **11**, 604–612 (2017)
12. S.-E. Molzahn, B.S. Kerner, H. Rehborn, J. Int. Transp. Syst. **24**, 569–584 (2020)
13. Next generation simulation programs (NGSIM), http://ops.fhwa.dot.gov/trafficanalysistools/ngsim.htm

14. π NEUMA large-scale field experiment, https://open-traffic.epfl.ch/
15. H. Rehborn, M. Koller, S. Kaufmann, *Data-Driven Traffic Engineering: Understanding of Traffic and Applications Based on Three-Phase Traffic Theory* (Elsevier, Amsterdam, 2021)
16. J. Treiterer, Investigation of traffic dynamics by aerial photogrammetry techniques, Ohio State University, Technical report PB 246 094 (Columbus, Ohio, 1975)
17. J. Treiterer, J.A. Myers, in *Proceedings of the 6th International Symposium on Transportation and Traffic Theory*, ed. by D.J. Buckley (A.H. & AW Reed, London, 1974), pp. 13–38
18. J. Treiterer, J.I. Taylor, Highw. Res. Rec. **142**, 1–12 (1966)
19. D. Wegerle, B.S. Kerner, M. Schreckenberg, S.L. Klenov, J. Int. Transp. Syst. **24**, 598–616 (2020)

Chapter 4
Why Does Traffic Breakdown Occur Mostly at a Bottleneck?

The importance of a bottleneck for traffic breakdown was empirically found and theoretically understood already in the first empirical and theoretical studies of vehicular traffic.[1] The question formulated in the title of this chapter can be answered after we consider the occurrence of a local speed decrease in free flow at the bottleneck.[2]

4.1 Local Speed Decrease in Free Flow at Road Bottlenecks

Each driver can drive with a desired or permitted speed on a highway, if there are no other drivers on the highway. This is because interactions between different vehicles cause speed decrease in comparison with the desired or permitted speed.

In particular, while approaching a slower moving preceding vehicle a driver should often decelerate before the driver can pass the preceding vehicle. This deceleration can lead to a time-limited speed decrease of the vehicle in comparison with the desired speed. A time-limited speed decrease of vehicles occurs at some road location and, therefore, it can also be considered a *local speed decrease* occurring during vehicle motion in free flow. Within the local speed decrease, the speed is lower than a desired or permitted speed of the vehicle in free flow.[3]

In real traffic, a bottleneck introduces a local speed decrease in free flow. The local speed decrease exists *permanently* in free flow at the bottleneck. The permanent

---

[1] See the classical work by Greenshields et al. [8] and references in [1–7, 9–12, 14–18].

[2] Discussions of the local speed decrease in free flow at the bottleneck is made in this chapter based on Sect. 5.3 of the book [13]. It should be mentioned that the term *local speed decrease in free flow at the bottleneck* used in this book was called in [13] "local perturbation in free flow at a bottleneck" (see Fig. 5.12a, b of [13]).

[3] Features of highway bottlenecks discussed in this Sect. 4.1 have been studied both empirically and theoretically in many papers of standard traffic science (see, e.g., reviews [3, 4, 10–14, 18]).

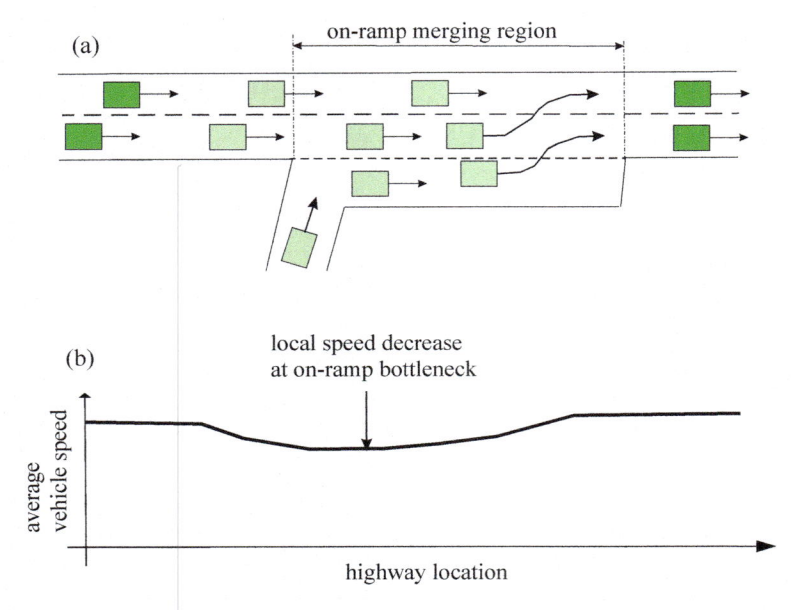

Fig. 4.1 Qualitative explanation of empirical results about features of local speed decrease in free flow at on-ramp bottleneck. **a** Qualitative explanation of merging of vehicles from the on-ramp lane onto the main road within the on-ramp merging region that causes the deceleration of vehicles; vehicles labeled by colored light-green move at a lower speed than vehicles labeled by colored green. **b** Qualitative space distribution of the speed averaged over time (averaged speed for short) within the local speed decrease

existence of such a local speed decrease in free flow at the bottleneck can explain why traffic breakdown in real traffic occurs mostly at the bottleneck: Traffic breakdown is much more probable to occur within the local speed decrease localized at the bottleneck than in free flow of a higher speed outside the bottleneck.

A bottleneck can be caused by road infrastructure. In this case, the bottleneck is called a road bottleneck. Examples of road bottlenecks are on- and off-ramps on highways, road works, traffic signals, road gradients and curves, the decrease in the number of road lanes, highway line constriction, etc.

4.1.1 On-Ramp Bottleneck

At an on-ramp bottleneck, vehicles should change from the on-ramp lane onto the main road (Fig. 4.1a). This vehicle merging occurs within a merging region of the on-ramp called *on-ramp merging region* (Fig. 4.1a).

When the flow rate is large enough, vehicles merging from the on-ramp onto the main road force the deceleration of vehicles (vehicles labeled by colored light-green in Fig. 4.1a) moving on the main road by their approaching the on-ramp merging

4.1 Local Speed Decrease in Free Flow at Road Bottlenecks

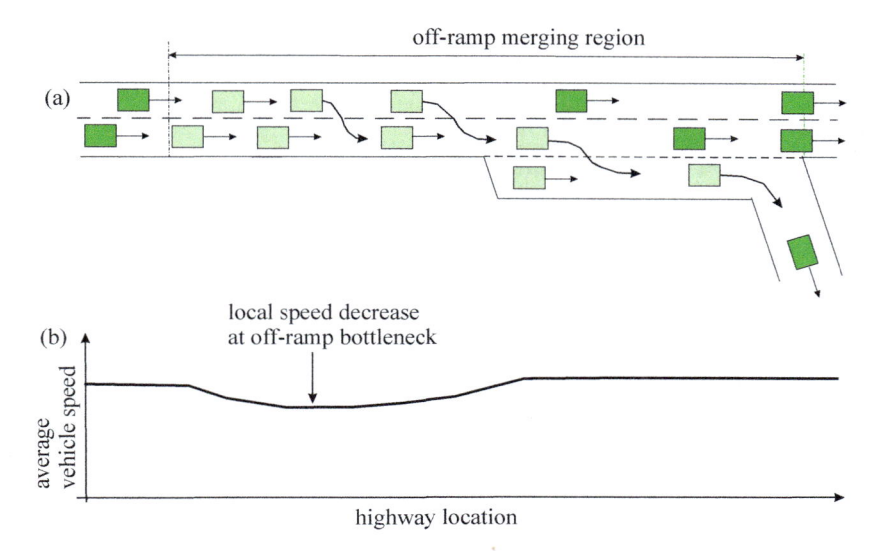

Fig. 4.2 Qualitative explanation of empirical results about features of local speed decrease in free flow at off-ramp bottleneck: **a** Qualitative explanation of lane changing upstream of the off-ramp lane caused by vehicles going to the off-ramp and the associated deceleration of vehicles on the main road; vehicles labeled by colored light-green move at a lower speed than vehicles labeled by colored green. **b** Qualitative space distribution of the averaged speed within the local speed decrease in free flow at off-ramp bottleneck

region. Some of the latter vehicles moving initially in the right lane change to the left lane to avoid a strong deceleration. The vehicle changing lanes can cause the deceleration of other vehicles moving in the left lane behind.

Due to vehicle merging from the on-ramp onto the main road and associated lane changing of vehicles on the main road a local decrease in the vehicle speed can occur in the vicinity of the on-ramp (Fig. 4.1). The local speed decrease is permanently localized in the vicinity of the on-ramp (labeled by "local speed decrease at on-ramp bottleneck" in Fig. 4.1b).

4.1.2 Off-Ramp Bottleneck

An off-ramp can also be the reason for a highway bottleneck (Fig. 4.2). In this case, vehicles going to the off-ramp must firstly change to the right lane.

Vehicles that want to leave the highway to the off-ramp moving initially in the left lane of the main road should decelerate while changing from the left lane to the right lane upstream of the off-ramp lane and in its vicinity. The lane changing of vehicles going to the off-ramp can force vehicles remaining to follow the highway reduce their speeds upstream of the off-ramp. In other words, the associated local speed decrease caused by the deceleration of vehicles going to the off-ramp appears usually upstream of the off-ramp (Fig. 4.2b). As for the on-ramp bottleneck, the local speed decrease is permanently localized in the vicinity of the off-ramp (labeled by "local speed decrease at off-ramp bottleneck" in Fig. 4.2b).

In general, at a large enough flow rate each highway bottleneck results in a local speed decrease localized in the vicinity of the bottleneck. Due to the local speed decrease at the bottleneck, there is the deceleration of vehicles passing the bottleneck.

4.2 Empirical Example of Local Speed Decrease in Free Flow at Road Bottleneck

An empirical example of a local speed decrease at an off-ramp bottleneck is shown in Fig. 4.3. We can see that probe vehicles leaving the main road to the off-ramp should reduce their speeds before they have left the highway (some of probe vehicles leaving to the off-ramp are labeled by "off-a", "off-b", "off-c", and "off-d" in Fig. 4.3b). Indeed, the decrease of the speed of probe vehicles "off-c" and "off-d" leaving the main road to the off-ramp is observed upstream of the beginning of the off-ramp lane at 15.7 km (Fig. 4.3c, d).

To see the effect of the speed decrease of vehicles leaving the main road to the off-ramp (Fig. 4.3c, d) on the motion of vehicles remaining on the main road, we consider the speeds of two such probe vehicles that remain on the main road (trajectories "n" and "p" in Fig. 4.3b, e, f). These two vehicles move in the vicinity of the off-ramp during approximately the same time interval as vehicles "off-c" and "off-d" (Fig. 4.3b). The speeds of probe vehicles "n" and "p" also decrease, when vehicles "n" and "p" are moving just upstream of the off-ramp (Fig. 4.3e, f). As a result, a local speed decrease appears just upstream of the off-ramp (labeled by "local speed decrease" in Fig. 4.3e, f). In probe vehicle data under consideration, there are no information about the lane in which a probe vehicle moves currently. However, the speed decrease of vehicles "off-c" and "off-d" is stronger than that of vehicles "n" and "p" (Fig. 4.3c–f). For this reason, we can assume that vehicles "n" and "p" move in other lanes (middle lane or left lane) than vehicles "off-c" and "off-d" moving in the right lane to leave the main road.

In general, empirical data measured on other highways and on different days confirm the common result that when vehicles leaving the main road to the off-ramp should reduce their speeds just upstream of the off-ramp lane, then a local speed decrease in free flow appears just upstream of the off-ramp bottleneck.

4.2 Empirical Example of Local Speed Decrease in Free Flow ...	43

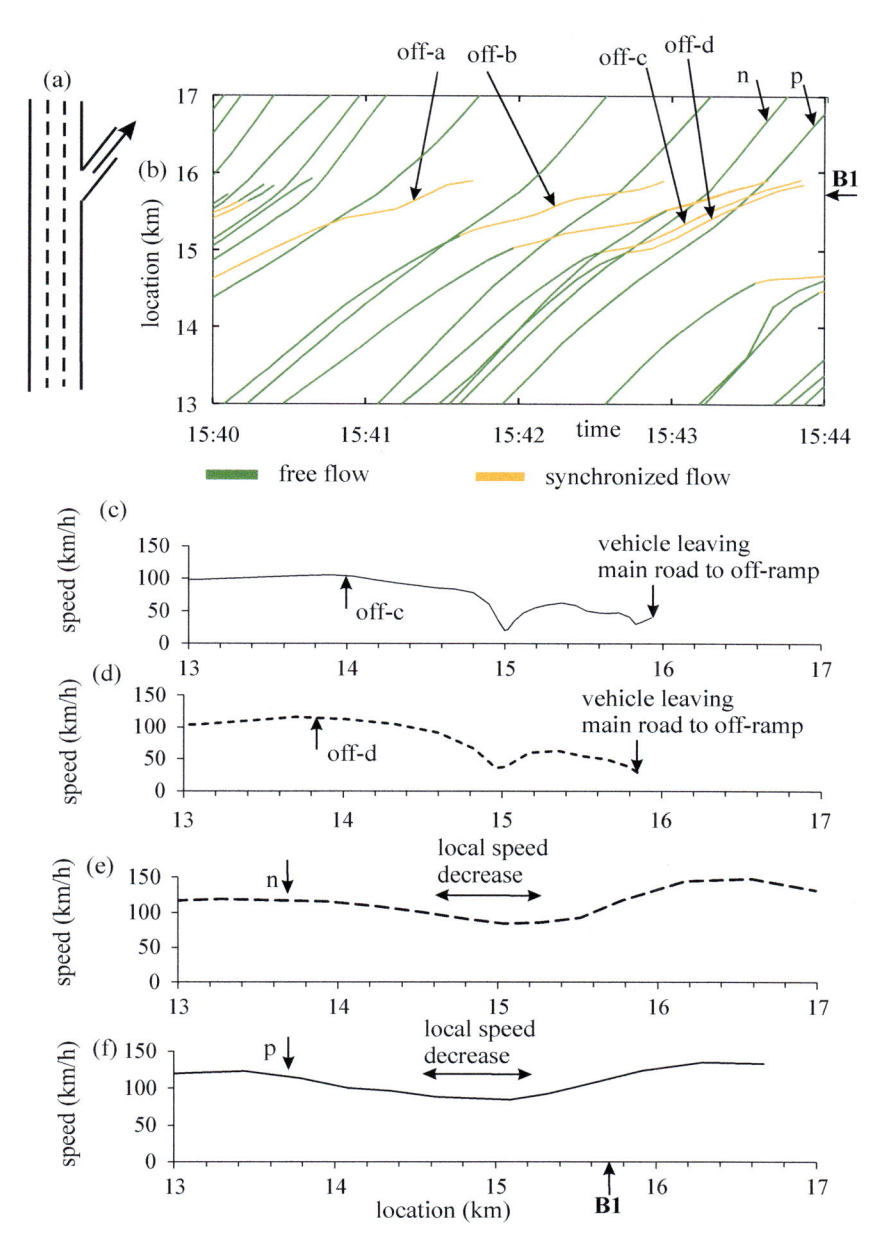

Fig. 4.3 Empirical local speed decrease in free flow at off-ramp bottleneck: **a, b** Schema of highway section (**a**) (fragment of Fig. 2.1a) and probe vehicle trajectories (**b**) (fragment of Fig. 2.1b). **c–f** Microscopic speeds of two probe vehicles leaving the main road to the off-ramp (**c, d**) and two probe vehicles that remain moving on the main road (**e, f**). In **c–f**, probe vehicles are labeled by the same numbers as those in **b**. Arrows B1 show the location of off-ramp bottleneck B1 (see caption to Fig. 2.1b)

4.3 Local Speed Decrease in Free Flow at Moving Bottleneck

In addition to road bottlenecks discussed in Sects. 4.1 and 4.2, in vehicular traffic there can be moving bottlenecks.[4] A moving bottleneck (MB) is caused by a slow vehicle or a platoon of slow vehicles moving in traffic flow. In free flow, the MB causes a local speed decrease that exists permanently at the MB; the local speed decrease is localized at the MB and, therefore, it moves downstream in free flow at the speed of the MB.

To explain the local speed decrease in free flow localized in the vicinity of the MB, we consider free flow on a two-lane road without road bottlenecks. The free flow consists of passenger vehicles and a truck moving in the right lane. We assume that the truck moves at a lower speed v_{MB} in free flow than the speed of the passenger vehicles (Fig. 4.4a). We denote the truck as "slow vehicle". The slow vehicle can be considered as the MB that moves at the speed of the slow vehicle v_{MB}.

While approaching the slow vehicle, passenger vehicles that move initially in the right lane change to the left lane to overtake the slow vehicle (Fig. 4.4a). This lane changing results in the increase in the vehicle density in the left lane in the vicinity of the slow vehicle (Fig. 4.4b). In accordance with the well-known empirical features of free flow (see Sect. 5.2 for more details), the flow rate is an increasing density function in free flow, whereas the average vehicle speed is a decreasing density function. Therefore, the flow rate increases (Fig. 4.4c) and the average speed of the passenger vehicles decreases (Fig. 4.4d) in the vicinity of the slow vehicle.

We assume that the slow vehicle moves at a time-independent speed v_{MB}. From the definition of a vehicle trajectory made in Sect. 3.1.1, we can make the conclusion that the trajectory of the slow moving vehicle in the road location–time plane is a line whose slope is equal to the speed of the slow vehicle v_{MB} (green line in Fig. 4.4e). In the vicinity of the slow vehicle, for free-flow conditions there are the local flow-rate increase and the local decrease in the average speed of the passenger vehicles (Fig. 4.4c, d). Therefore, the slow vehicle causes the wave of the flow-rate increase (blue wave in Fig. 4.4f) and the wave of the local decrease in the average speed of the passenger vehicles (gray wave in Fig. 4.4g): Both waves move at the slow vehicle speed v_{MB} (Fig. 4.4e–g).

The local decrease in the average speed in free flow occurring at the location of the slow vehicle (Fig. 4.4d) is qualitatively the same as the occurrence of the local speed decrease in free flow at a road bottleneck (Fig. 4.1b). Therefore, the slow vehicle can be considered a moving bottleneck (MB). Respectively, the local decrease of the average vehicle speed at the location of the MB (Fig. 4.4d) can indeed be considered the local speed decrease in free flow at the MB.

---

[4] To the author's knowledge, a possibility of a moving bottleneck (MB) was predicted in the works by Gazis and Herman [5, 6]. Newell [16, 17] revealed that in a system coordinate moving at the MB speed traffic breakdown at the MB should exhibit qualitatively the same features as those at road bottlenecks.

4.3 Local Speed Decrease in Free Flow at Moving Bottleneck 45

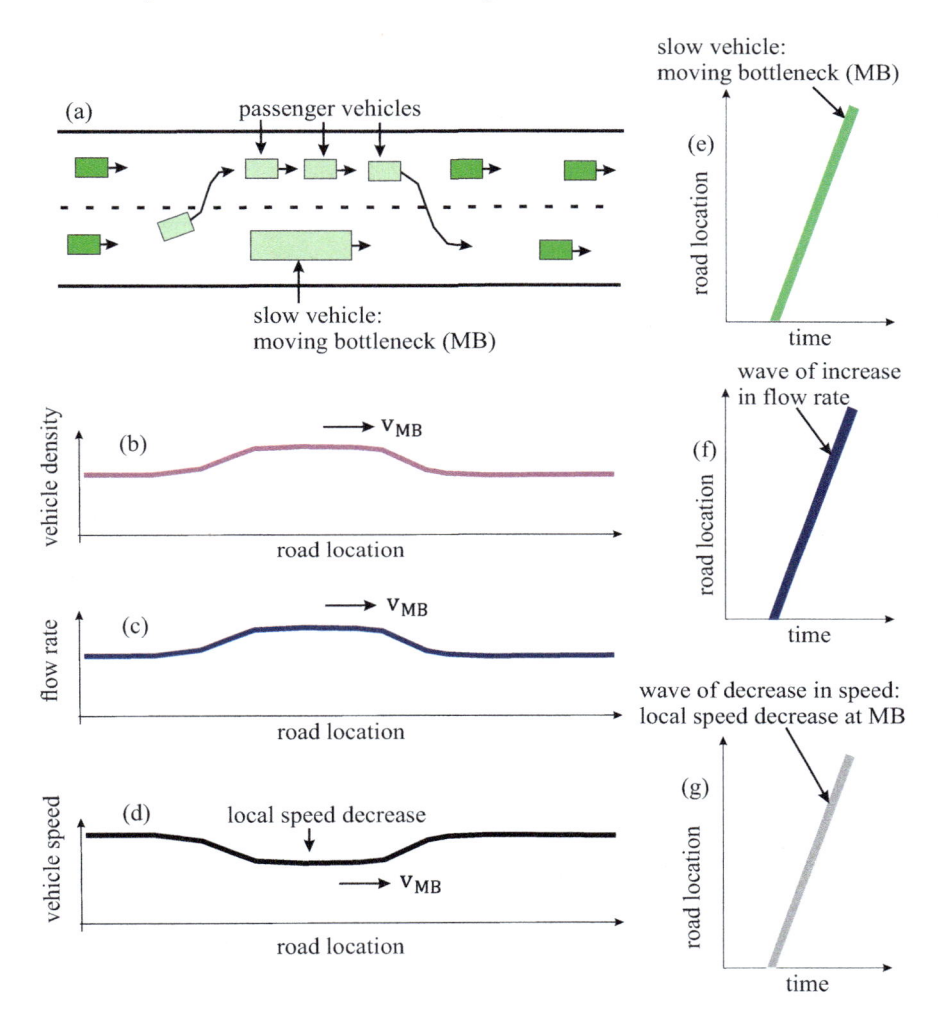

Fig. 4.4 Qualitative explanation of the occurrence of local speed decrease in free flow at a moving bottleneck (MB): **a** Two-lane road with passenger vehicles that pass a slow vehicle (MB) at a given time instant; vehicles labeled by colored light-green move at a lower speed than vehicles labeled by colored green. **b–d** The density (**b**) and flow rate (**c**) increases whereas the average speed decrease (**d**) at the location of MB; horizontal arrows v_{MB} show the downstream motion of the density (**b**) and flow rate (**c**) increase as well as the average speed decreases (**d**) at the speed of the MB denoted by v_{MB}. **e–g** Qualitative explanation of waves in free flow caused by MB in space and time: MB (**e**) causes a wave of the local increase in the flow rate (**f**) and a wave of the local decrease in the average vehicle speed (**g**) occurring at the location of the slow moving vehicle and propagating at the speed v_{MB} of the MB

The difference between local speed decreases, respectively, at a road bottleneck and at an MB is as follows: For road bottlenecks, the local speed decrease is localized at the bottleneck; therefore, it is on average motionless. Contrarily, the local speed decrease in free flow at the MB moves at the speed of the MB $v_{\rm MB}$; therefore, the wave of the local decrease in the average speed of the passenger vehicles (gray wave in Fig. 4.4g) is a synonym of the local speed decrease in free flow at the MB.

> The local speed decrease in free flow at the MB is the wave of the decrease in the free flow speed (gray wave moving at the MB speed in Fig. 4.4g).

References

1. W.D. Ashton, *The Theory of Traffic Flow* (Methuen & Co., London, Wiley, New York, 1966)
2. D.R. Drew, *Traffic Flow Theory and Control* (McGraw Hill, New York, 1968)
3. L. Elefteriadou, *An Introduction to Traffic Flow Theory* (Springer, Berlin, 2014)
4. N.H. Gartner, C.J. Messer, A.K. Rathi (eds.), *Traffic Flow Theory: a State-of-the-Art Report* (Transportation Research Board, Washington, 2001)
5. D.C. Gazis, R. Herman, *The Moving and 'Phantom' Bottlenecks* (IBM Thomas J. Watson Research Division, 1990)
6. D.C. Gazis, R. Herman, Transp. Sci. **26**(3), 223–229 (1992)
7. D.L. Gerlough, M.J. Huber, *Traffic Flow Theory Special Report 165* (Transportation Research Board, Washington, 1975)
8. B.D. Greenshields, J.R. Bibbins, W.S. Channing, H.H. Miller, Highw. Res. Board Proc. **14**, 448–477 (1935)
9. F.A. Haight, *Mathematical Theories of Traffic Flow* (Academic, New York, 1963)
10. F.L. Hall, in [4], pp. 2-1–2-36
11. *Highway Capacity Manual 2000* (National Research Council, Transportation Research Board, Washington, 2000)
12. *Highway Capacity Manual 2016*, 6th edn. (National Research Council, Transportation Research Board, Washington, 2016)
13. B.S. Kerner, *The Physics of Traffic* (Springer, Berlin, 2004)
14. A.D. May, *Traffic Flow Fundamentals* (Prentice-Hall Inc., New Jersey, 1990)
15. G.F. Newell, *Applications of Queuing Theory* (Chapman Hall, London, 1982)
16. G.F. Newell, A moving bottleneck, Technical report UCB-ITS-RR-93-3 (Institute of Transportation Studies, University of California, Berkeley, 1993)
17. G.F. Newell, Transp. Res. B **32**(8), 531–537 (1998)
18. R.P. Roess, E.S. Prassas, *The Highway Capacity Manual: a Conceptual and Research History* (Springer, Berlin, 2014)

Chapter 5
Empirical Spontaneous Traffic Breakdown—Fundamental Problem for Understanding Real Traffic

In the three-phase traffic theory, there is the distinguishing between the *empirical spontaneous* traffic breakdown (Sect. 2.2) and the *empirical induced* traffic breakdown at a highway bottleneck (Sect. 2.3).

Contrary to the three-phase traffic theory, there is *no* distinguishing between empirical spontaneous and induced traffic breakdowns in standard traffic and transportation science[1]: The terms *spontaneous* and *induced* have not been used in the application to real traffic breakdown. In all publications of the standard traffic research devoted to empirical studies of traffic breakdown, which the author knows, traffic breakdown at a highway bottleneck is the empirical traffic phenomenon that in the three-phase traffic theory as well as in this book we call empirical *spontaneous* traffic breakdown at the bottleneck. In other words, the term *traffic breakdown* used in the standard methodologies of studies of empirical traffic phenomena and the term *spontaneous traffic breakdown* used in the three-phase traffic theory can be considered synonymous: Traffic researchers that have worked within the framework of the standard methodologies have not appreciated the empirical traffic phenomenon "empirical induced traffic breakdown (F→S transition)" at the bottleneck.

However, why should a study of empirical induced traffic breakdown at a highway bottleneck exhibit the fundamental importance for understanding real traffic? To answer this question in Chap. 6, in this chapter, we should firstly show that there is a fundamental problem for understanding real traffic associated with the use of the standard methodologies of studies of empirical traffic phenomena.

---

[1] The standard methodologies for studies of empirical traffic breakdown can be found, for example, in papers [2–4, 9–16, 22, 29, 30, 39, 42–49, 72, 82–85] as well as reviews and books [1, 5, 8, 21, 28, 31, 34, 35, 40, 41, 52, 53, 61, 73, 95, 101].

> The fundamental problem for understanding real traffic is as follows: A definitive conclusion about the nature of empirical traffic breakdown based solely on the studies of empirical *spontaneous* traffic breakdown (F→S transition) is not possible.

> The ignoring of the traffic phenomenon "empirical induced traffic breakdown (induced F→S transition) at a bottleneck" made in standard traffic and transportation science is the origin of the failure of applications of the standard traffic theories in the real world.

Traffic breakdown limits highway capacity of free flow at a bottleneck. The meaning of highway capacity of free flow is the basis for the development of any method for traffic control, management, and organization of a traffic network as well as for many other ITS applications. Therefore, the following question arises:

- How does the ignoring of the traffic phenomenon "empirical induced traffic breakdown (F→S transition) at the bottleneck" effect on the perception of highway capacity in standard traffic and transportation science?

5.1 Perception of Highway Capacity in Standard Traffic and Transportation Science

Since 1920s–1930s the perception of highway capacity in the standard traffic and transportation research was as follows:

- Highway capacity is equal to the maximum flow rate in free flow at which free flow still persists at a bottleneck. This means that if the flow rate at the bottleneck is less than highway capacity, no traffic breakdown can occur at the bottleneck. Contrarily, when the flow rate in free flow upstream of the bottleneck exceeds highway capacity of free flow at the bottleneck, traffic breakdown must occur at the bottleneck.

This standard understanding of highway capacity as some maximum flow rate in free flow at which free flow still persists at a bottleneck is generally accepted in stan-

5.1 Perception of Highway Capacity in Standard Traffic and Transportation ...

dard traffic and transportation science.[2] Therefore, in "Highway Capacity Manual",[3] which is used as a textbook for most students in transportation science, there is a consideration of diverse methods that can be used for measurements of the maximum flow rate in free flow at which free flow still persists at a bottleneck, i.e., the methods for the estimation of highway capacity of free flow at the bottleneck.

In 1995, it was found that empirical traffic breakdown at a bottleneck exhibits the probabilistic (stochastic) behavior[4]: At a given flow rate in free flow at the bottleneck, traffic breakdown can occur but it should not necessarily occur. Thus, on one day traffic breakdown occurs, however, on another day at the same flow rates traffic breakdown is not observed. Therefore, highway capacity should exhibit a stochastic behavior. In the standard traffic and transportation research, the stochastic highway capacity is often explained through a strongly heterogeneous traffic consisting of both personal cars and trucks. The share of trucks as well as the share of vehicles with different vehicle and driver characteristics are usually stochastic time-functions in real traffic flow. Therefore, highway capacity that should depend on vehicle and driver characteristics should also be a stochastic time-function.

Stochastic highway capacity in the standard traffic and transportation research is considered within the standard perception of highway capacity[5]:

- *At a time instant* stochastic highway capacity is equal to the maximum flow rate in free flow at which free flow still persists at a bottleneck at this time instant.
- If at a time instant the flow rate at the bottleneck is less than the value of stochastic highway capacity at this time instant, no traffic breakdown can occur at the bottleneck.
- Contrarily, when at another time instant the flow rate in free flow at the bottleneck exceeds the value of stochastic highway capacity at this time instant traffic breakdown must occur at the bottleneck.

In other words, in the standard understanding of stochastic highway capacity it is assumed that *at any time instant*, there is a particular value of highway capacity of free flow at a bottleneck; the value of highway capacity is a stochastic time-function.

It must be emphasized that the standard understanding of stochastic highway capacity does not contradict features of the empirical spontaneous traffic breakdown at a bottleneck. Indeed, we might assume that empirical spontaneous traffic breakdown at an off-ramp bottleneck shown in Fig. 2.2 has occurred because at the time

---

[2] See, e.g., papers [2–4, 9–16, 22, 29, 30, 39, 42–49, 72, 82–85] as well as reviews and books [1, 5–8, 18, 20, 21, 28, 31–35, 38, 40, 41, 50, 52–55, 61, 68, 70, 71, 73, 79–81, 86, 89, 90, 95–97, 99–101].

[3] See Refs. [52, 53]. It should also be noted that the standard perception of highway capacity is consistent with applications of the well-known *queuing theory* and the associated queuing models of vehicular traffic; see references, e.g., in the book [79].

[4] The probabilistic behavior of empirical spontaneous traffic breakdown was observed by Elefteriadou et al. [30]. It was also found that the probability of empirical spontaneous traffic breakdown at a bottleneck is a sharp increasing function of the flow rate. The probability of empirical spontaneous traffic breakdown was firstly found by Persaud et al. [84] in 1998.

[5] The standard understanding of stochastic highway capacity has been a subject of many empirical and theoretical studies; see, e.g., [9–13, 28–30, 53, 84, 85].

instant of traffic breakdown (time about 16:19 in Fig. 2.2) the flow rate in free flow at bottleneck B1 exceeds the value of stochastic highway capacity at this time instant and, therefore, in accordance with the standard understanding of stochastic highway capacity traffic breakdown must occur at bottleneck B1.[6]

However, this seeming consistence of empirical features of spontaneous traffic breakdown with the standard perception of highway capacity considered above is invalid. Indeed, as we will show in Chaps. 6 and 7, the standard perception of stochastic highway capacity contradicts basically the evidence of the empirical induced traffic breakdown at a bottleneck. As we will show in Chap. 8, traffic breakdown exhibits the nucleation nature independent of whether spontaneous or induced traffic breakdown is considered. The empirical nucleation nature of real traffic breakdown contradicts fundamentally the standard perception of highway capacity.[7]

There are the following consequences of the seeming consistence of empirical features of spontaneous traffic breakdown with the standard perception of stochastic highway capacity:

- This might be the reason for the ignoring of the evidence of the empirical induced traffic breakdown at a bottleneck in standard traffic and transportation science. For this reason, the term *spontaneous* is not usually used in the standard traffic and transportation research, when traffic breakdown is studied.
- This explains the failure of applications of standard traffic and transportation theories in the real world.
- This explains also why a long cumulative process of empirical and theoretical studies of traffic breakdown and resulting traffic congestion made in the standard traffic and transportation research cannot lead to understanding real traffic.

The ignoring of the empirical induced traffic breakdown at a highway bottleneck in the standard traffic and transportation research causes the fundamental problem for understanding real traffic formulated above: The definitive conclusion about the nature of empirical traffic breakdown and the meaning of stochastic highway capacity based solely on studies of empirical spontaneous traffic breakdown is not possible.

---

[6] In a larger scale, the same empirical spontaneous traffic breakdown as that in Fig. 2.2 is shown in Fig. 5.5 below.

[7] In more details, the criticism of the standard understanding of stochastic highway capacity can be found in Sect. 7.3 as well as in Appendix C.

5.2 Empirical Fundamental Diagram of Traffic Flow

In standard traffic science, empirical characteristics of traffic breakdown at a bottleneck are often studied with the use of the empirical fundamental diagram of traffic flow.[8] Here and in next Sect. 5.3, we try to answer the following question:

- Is the understanding of the nature of real traffic breakdown from a sole study of empirical traffic data with the use of the empirical fundamental diagram possible?

To answer this question, we consider empirical data measured through road detectors located at an on-ramp bottleneck (Fig. 5.1a, b). At the bottleneck location ($x = 6.4$ km in Fig. 5.1b), the speed drops sharply during traffic breakdown (arrow labeled by "spontaneous traffic breakdown" in Fig. 5.1c, left). Contrary to this speed behavior, the flow rate in synchronized flow can remain on average the same as that in an initial free flow (Fig. 5.1c, right). This is usual result for other data measured at the location of the local speed decrease at the bottleneck.

The empirical fundamental diagram is a flow rate–density relationship between the vehicle density and the flow rate presented in the flow–density plane (Fig. 5.2a). Each point in the empirical fundamental diagram presented in Fig. 5.2a is related to empirical data of the flow rate measured by the road detector and the density estimated from the flow rate and speed measurements shown in Fig. 5.1c. To estimate the vehicle density, we take into account that the flow rate denoted by q is the product of the vehicle density denoted by ρ and the vehicle speed v:

$$q = \rho v. \qquad (5.1)$$

From (5.1), the vehicle density can be estimated through formula[9]

$$\rho = q/v. \qquad (5.2)$$

Additionally with the fundamental diagram (Fig. 5.2a), traffic researchers and engineers use often equivalent data presentations in the speed–density plane (Fig. 5.2b) and speed–flow plane (Fig. 5.2c). With arrows F→S in Fig. 5.2a–c we have symbolically shown changes in the flow rate, the average speed, and the density occurring during spontaneous traffic breakdown (F→S transition) shown in Fig. 5.1b, c. As follows from empirical data (Fig. 5.2), in free flow, the average vehicle speed is a decreasing flow-rate function (Fig. 5.2c).

A qualitative presentation of these well-known empirical features of free flow at the bottleneck (Fig. 5.2) is shown in Fig. 5.3. We recall that there is a local

---

[8] See, for example, papers [2–4, 9–16, 22, 29, 30, 39, 42–49, 72, 82–85] as well as reviews and books [1, 5, 8, 21, 28, 31, 34, 35, 41, 52, 53, 61, 73, 95].

[9] It should be noted that the density estimation through formula (5.2) can only be made, if the speed in the empirical data is not equal to zero. This is the case for the traffic data presented in Fig. 5.2 that have been measured at the bottleneck location $x = 6.4$ km (Fig. 5.1c). We note also that, at low-speed values, there can be a large systematic error in the density estimation through formula (5.2) (see explanations on page 411 of the review [59]).

52   5 Empirical Spontaneous Traffic Breakdown—Fundamental Problem …

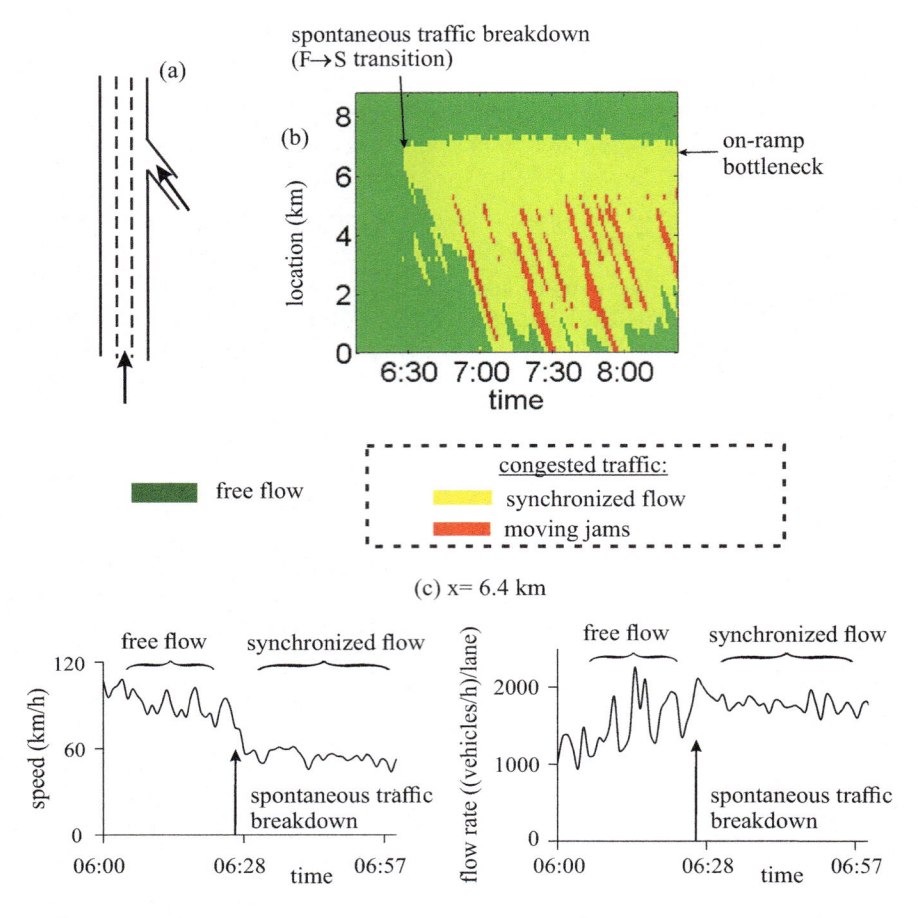

Fig. 5.1 Empirical spatiotemporal traffic dynamics reconstructed through the use of road detector data: **a** Sketch of highway section with on-ramp bottleneck. **b** Empirical speed data presented in space and time with the use of averaging method described in Sect. C.2 of [60]. **c** Time functions of the average speed (left) and the flow rate averaged per road lane (right) related to location $x = 6.4$ km of empirical spontaneous traffic breakdown (F→S transition). Empirical example of a congested traffic pattern measured on three-lane highway A5-South in Germany on January 13, 1997; data are averaged during 1-min time intervals. Adapted from [56]

speed decrease in free flow at the bottleneck location (Sect. 4.1). Traffic data shown in Fig. 5.2 are related to the measurement of the *average speed* at the bottleneck. Therefore, we consider some qualitative features of *the average local speed decrease* in free flow at the bottleneck. We denote the minimum speed within the average local speed decrease at the bottleneck by $v_{\rm free}^{\rm (B)}$ (Fig. 5.3a). The flow-rate function of the speed $v_{\rm free}^{\rm (B)}$ shown in Fig. 5.3b is qualitatively consistent with empirical traffic data

5.2 Empirical Fundamental Diagram of Traffic Flow

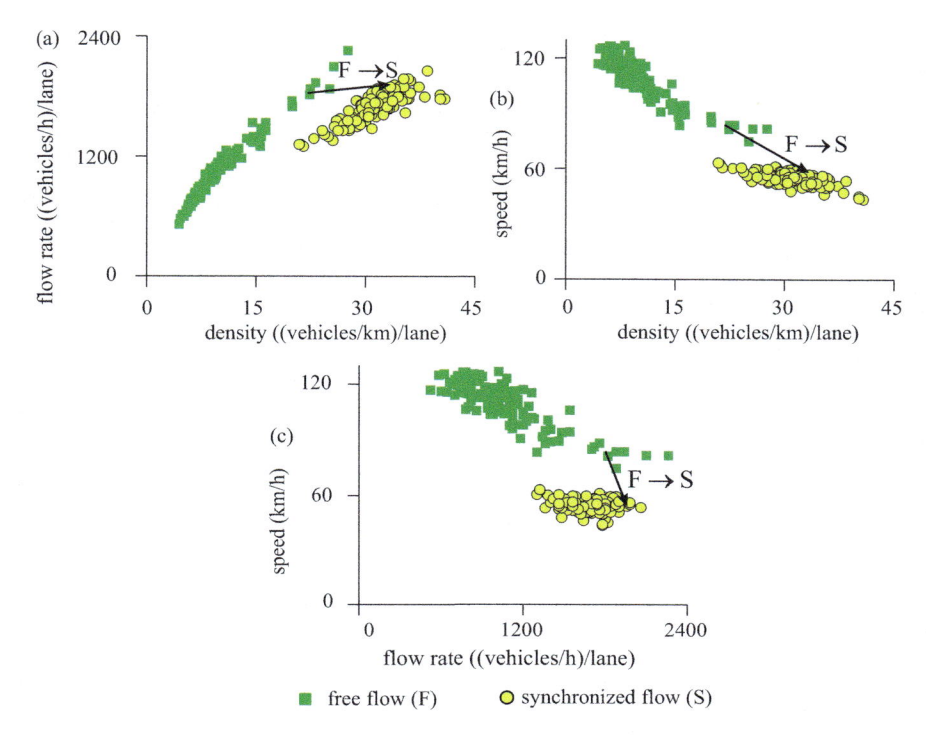

Fig. 5.2 Continuation of Fig. 5.1. **a** Empirical fundamental diagram of traffic flow measured at the bottleneck location at which spontaneous traffic breakdown (F→S transition) has occurred ($x = 6.4$ km); data averaged during 1-min time intervals are related to congested pattern in Fig. 5.1b. **b, c** The same empirical data as those in (**a**) presented in the speed–density plane (**b**) and speed–flow plane (**c**). Empirical data related to free flow at the bottleneck (green squares) have been measured before traffic breakdown has occurred, whereas empirical data related to congested traffic (synchronized flow) at the bottleneck (yellow circles) have been measured after traffic breakdown has occurred. Qualitatively, the same empirical results have been found in many different empirical traffic data (see, e.g., [43, 73])

shown in Fig. 5.2c.[10] When the flow rate q at the bottleneck is small, the speed $v_{\rm free}^{(B)}$ is equal to some maximum value denoted by $v_{\rm free}^{(\max)}$; when the flow rate q reaches some maximum possible value in free flow, the speed $v_{\rm free}^{(B)}$ is equal to some minimum speed in free flow denoted by $v_{\rm free}^{(\min)}$.

---

[10] Indeed, traffic data shown in Fig. 5.2 are related to measurements of the average speed by a road detector located at the bottleneck. Empirical data measured at bottleneck locations in many countries show that qualitative characteristics of the fundamental diagram do not depend on the exact location of the detector in the bottleneck vicinity: We find that when the flow rate is very small, the average speed corresponds to some maximum speed in free flow; when the flow rate reaches some maximum possible value in free flow, the average speed is equal to some minimum speed in free flow. Thus, we can assume that the *minimum speed* $v_{\rm free}^{(B)}$ within the average local speed decrease at the bottleneck satisfies the above-mentioned empirical features of free flow.

Fig. 5.3 Qualitative characteristics of free flow at road bottlenecks: **a** Qualitative space dependence of the speed within the average local speed decrease at a road bottleneck for a given flow rate q in free flow; $v_{\text{free}}^{(B)}$ is the minimum speed within the average local speed decrease at the bottleneck. **b** Qualitative dependence of the speed $v_{\text{free}}^{(B)}$ on the flow rate q in free flow at the bottleneck

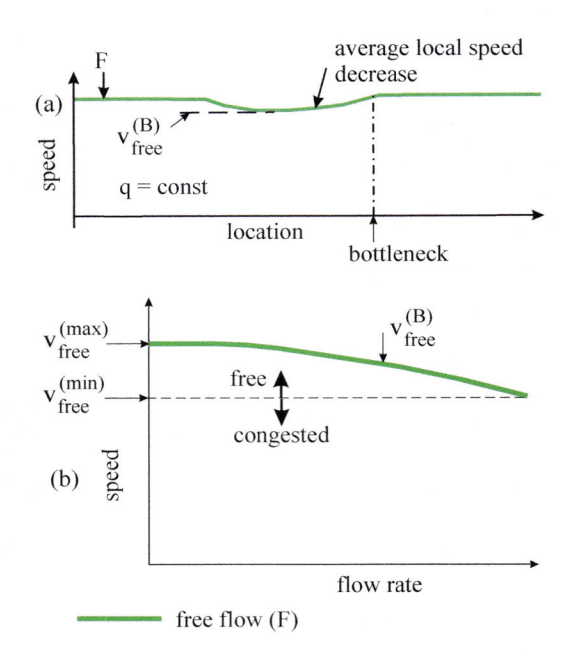

In this qualitative consideration of macroscopic characteristics of free flow at the bottleneck, the average speed $v_{\text{free}}^{(\min)}$ is related to the limit point of free flow: The average speed $v \geq v_{\text{free}}^{(\min)}$ correspond to states of free flow (green curve for the flow-rate dependence of the speed $v_{\text{free}}^{(B)}$ in Fig. 5.3b). The average speed $v < v_{\text{free}}^{(\min)}$ correspond to states of congested traffic. Arrows labeled by "free" and "congested" in Fig. 5.3b represent qualitatively this well-known empirical result. This result is often used to distinguish between free flow and congested traffic as well as for the following definition of congested traffic: In congested traffic, the average vehicle speed is less than the minimum average speed $v_{\text{free}}^{(\min)}$ that is still possible in free flow (Fig. 5.3b).

It should be emphasized that the development of traffic breakdown (F→S transition) at the bottleneck occurs in space and time (Fig. 5.1b). However, many *spatiotemporal features* of traffic breakdown are lost in the empirical fundamental diagram (Fig. 5.2a).[11] We can make the conclusion:

- For a correct understanding of empirical traffic data, the empirical fundamental diagram (Fig. 5.2a) should be used together with spatiotemporal empirical data that contain the full information about the development of traffic breakdown (F→S transition) in space and time.

---

[11] An example of empirical spatiotemporal features of traffic breakdown that cannot be resolved with the sole use of the empirical fundamental diagram are the empirical transitions from free flow to synchronized flow and backward (F→S→F transitions) that will be studied in Sect. 9.2.

5.3 Empirical Hysteresis Effect

While discussing empirical data shown in Figs. 5.1 and 5.2, we have limited by a consideration of spontaneous traffic breakdown (F→S transition) at the bottleneck. It is clear that traffic congestion at the bottleneck has a limited duration: When the flow rate at the bottleneck decreases over time considerably, a return S→F transition should occur at the bottleneck and, therefore, free flow returns at the bottleneck. Such an empirical example is shown in Fig. 5.4. The F→S transition at the on-ramp bottleneck shown in Fig. 5.4b has qualitatively the same features as those discussed above for the empirical example in Figs. 5.1 and 5.2. However, the empirical data in Fig. 5.4b are shown for a longer time interval. It turns out that the flow rate at the bottleneck in Fig. 5.4b decreases at $t \approx 10{:}30$. This can explain the return S→F transition at the bottleneck (labeled by "S→F transition" in Fig. 5.4b).

The F→S transition together with the return S→F transition shown in Fig. 5.4b at the bottleneck location are presented, respectively, by arrows F→S and S→F in the flow–density plane (the fundamental diagram) as well as in the speed–density and speed–flow planes in Fig. 5.4c–e. The F→S transition together with the return S→F transition are built a so-called hysteresis loop that is usual for a well-known hysteresis effect. The hysteresis effect is the dependence of the state of a traffic system on its history. Indeed, the return S→F transition at the bottleneck can only be observed if the F→S transition occurred in the past at this bottleneck.

As in Fig. 5.2a, many spatiotemporal features of the F→S transition and return S→F transition are lost in the empirical fundamental diagram shown in Fig. 5.4c. Traffic is a dynamic process that occurs in space and time. Therefore, a question arises: What is the sense of the empirical fundamental diagram, if for a correct understanding of empirical traffic data the empirical fundamental diagram (Figs. 5.2a and 5.4c) should be used together with spatiotemporal empirical data (Figs. 5.1b and 5.4b)? One of the answers is that spatiotemporal empirical data (Figs. 5.1b and 5.4b) are too complex for the illustration of some traffic characteristics, like hysteresis effect mentioned above. In other words, the empirical fundamental diagram is very useful for illustrations of some macroscopic traffic characteristics. However, a consideration of the empirical fundamental diagram without relation to the spatiotemporal empirical data can lead and it does very often lead to incorrect conclusions about traffic breakdown.[12]

---

[12] This criticism related to many standard empirical traffic data studies will be considered in Sect. C.4 of Appendix C. In particular, we will explain that a diverse variety of traffic hysteresis effects can be observed due to a huge number of qualitatively different effects associated with the three traffic phases (F, S, and J) and transitions between the phases. Therefore, the observation of a traffic hysteresis in a traffic data study in the flow–density plane (or speed–density or else speed–flow planes) should not necessarily be associated with an F→S transition (traffic breakdown) together with a return S→F transition at a bottleneck. This conclusion is related, for example, to a well-known Treiterer's hysteresis effect [102–104] that will be discussed in Sect. B.3 of Appendix B. In general, to understand the meaning of any particular traffic hysteresis effect, a spatiotemporal data analysis with the use of the related empirical traffic data measured in space and time should be performed.

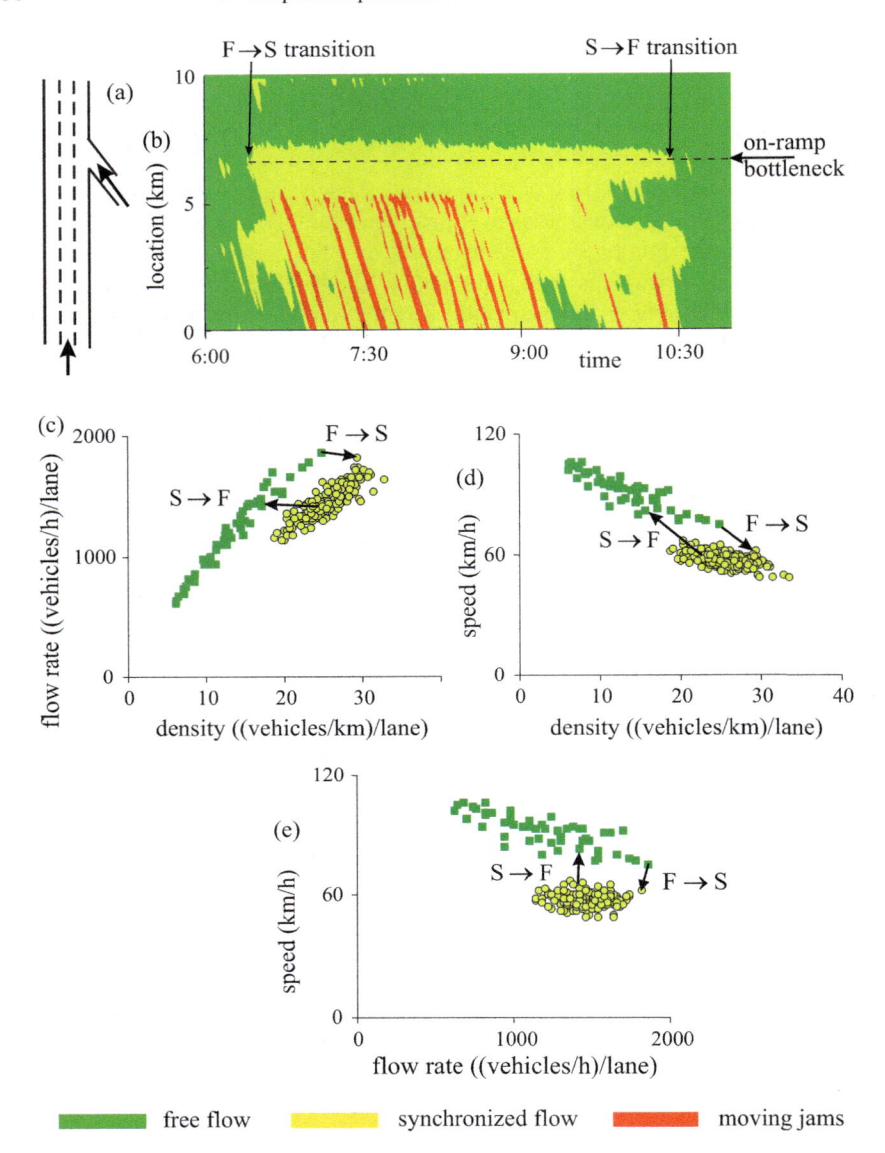

Fig. 5.4 Empirical spatiotemporal traffic dynamics reconstructed through the use of road detector data: **a** Sketch of highway section with on-ramp bottleneck that is the same as that in Fig. 5.1a. **b** Empirical speed data presented in space and time for data measured on three-lane highway A5-South in Germany on March 23, 1998. **c** Empirical fundamental diagram of traffic flow measured at the bottleneck location $x = 6.4$ km in (**b**) at which traffic breakdown has occurred. **d**, **e** The same empirical data as those in (**c**) presented in the speed–density plane (**d**) and speed–flow plane (**e**). Arrows F→S and S→F in (**c**–**e**) show changes in traffic variables during spontaneous traffic breakdown (F→S transition) and return S→F transition, respectively. Qualitatively, the same empirical results have been found in many studies (see, e.g., [43, 73]). Adapted from [57]

5.3 Empirical Hysteresis Effect   57

> From a sole study of empirical traffic data with the use of the empirical fundamental diagram, the understanding of the nature of real traffic breakdown is not possible. To understand real traffic breakdown, traffic data measured in space and time should be used.

5.4 Microscopic Spatiotemporal Features of Empirical Spontaneous Traffic Breakdown at Bottleneck

Some of the empirical spatiotemporal features of spontaneous traffic breakdown at off-ramp bottleneck B1 (Fig. 2.2) can be found from a microscopic analysis of the probe vehicle data presented in Fig. 5.5. Probe vehicles labeled by numbers 1–4 remain on the main road, whereas probe vehicles labeled by "off-1" and "off-2" leave the main road to the off-ramp (Fig. 5.5b). Vehicles "off-1" and "off-2" going to the off-ramp maintain the local speed decrease at the off-ramp bottleneck (labeled by "local speed decrease" in Fig. 5.5c), as already explained in Sect. 4.2.[13] Independent of the local speed decrease, vehicle 1 moves with a free flow speed: No traffic breakdown has occurred until $t = 16{:}18$.

Although the behavior of the speed decrease of probe vehicles leaving the main road to the off-ramp remains the same, suddenly traffic breakdown occurs spontaneously on the main road at $t \approx 16{:}19$[14]: The speed of vehicle 2 remaining on the main road drops to a lower speed related to synchronized flow (congested traffic) (labeled by "spontaneous traffic breakdown" in Fig. 5.5d). This speed drop at the bottleneck is also realized for the following probe vehicles remaining on the main road (vehicles 3 and 4 in Figs. 5.5b and 5.6). We can make the conclusion:

- No particular event has been detected that has caused empirical traffic breakdown (Fig. 5.5d). This explains the use of the word *spontaneous* in the definition of this empirical traffic breakdown (F→S transition) made in the three-phase traffic theory.

After traffic breakdown has occurred, the downstream front of synchronized flow at which vehicles 2, 3, and 4 accelerate to free flow is localized at the bottleneck

---

[13] We can make this conclusion from a comparison of the local speed decrease shown in Fig. 4.3e, f caused by probe vehicles "off-c" and "off-d" (Fig. 4.3c, d) with the local speed decrease shown in Fig. 5.5c caused by vehicle "off-1".

[14] It should be emphasized that in Fig. 5.5b trajectories of only probe vehicles are shown. We do not know how many other vehicles move between the probe vehicles. For example, we can state that vehicles "off-1" and 1 are different probe vehicles that are close to each other, however, we do not know whether there are other vehicles between them. Moreover, we cannot state whether traffic breakdown has occurred just before vehicle "off-2" or after this vehicle. For this reason, we can only approximately determine the time instant of traffic breakdown.

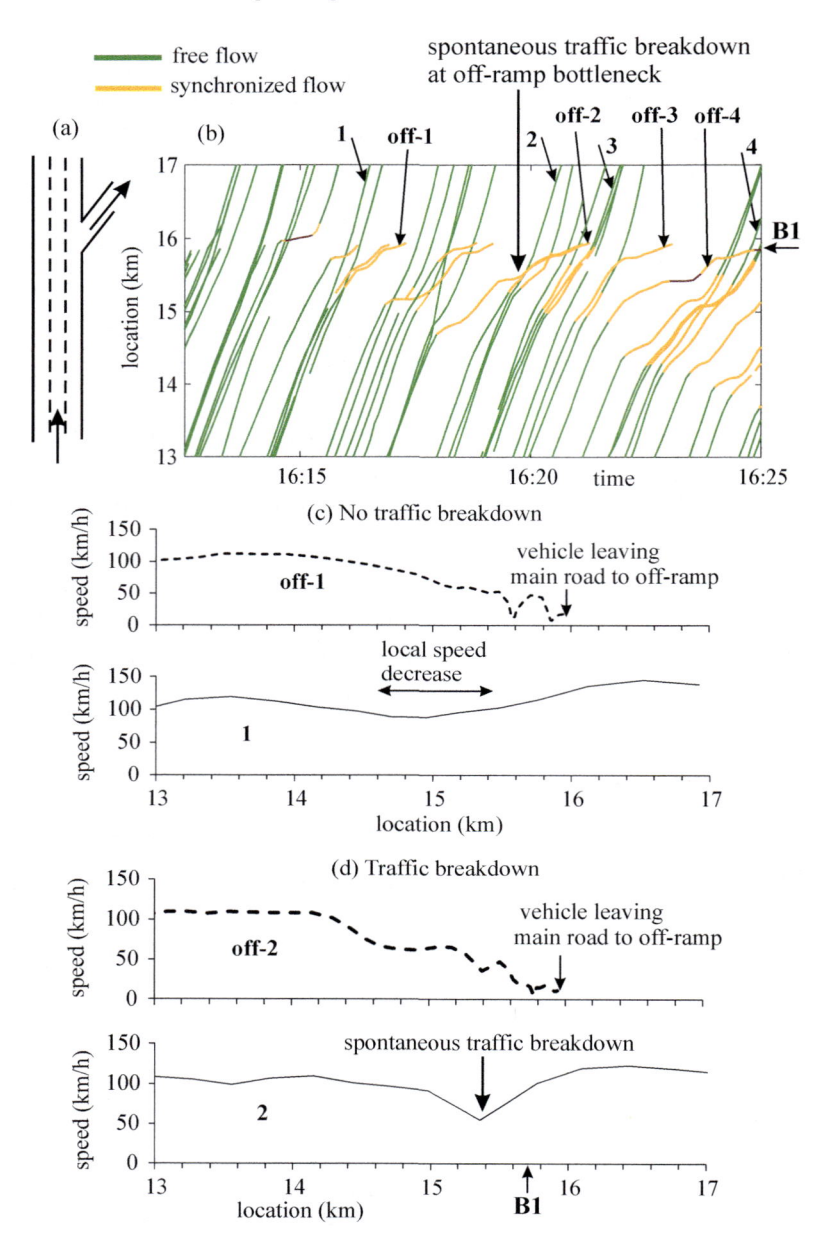

Fig. 5.5 Empirical spontaneous traffic breakdown at off-ramp bottleneck: **a** Schema of highway section taken from Fig. 4.3a. **b** Empirical trajectories of probe vehicles in the space–time plane (fragment of Fig. 2.2). **c** No traffic breakdown: Road-location functions of speeds on trajectories of vehicles "off-1" and 1 labeled as those in (**b**). **d** Traffic breakdown: Road-location functions of speeds on trajectories of vehicles "off-2" and 2 labeled as those in (**b**). Arrows B1 show the location of off-ramp bottleneck B1 (see caption to Fig. 2.1b)

5.4 Microscopic Spatiotemporal Features of Empirical Spontaneous ...

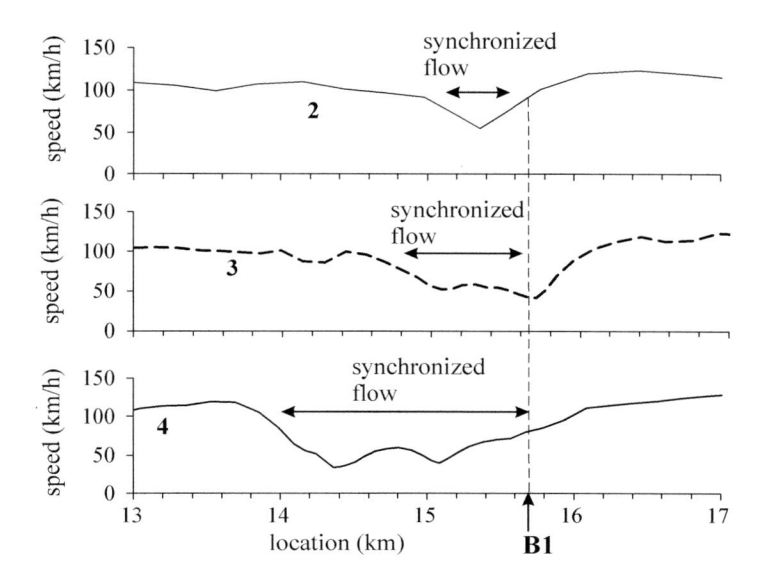

Fig. 5.6 Continuation of Fig. 5.5. Development of synchronized flow on the road after traffic breakdown has occurred: Road-location functions of speeds on trajectories of vehicles 2, 3, and 4 labeled in Fig. 5.5b by the same numbers; speed of vehicle 2 is taken from Fig. 5.5d. Arrow B1 and dashed vertical line show the location of off-ramp bottleneck (see caption to Fig. 2.1b)

(Figs. 5.5b and 5.6). The effect of the localization of the downstream front of traffic congestion remains for other probe vehicles that do not leave the main road to the off-ramp (Fig. 2.2). In other words, congested traffic resulting from traffic breakdown belongs to the synchronized flow traffic phase.

The origin of the localization of the downstream front of synchronized flow at the bottleneck (Figs. 2.2 and 5.6) is qualitatively the same as the origin of the localization of the local speed decrease at the bottleneck: As explained in Sect. 4.1, slow vehicles "off-c" and "off-d" in Fig. 4.3c, d as well as vehicle "off-1" in Fig. 5.5c that go to the off-ramp maintain the local speed decrease at the bottleneck before traffic breakdown has occurred. The localization of the downstream front of synchronized flow at the bottleneck can be considered the *self-maintaining* of synchronized flow at the bottleneck: After traffic breakdown has occurred, due to slow vehicles going to the off-ramp (for example, vehicles "off-3" and "off-4" marked in Fig. 5.5b),[15] synchronized flow is self-maintained at the bottleneck.

The self-maintaining of synchronized flow is also observed at an on-ramp bottleneck. In this case, slow vehicles merging from the on-ramp onto the main road maintain both the local speed decrease before traffic breakdown has occurred at the bottleneck and the localization of the downstream front of synchronized flow at the bottleneck after traffic breakdown has already occurred.

---

[15] Space dependencies of the speeds of vehicles "off-3" and "off-4" (not shown in Fig. 5.5) are qualitatively the same as that for vehicle "off-2" in Fig. 5.5d.

At the upstream front of synchronized flow, probe vehicles decelerate while approaching the bottleneck (see trajectories of vehicles 3 and 4 in Figs. 5.5b and 5.6). Contrary to the downstream front of the synchronized flow that is localized at the bottleneck, the upstream front of synchronized flow moves in the upstream direction. This means that the synchronized flow region is widening in the upstream direction from the bottleneck location.

There is a huge number of empirical probe vehicle data in which empirical spontaneous traffic breakdown (F→S transition) is observed at different kinds of highway bottlenecks on different days and highways. All empirical data show qualitatively the same features of empirical spontaneous traffic breakdown as those discussed in the empirical example presented in Figs. 5.5 and 5.6. With the use of Fig. 5.7, we can qualitatively summarized these features as follows[16]:

1. Before spontaneous traffic breakdown (F→S transition) occurs ($t < t_0$ in Fig. 5.7a), there is free flow both downstream and upstream of the bottleneck as well as at the bottleneck location (labeled by "free flow" in Fig. 5.7a).
2. Before traffic breakdown occurs ($t < t_0$), there is a local speed decrease in free flow at the bottleneck location (labeled by "local speed decrease" in Fig. 5.7b).
3. Traffic breakdown occurs at the location of the local speed decrease in free flow at the bottleneck.
4. Traffic breakdown is a sudden speed drop (call also as *speed breakdown*) at the location of the local speed decrease leading to the formation of synchronized flow at the bottleneck. During traffic breakdown, the upstream front of synchronized flow propagates upstream, whereas the downstream front of synchronized flow is localized at the bottleneck (labeled by "synchronized flow" in Fig. 5.7c, d).
5. After empirical spontaneous traffic breakdown has just occurred at the bottleneck, the flow rate in the emergent synchronized flow can remain on average the same as that in an initial free flow.[17]
6. Traffic breakdown (F→S transition) together with a return S→F transition at the bottleneck location are usually built a hysteresis loop in the flow–density plane (as well as in the speed–density and speed–flow planes).

---

[16] These empirical features of spontaneous traffic breakdown have been found in the standard methodologies for studies of empirical traffic breakdown (see, for example, papers [2–4, 9–16, 22, 29, 30, 39, 42–49, 72, 82–85] as well as reviews and books [1, 5, 8, 21, 28, 31, 34, 35, 41, 52, 53, 61, 73, 95]).

[17] In the particular empirical example shown in Fig. 5.2a–c, the flow rate increases slightly during the F→S transition (traffic breakdown). In many other empirical data, the flow rate can slightly decrease during the F→S transition. This behavior of the flow rate during the F→S transition (traffic breakdown) is well known from many empirical observations of spontaneous traffic breakdown [43, 44] (see a discussion about either a drop or a jump in the flow rate during traffic breakdown at highway bottlenecks in Sect. 2.2.2 of the book [57]).

5.4 Microscopic Spatiotemporal Features of Empirical Spontaneous … 61

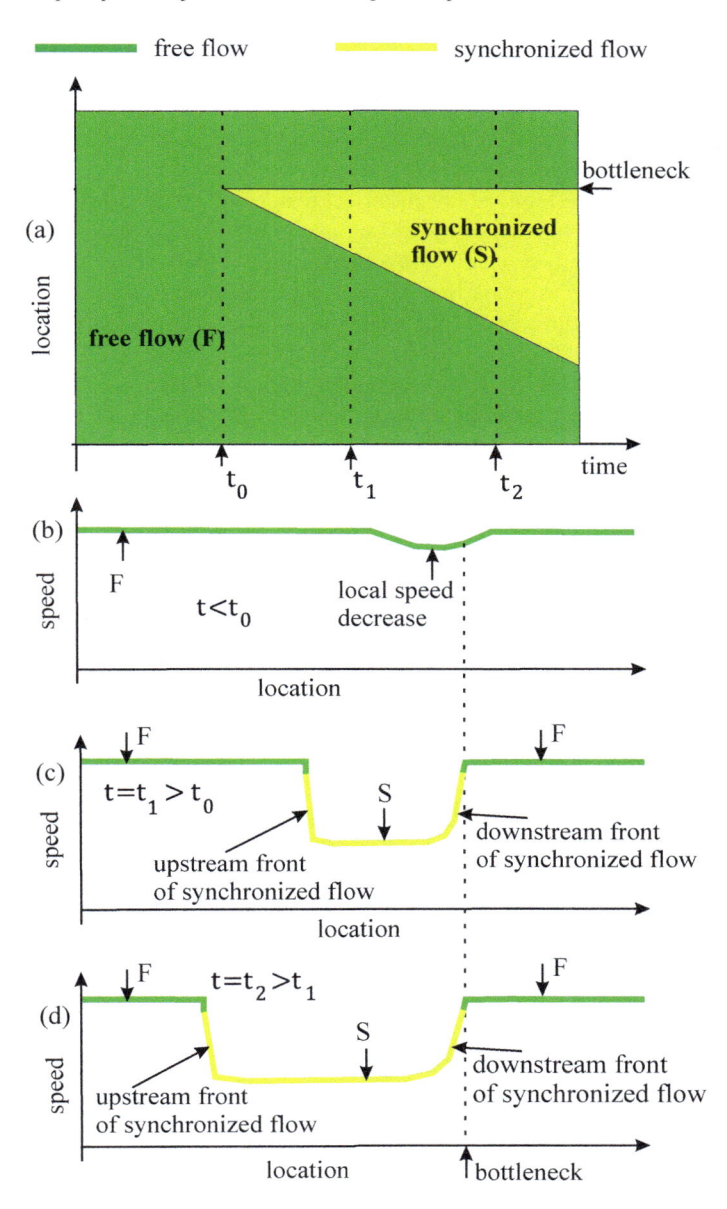

Fig. 5.7 Qualitative explanations of empirical spontaneous traffic breakdown (F→S transition) at bottleneck: **a** Free flow (green) and synchronized flow (yellow) in the time–road-location plane; at $t = t_0$ spontaneous traffic breakdown occurs at the bottleneck resulting in the occurrence of synchronized flow (yellow) propagating upstream. **b–d** Qualitative space dependencies of speed at $t < t_0$ in free flow, i.e., before traffic breakdown has occurred (**b**) as well as at two subsequent time instants $t_1 > t_0$ (**c**) and $t_2 > t_1$ (**d**) after traffic breakdown has occurred

5.5 Ignoring of Phenomenon "Empirical Induced Traffic Breakdown"—Emergence of a Diverse Variety of Traffic Theories

Traffic researchers have found many empirical features of spontaneous traffic breakdown.[18] Some of the features of empirical spontaneous traffic breakdown (F→S transition) found in the standard traffic and transportation research have been explained in Sects. 5.2–5.4 above. Another important empirical feature of traffic breakdown is the probabilistic behavior of empirical spontaneous traffic breakdown mentioned in Sect. 5.1.

These achievements of the empirical study of traffic breakdown made by traffic researchers have had a considerable impact on the development of traffic science. Nevertheless, the empirical nature of traffic breakdown could not be *definitely* understood from these empirical studies made in the standard traffic and transportation research.[19]

> Because the empirical nature of traffic breakdown (F→S transition) has not been understood in standard traffic and transportation science, a diverse variety of traffic and transportation theories have been introduced and developed.

The fundamentals of these qualitatively different standard traffic and transportation theories[20] were introduced in the 1950s–1960s.[21] There are many important contributions of the standard traffic theories for the understanding of the driver behaviors and real traffic phenomena that are also used in the three-phase traffic theory and in this book:

- Methodologies for the measurement of real traffic data (Chap. 3).
- The importance of a highway bottleneck for the occurrence of spontaneous traffic breakdown (F→S transition) (Chap. 4).

---

[18] See, e.g., [2–4, 9–16, 22, 29, 30, 39, 42–49, 72, 82–85] as well as reviews and books [1, 5, 8, 21, 28, 31, 34, 35, 41, 52, 53, 61, 73, 95].

[19] Because the empirical nature of traffic breakdown (F→S transition) could not been understood, there have been introduced many invalid terms in standard traffic and transportation science: The meaning of the terms contradicts basically the empirical nucleation nature of traffic breakdown. A prominent example of such invalid terms is "capacity drop". Unfortunately, these and many other invalid terms are currently well accepted in the state-of-the-art of the standard traffic and transportation research. A detailed explanation of this criticism can be found in Sect. 4.11 of the book [57].

[20] Reviews of the qualitatively different standard traffic and transportation theories can be found in [1, 5, 7, 8, 19, 21, 28, 31, 34, 35, 40, 41, 50, 61, 69, 73, 75, 76, 95, 98, 101].

[21] The generally accepted fundamentals of standard traffic theories were introduced, in particular, in the works by Lighthill and Whitham [67], Richards [94], Prigogine [88], Reuschel [91–93], Pipes [87], Kometani and Sasaki [62–65], Herman, Gazis, Rothery, Montroll, Chandler, and Potts [17, 36, 37, 51], Edie et al. [23–27], Moskowitz [74], Newell [77, 78].

5.5 Ignoring of Phenomenon "Empirical Induced Traffic ..."

- Traffic flow characteristics (for example, the hysteresis effect) resulting from empirical spontaneous traffic breakdown at the bottleneck (Sects. 5.2–5.4).
- The probabilistic behavior of traffic breakdown (Sect. 5.1).[22]
- The existence of a driver reaction time (Chap. 10).
- The existence of classical traffic flow instability that is responsible for the emergence of moving jams (stop-and-go-traffic) (Chap. 10).

Because the standard traffic theories have considerably contributed in the development of traffic science, the following question arises: Which of the standard traffic flow theories can be used for understanding real traffic?

> The answer to this question is that *none* of the standard traffic flow theories can be used for understanding real traffic.

This is because, as we will see in Chap. 6, traffic breakdown (F→S transition) at a highway bottleneck exhibits the empirical nucleation nature. However, *none* of the standard traffic flow theories can explain the empirical nucleation nature of traffic breakdown (F→S transition) at the bottleneck. In more details, the failure of standard traffic flow theories for understanding real traffic will be considered in Appendix C.

The failure of standard traffic flow theories for understanding real traffic might be explained by a very long time interval between the development of the classical traffic theories made in 1950s–1960s and the understanding of the empirical nucleation nature of traffic breakdown at highway bottlenecks made by the author at the end of 1990s.[23] During this long time interval, several generations of traffic researches developed a huge number of traffic flow models based on the standard approaches. Thus, in accordance with historical analysis of scientific revolutions in other fields of science made by Kuhn,[24] we may assume that it is very difficult for traffic researchers to see and accept that there is the empirical traffic phenomenon that contradicts basically all fundamentals of the standard traffic flow theories and methodologies.

---

[22] To describe the probabilistic behavior of spontaneous traffic breakdown at a bottleneck, a study of the probability of spontaneous traffic breakdown is required that is out of the scope of this book devoted for nonspecialists (see a detailed consideration of this subject in Chap. 5 of the book [57]).

[23] See explanations of the empirical traffic phenomenon "the empirical nucleation nature of traffic breakdown at a highway bottleneck" and the associated references in Chaps. 6 and 7.

[24] See Kuhn's book [66] as well as Sect. 12.1 for a more detailed consideration of the application of Kuhn's theory to transportation science.

5.6 Ignoring of Phenomenon "Empirical Induced Traffic Breakdown"—Consequences for Transportation Science

As emphasized in Sect. 5.5, none of the standard traffic flow theories can explain the empirical nucleation nature of traffic breakdown at a bottleneck. This critical conclusion allows us to draw the following consequences of the ignoring of the traffic phenomenon "empirical induced traffic breakdown (F→S transition) at the bottleneck"[25]:

- Understanding real traffic is not possible.
- Many applications of empirical fundamental diagram are invalid (Sect. C.4).
- The application of the empirical probabilistic behavior of spontaneous traffic breakdown at a highway bottleneck for the standard definition of stochastic highway capacity is invalid for real traffic (Sects. 7.3 and C.1).
- None of the standard traffic theories can explain the empirical nucleation nature of traffic breakdown (Appendix C).
- The evaluation of the effect of intelligent transportation systems (ITS) on traffic breakdown with the use of standard traffic models leads to invalid results (Sect. C.6).[26]
- The standard traffic theories failed by their ITS applications.[27]

> For understanding real traffic, a study of the traffic phenomenon "empirical induced traffic breakdown at a highway bottleneck" is needed.

---

[25] The ignoring of the traffic phenomenon "empirical induced traffic breakdown (F→S transition) at the bottleneck" is the state-of-the-art of the most works of the standard traffic and transportation research up to now (see, e.g., reviews and books [5, 28, 31, 53, 61, 95, 101]).

[26] Results of the evaluation of ITS applications with the standard traffic theories, models, and simulation tools can be found, e.g., in [1, 5, 7, 8, 20, 21, 28, 31, 34, 38, 40, 41, 50, 52–55, 61, 68, 79, 80, 86, 96, 100, 101]. Prove of the criticism of the standard traffic flow models used for the evaluation of ITS applications can be found in [58].

[27] It should be emphasized that the standard understanding of stochastic highway capacity or/and the standard traffic flow theories are the theoretical basis of most of the standard methods for traffic control, management, and organization of a traffic network as well as of many other ITS applications (see, e.g., reviews and books [1, 5–8, 18, 20, 21, 28, 31–34, 38, 40, 41, 50, 52–55, 61, 68, 70, 71, 73, 79–81, 86, 89, 90, 95–97, 99–101]). Criticisms of ITS applications based on the standard traffic theories can be found in [57, 58].

References

1. W.D. Ashton, *The Theory of Traffic Flow* (Methuen & Co. London; Wiley, New York, 1966)
2. J.H. Banks, Transp. Res. Rec. **1510**, 1–10 (1995)
3. J.H. Banks, Transp. Res. Rec. **1678**, 128–134 (1999)
4. J.H. Banks, Transp. Res. Rec. **2099**, 14–21 (2009)
5. J. Barceló (ed.), *Fundamentals of Traffic Simulation* (Springer, Berlin, 2010)
6. M.G.H. Bell, Y. Iida, *Transportation Network Analysis* (Wiley, Incorporated, Hoboken, NJ 07030-6000 USA, 1997)
7. N. Bellomo, V. Coscia, M. Delitala, Math. Mod. Meth. App. Sc. **12**, 1801–1843 (2002)
8. M. Brackstone, M. McDonald, Transp. Res. F **2**, 181 (1998)
9. W. Brilon, Transp. Res. Record **2483**, 57–65 (2015)
10. W. Brilon, J. Geistefeldt, M. Regler, in *Traffic and Transportation Theory*, ed. by H.S. Mahmassani (Elsevier Science, Amsterdam, 2005), pp. 125–144
11. W. Brilon, J. Geistefeld, H. Zurlinden, Transp. Res. Record **2027**, 1–8 (2007)
12. W. Brilon, M. Regler, J. Geistefeld, Straßenverkehrstechnik. Heft **3**, 136 (2005)
13. W. Brilon, H. Zurlinden, Straßenverkehrstechnik. Heft **4**, 164 (2004)
14. M.J. Cassidy, S. Ahn, Transp. Res. Rec. **1934**, 140–147 (2005)
15. M.J. Cassidy, R.L. Bertini, Transp. Res. B **33**, 25–42 (1999)
16. M.J. Cassidy, B. Coifman, Transp. Res. Rec. **1591**, 1–6 (1997)
17. R.E. Chandler, R. Herman, E.W. Montroll, Oper. Res. **6**, 165–184 (1958)
18. Y.-C. Chiu, J. Bottom, M. Mahut, A. Paz, R. Balakrishna, T. Waller, J. Hicks, Dynamic traffic assignment, a primer. Trans. Res. Circular E-C153 (2011), http://onlinepubs.trb.org/onlinepubs/circulars/ec153.pdf
19. D. Chowdhury, L. Santen, A. Schadschneider, Phys. Rep. **329**, 199 (2000)
20. C.F. Daganzo, *Fundamentals of Transportation and Traffic Operations* (Elsevier Science Inc., New York, 1997)
21. D.R. Drew, *Traffic Flow Theory and Control* (McGraw Hill, New York, NY, 1968)
22. S.M. Easa, A.D. May, Transp. Res. Rec. **772**, 24–37 (1980)
23. L.C. Edie, Oper. Res. **2**, 107–138 (1954)
24. L.C. Edie, Oper. Res. **9**, 66–77 (1961)
25. L.C. Edie, R.S. Foote, in *Highway Research Board Proceedings*, vol. 37 (HRB, National Research Council, Washington, D.C., 1958), pp. 334–344
26. L.C. Edie, R.S. Foote, in *Highway Research Board Proceedings*, vol. 39 (HRB, National Research Council, Washington, D.C., 1960), pp. 492–505
27. L.C. Edie, R. Herman, T.N. Lam, Transp. Sci. **14**, 55–76 (1980)
28. L. Elefteriadou, *An Introduction to Traffic Flow Theory* (Springer, Berlin, 2014)
29. L. Elefteriadou, A. Kondyli, W. Brilon, F.L. Hall, B. Persaud, S. Washburn, J. Transp. Eng. **140**, 04014003 (2014)
30. L. Elefteriadou, R.P. Roess, W.R. McShane, Transp. Res. Rec. **1484**, 80–89 (1995)
31. A. Ferrara, S. Sacone, S. Siri, *Freeway Traffic Modelling and Control* (Springer, Berlin, 2018)
32. T.L. Friesz, D. Bernstein, *Foundations of Network Optimization and Games, Complex Networks and Dynamic Systems*, vol. 3 (Springer, New York, Berlin, 2016)
33. N.H. Gartner, C.J. Messer, A.K. Rathi (eds.), *Special Report 165: Revised Monograph on Traffic Flow Theory* (Trans. Res. Board, Washington, DC, 1997)
34. N.H. Gartner, C.J. Messer, A.K. Rathi (eds.), *Traffic Flow Theory: A State-of-the-Art Report* (Transportation Research Board, Washington DC, 2001)
35. D.C. Gazis, *Traffic Theory* (Springer, Berlin, 2002)
36. D.C. Gazis, R. Herman, R.B. Potts, Oper. Res. **7**, 499–505 (1959)
37. D.C. Gazis, R. Herman, R.W. Rothery, Oper. Res. **9**, 545–567 (1961)
38. D.L. Gerlough, M.J. Huber, *Traffic Flow Theory Special Report 165* (Transp. Res. Board, Washington D.C., 1975)
39. B.D. Greenshields, J.R. Bibbins, W.S. Channing, H.H. Miller, in Highway Res. Board Proc. **14**, 448–477 (1935)

40. M. Guerrieri, R. Mauro, *A Concise Introduction to Traffic Engineering* (Springer, Berlin, 2021)
41. F.A. Haight, *Mathematical Theories of Traffic Flow* (Academic Press, New York, 1963)
42. F.L. Hall, Transp. Res. A **21**, 191–201 (1987)
43. F.L. Hall, in [34], pp. 2-1–2-36
44. F.L. Hall, K. Agyemang-Duah, Transp. Res. Rec. **1320**, 91–98 (1991)
45. F.L. Hall, W. Brilon, Transp. Res. Rec. **1457**, 35–42 (1994)
46. F.L. Hall, M.A. Gunter, Transp. Res. Rec. **1091**, 1–9 (1986)
47. F.L. Hall, V.F. Hurdle, J.H. Banks, Transp. Res. Rec. **1365**, 12–18 (1992)
48. F.L. Hall, B.N. Persaud, Transp. Res. Rec. **1232**, 9–16 (1989)
49. F.L. Hall, A. Pushkar, Y. Shi, Transp. Res. Rec. **1398**, 24–30 (1993)
50. D. Helbing, Rev. Mod. Phys. **73**, 1067–1141 (2001)
51. R. Herman, E.W. Montroll, R.B. Potts, R.W. Rothery, Oper. Res. **7**, 86–106 (1959)
52. *Highway Capacity Manual 2000*, (National Research Council, Transportation Research Board, Washington, DC, 2000)
53. *Highway Capacity Manual 2016*, 6th edn. (National Research Council, Transportation Research Board, Washington, DC, 2016)
54. A. Horni, K. Nagel, K.W. Axhausen (eds.), *The Multi-Agent Transport Simulation MATSim* (Ubiquity, London, 2016), http://matsim.org/the-book. https://doi.org/10.5334/baw
55. P. Kachroo, K.M.A. Özbay, *Feedback Control Theory for Dynamic Traffic Assignment* (Springer, Berlin, 2018)
56. B.S. Kerner, *The Physics of Traffic* (Springer, Berlin, Heidelberg, New York, 2004)
57. B.S. Kerner, *Breakdown in Traffic Networks* (Springer, Berlin, New York, 2017)
58. B.S. Kerner, in *Complex Dynamics of Traffic Management*, ed. by B.S. Kerner, Encyclopedia of Complexity and Systems Science Series (Springer, New York, NY, 2019), pp. 195–283
59. B.S. Kerner, in *Complex Dynamics of Traffic Management*, ed. by B.S. Kerner, Encyclopedia of Complexity and Systems Science Series (Springer, New York, NY, 2019), pp. 387–500
60. B.S. Kerner, H. Rehborn, R.-P. Schäfer, S.L. Klenov, J. Palmer, S. Lorkowski, N. Witte, Physica A **392**, 221–251 (2013)
61. F. Kessels, *Traffic Flow Modelling* (Springer, Berlin, 2019)
62. E. Kometani, T. Sasaki, J. Oper. Res. Soc. Jap. **2**, 11–26 (1958)
63. E. Kometani, T. Sasaki, Oper. Res. **7**, 704–720 (1959)
64. E. Kometani, T. Sasaki, Oper. Res. Soc. Jap. **3**, 176–190 (1961)
65. E. Kometani, T. Sasaki, in *Theory of Traffic Flow*, ed. by R. Herman (Elsevier, Amsterdam, 1961), pp. 105–119
66. T.S. Kuhn, *The Structure of Scientific Revolutions*, 4th edn. (The University of Chicago Press, Chicago, London, 2012)
67. M.J. Lighthill, G.B. Whitham, Proc. Roy. Soc. A **229**, 281–345 (1955)
68. W. Leutzbach, *Introduction to the Theory of Traffic Flow* (Springer, Berlin, 1988)
69. S. Maerivoet, B. De Moor, Phys. Rep. **419**, 1–64 (2005)
70. H.S. Mahmassani, Networks and Spatial Economics **1**, 267–292 (2001)
71. F.L. Mannering, W.P. Kilareski, *Principles of Highway Engineering and Traffic Analysis*, 2nd edn. (Wiley, New York, 1998)
72. A.D. May, Highway Res. Rec. **59**, 9–38 (1964)
73. A.D. May, *Traffic Flow Fundamentals* (Prentice-Hall Inc., New Jersey, 1990)
74. K. Moskowitz, Proc. Highw. Res. Board **33**, 385–395 (1954)
75. T. Nagatani, Rep. Prog. Phys. **65**, 1331–1386 (2002)
76. K. Nagel, P. Wagner, R. Woesler, Oper. Res. **51**, 681–716 (2003)
77. G.F. Newell, Oper. Res. **9**, 209–229 (1961)
78. G.F. Newell, in *Proceedings of the Second International Symposium on Traffic Road Traffic Flow* (OECD, London, 1963), pp. 73–83
79. G.F. Newell, *Applications of Queuing Theory* (Chapman Hall, London, 1982)
80. M. Papageorgiou, I. Papamichail, Transp. Res. Rec. **2047**, 28–36 (2008)
81. S. Peeta, A.K. Ziliaskopoulos, Netw. Spat. Econ. **1**, 233–265 (2001)

References

82. B.N. Persaud, F.L. Hall, Transp. Res. A **23**, 103–113 (1989)
83. B.N. Persaud, V.F. Hurdle, Transp. Res. Rec. **1194**, 191–198 (1988)
84. B.N. Persaud, S. Yagar, R. Brownlee, Transp. Res. Rec. **1634**, 64–69 (1998)
85. B.N. Persaud, S. Yagar, D. Tsui, H. Look, Transp. Res. Rec. **1748**, 110–115 (2001)
86. B. Piccoli, A. Tosin, in *Encyclopedia of Complexity and System Science*, ed. by R.A. Meyers (Springer, Berlin, 2009), pp. 9727–9749
87. L.A. Pipes, J. Appl. Phys. **24**, 274–287 (1953)
88. I. Prigogine, in *Theory of Traffic Flow*, ed. by R. Herman (Elsevier, Amsterdam, 1961), pp. 158–164
89. I. Prigogine, R. Herman, *Kinetic Theory of Vehicular Traffic* (American Elsevier, New York, 1971)
90. B. Ran, D. Boyce, *Modeling Dynamic Transportation Networks* (Springer, Berlin, 1996)
91. A. Reuschel, Österreichisches Ingenieur-Archiv **4**(3/4), 193–215 (1950)
92. A. Reuschel, Z. Österr, Ing.-Archit.-Ver. **95**, 59–62 (1950)
93. A. Reuschel, Z. Österr, Ing.-Archit.-Ver. **95**, 73–77 (1950)
94. P.I. Richards, Oper. Res. **4**, 42–51 (1956)
95. R.P. Roess, E.S. Prassas, *The Highway Capacity Manual: A Conceptual and Research History* (Springer, Berlin, 2014)
96. M. Saifuzzaman, Z. Zheng, Transp. Res. C **48**, 379–403 (2014)
97. T. Seo, A.M. Bayen, T. Kusakabe, Y. Asakura, Annu. Rev. Cont. **43**, 128–151 (2017)
98. A. Schadschneider, D. Chowdhury, K. Nishinari, *Stochastic Transport in Complex Systems* (Elsevier Science Inc., New York, 2011)
99. Y. Sheffi, *Urban Transportation Networks: Equilibrium Analysis with Mathematical Programming Methods* (Prentice-Hall, New Jersey, 1984)
100. TRANSIMS (U.S. Department of Transportation, Federal Highway Administration, Washington, DC, 2017), https://www.fhwa.dot.gov/planning/tmip/resources/transims/
101. M. Treiber, A. Kesting, *Traffic Flow Dynamics* (Springer, Berlin, 2013)
102. J. Treiterer, *Investigation of Traffic Dynamics by Aerial Photogrammetry Techniques*, Ohio State University Technical Report PB 246 094 (Columbus, Ohio, 1975)
103. J. Treiterer, J.A. Myers, in *Proceedings of the 6th International Symposium on Transportation and Traffic Theory*, ed. by D.J. Buckley (A.H. & AW Reed, London, 1974), pp. 13–38
104. J. Treiterer, J.I. Taylor, Highway Res. Rec. **142**, 1–12 (1966)

Chapter 6
Empirical Induced Traffic Breakdown—Nucleation Nature of Traffic Breakdown

In this chapter,[1] we explain that the crucial empirical traffic phenomenon that has led to understanding real traffic is the empirical induced traffic breakdown (induced F→S transition) at a bottleneck.

6.1 Features of Empirical Induced Traffic Breakdown at Bottleneck

6.1.1 Microscopic Characteristics of Empirical Induced Traffic Breakdown

To understand the empirical induced traffic breakdown defined in Sect. 2.3, we consider Fig. 2.2 in more details (Fig. 6.1). In the empirical example under consideration (Fig. 6.1b) there are wide moving traffic jams that have emerged within synchronized flow[2] upstream of off-ramp bottleneck B1 shown in Fig. 2.2.

We consider one of the wide moving jams (labeled by "wide moving jam (J)" in Fig. 6.1b). After the moving jam has emerged within synchronized flow upstream of bottleneck B1 (see Fig. 2.2 in which this wide moving jam has been labeled by "wide moving jam"), the wide moving jam propagates upstream through the highway section. Propagating through on-ramp bottleneck B3, the wide moving jam induces traffic breakdown at bottleneck B3 (Fig. 6.1b).

Indeed, before the jam has reached bottleneck B3, free flow has been at this bottleneck (we can see a free flow speed in the vicinity of the bottleneck location (about

---

[1] A consideration of the empirical nucleation nature of traffic breakdown presented in this chapter is based on results of papers and reviews [1–21].

[2] Explanations of the wide moving jam emergence will be done in Chap. 10.

© The Author(s), under exclusive license to Springer Nature Switzerland AG 2021
B. S. Kerner, *Understanding Real Traffic*,
https://doi.org/10.1007/978-3-030-79602-0_6

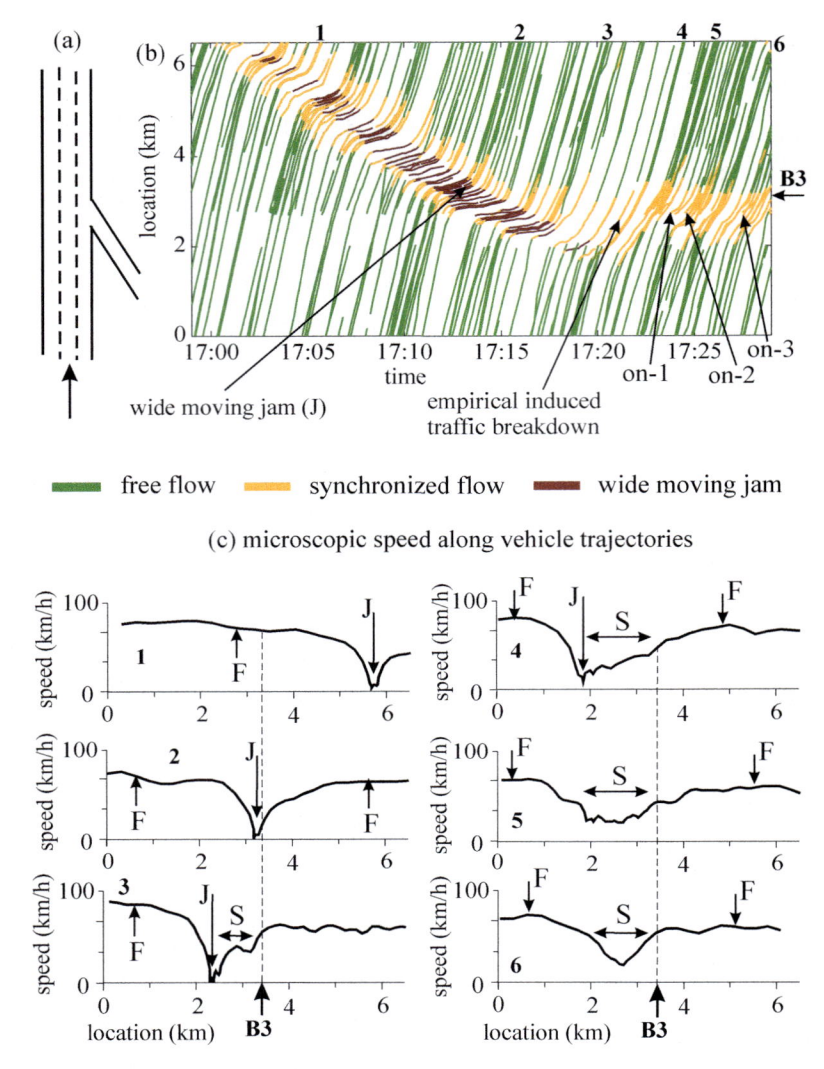

Fig. 6.1 Microscopic development of empirical induced traffic breakdown: **a** Fragment of simplified schema of highway section taken from Fig. 2.1a (bottleneck B3 is the same on-ramp bottleneck as bottleneck B3 shown in Fig. 2.1a). **b** Fragment of trajectories of probe vehicles in space and time taken from Fig. 2.2; some of the slow vehicles merging from the on-ramp onto the main road that maintain synchronized flow induced by the wide moving jam at on-ramp bottleneck B3 are labeled by "on-1", "on-2", and "on-3". **c** Microscopic speed along empirical vehicle trajectories as location functions for probe vehicles 1–6 whose numbers are, respectively, the same as those shown in the space–time plane in (**b**); dashed vertical lines show the location of bottleneck B3. F—free flow, S—synchronized flow, J—wide moving jam

6.1 Features of Empirical Induced Traffic Breakdown at Bottleneck 71

3.4 km) at time interval $t < 17{:}10$ in Fig. 6.1b as well as on vehicle trajectory 1 in Fig. 6.1c). From the space dependencies of the speed of probe vehicles 2 and 3 (Fig. 6.1b, c), we can see that when the wide moving jam reaches on-ramp bottleneck B3, the speed in the bottleneck vicinity decreases to values that are as low as zero. This strong speed decrease caused by the jam propagation through on-ramp bottleneck B3 causes the emergence of synchronized flow that is self-maintained after the jam is upstream from bottleneck B3 (probe vehicles 3 and 4 in Fig. 6.1b, c): The synchronized flow is induced by the jam propagation at the bottleneck. Indeed, synchronized flow remains at on-ramp bottleneck B3 during a long enough interval after the wide moving jam does not influence on the speed at on-ramp bottleneck B3 any more (trajectories of vehicles 5 and 6 in Fig. 6.1b, c).

To see the effect of the empirical induced synchronized flow at on-ramp bottleneck B3 (arrows labeled by "S" on trajectories of vehicles 5 and 6 in Fig. 6.1c) in more details, we consider several different time intervals of the propagation of the wide moving jam. The location of the wide moving jam on trajectory 1 is still downstream of on-ramp bottleneck B3. As already mentioned, before the wide moving jam has reached bottleneck B3 there has been free flow at bottleneck B3 (free flow labeled by "F" on trajectory 1 in Fig. 6.1c). Later, the wide moving jam reaches bottleneck B3 (Fig. 6.1b, c). During the time interval of the jam duration (about 2 min as it can be seen on trajectory 2 in Fig. 6.1b), there is congested traffic at bottleneck B3. It is traffic congestion within the wide moving jam: This congested traffic results from the propagation of the wide moving jam to the location of bottleneck B3.

Due to the subsequent upstream propagation of the wide moving jam, the jam has passed bottleneck B3 (time interval $t > 17{:}18$ in Fig. 6.1b). At $t > 17{:}18$, after the wide moving jam has passed bottleneck B3, there is no downstream congestion that could affect bottleneck B3 (Fig. 6.1b). Nevertheless, free flow does not return at bottleneck B3. Instead, synchronized flow persists at bottleneck B3 (arrows labeled by "S" on trajectories of vehicles 5 and 6 in Fig. 6.1c). Thus, this synchronized flow persisting at bottleneck B3 has been indeed induced due to the jam propagation through bottleneck B3: This induced traffic breakdown is empirical induced F→S transition (labeled by "empirical induced traffic breakdown" in Fig. 6.1b).

In the example of the empirical induced traffic breakdown at on-ramp bottleneck B3 (Fig. 6.1b), a synchronized flow pattern (SP) emerges at bottleneck B3. An SP is a congested traffic pattern in which congested traffic consists of the synchronized flow phase only. It should be noted that synchronized flow within the SP induced at on-ramp bottleneck B3 propagates only about 1.3–1.7 km upstream of bottleneck B3. In other words, in the case under consideration the SP is a localized synchronized flow pattern (LSP) at bottleneck B3: As for many other SPs at a bottleneck, the downstream front of the LSP is fixed at bottleneck B3. However, the upstream front of the LSP that separates synchronized flow and free flow upstream of the LSP does not propagates continuously upstream.

The reason for the LSP formation is as follows: On the highway section under consideration, there is an off-ramp at highway locations 1.53–2 km (this off-ramp that is about 1.5 km upstream of on-ramp bottleneck B3 is not shown in Fig. 6.1). Due to vehicles leaving the main road to this off-ramp (probe vehicles leaving the main road to the off-ramp at location about 2 km can be clearly seen in Fig. 6.1b), the flow-rate upstream on-ramp bottleneck B3 reduces. This flow-rate decrease prevents upstream propagation of the synchronized flow induced at bottleneck B3.

> The importance of the evidence of the empirical induced traffic breakdown (induced F→S transition) at a bottleneck is as follows: This empirical evidence leads to the conclusion about the *empirical nucleation nature of traffic breakdown*. The empirical nucleation nature of traffic breakdown changes basically understanding real traffic.

> The empirical nucleation nature of traffic breakdown at a bottleneck is as follows. A small enough local speed decrease in free flow at the bottleneck does not lead to traffic breakdown. However, a large enough short-time local speed decrease in free flow at the bottleneck does lead to traffic breakdown (F→S transition) at the bottleneck.

> A large enough short-time speed decrease in free flow at a bottleneck that causes traffic breakdown at the bottleneck can be considered a *nucleus* for traffic breakdown (nucleus for F→S transition) at the bottleneck.

In Fig. 6.1, the large enough short-time speed decrease at bottleneck B3 initiating empirical induced traffic breakdown (F→S transition) is caused by the wide moving jam propagating through bottleneck B3. In other words, in this particular empirical example, the wide moving jam can be considered a nucleus for traffic breakdown (F→S transition) at bottleneck B3 (Fig. 6.1b). From Figs. 2.2 and 6.1 we can see that to observe an empirical induced traffic breakdown (induced F→S transition) at a bottleneck, simultaneous traffic measurements on a long enough road section containing several neighborhood bottlenecks are usually required.

> The nucleation nature of traffic breakdown at a bottleneck is basically a spatiotemporal traffic phenomenon: To understand the empirical nucleation nature of traffic breakdown at the bottleneck, empirical traffic data should simultaneously be measured in space and time.

6.1 Features of Empirical Induced Traffic Breakdown at Bottleneck 73

6.1.2 Macroscopic Characteristics of Empirical Induced Traffic Breakdown

To support conclusions made from the above consideration of Fig. 6.1, we consider another empirical example of induced traffic breakdown shown in Fig. 6.2a that has already been briefly discussed in Sect. 2.3. After the wide moving jam has passed on-ramp bottleneck B2, there is no downstream congestion that could affect bottleneck B2 (Fig. 6.2a). Moreover, when the wide moving jam is far away upstream of bottleneck B2, the jam propagates through free flow as it can be seen from Fig. 6.2a, c. Nevertheless, as in the empirical example shown in Fig. 6.1, free flow does not return at on-ramp bottleneck B2 in Fig. 6.2a. Instead, synchronized flow persists at bottleneck B2. Thus, this synchronized flow persisting at bottleneck B2 (Fig. 6.2a, b) has been indeed induced due to the jam propagation through bottleneck B2. In other words, the wide moving jam can be considered a nucleus for traffic breakdown (F→S transition) at bottleneck B2 (Fig. 6.2a).

Synchronized flow induced at on-ramp bottleneck B2 in Fig. 6.2a propagates only about 3–4 km upstream of bottleneck B2. In other words, an LSP occurs at bottleneck B2: The upstream front of the LSP that separates synchronized flow and free flow upstream of the LSP does not propagates continuously upstream. The reason for the LSP formation is probably the same one as that in the example shown in Fig. 6.1: On the highway section, there is an off-ramp[3] upstream of on-ramp bottleneck B2. Due to vehicles leaving the main road to this off-ramp, the flow-rate upstream on-ramp bottleneck B2 reduces. This flow-rate decrease prevents continuous upstream propagation of the synchronized flow induced at bottleneck B2.

6.1.3 Common Empirical Features of Synchronized Flow Resulting from Spontaneous and Induced Traffic Breakdowns

As mentioned in Sect. 3.2, a road detector measures both the speed of all vehicles passing the detector and the flow rate. With the use of road detector data it has been found that synchronized flows resulting from empirical spontaneous and empirical induced traffic breakdowns exhibit qualitatively the same empirical features (Fig. 6.2). In particular, after empirical induced traffic breakdown has occurred at a bottleneck, the flow rate in synchronized flow can remain on average the same as that in an initial free flow (Fig. 6.2 (b, right)). This conclusion is independent of whether the empirical induced traffic breakdown is the cause of synchronized flow at the bottleneck (Fig. 6.2a), or synchronized flow at the bottleneck has occurred due to empirical spontaneous traffic breakdown (Figs. 5.1b and (c, right)). Contrary to

---

[3] The off-ramp is not shown in Fig. 6.2a. Readers can see this off-ramp within intersection I2 on Fig. 2.1 of the book [13].

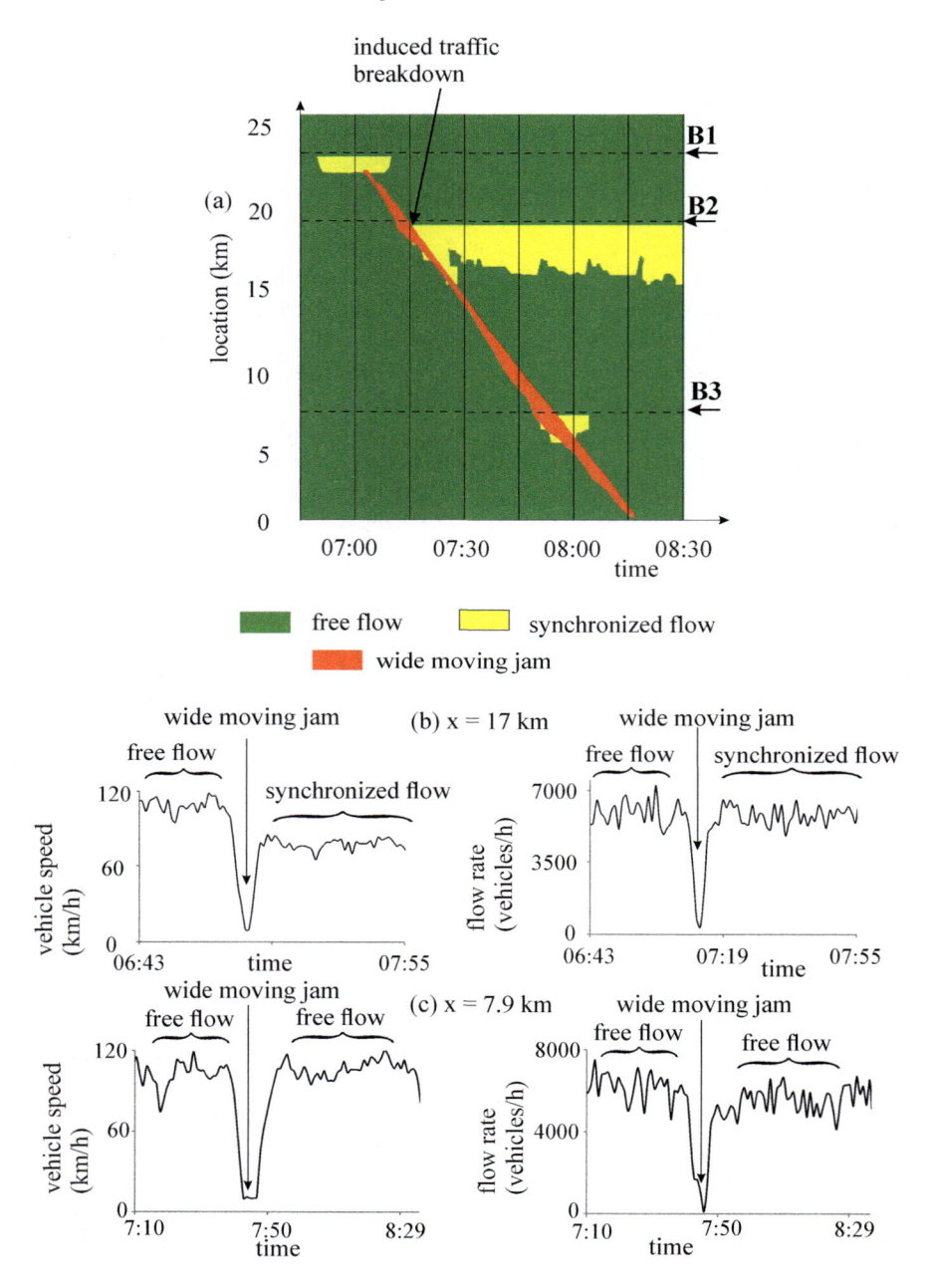

Fig. 6.2 Empirical induced traffic breakdown in macroscopic traffic data: **a** Empirical speed data in space and time taken from Fig. 2.3b. **b**, **c** Empirical average speed (left column) and the total flow rate averaged across the highway (right column) at two different road locations: **b** $x = 17$ km is related to the location of on-ramp bottleneck B2, **c** $x = 7.9$ km is related to a road location that is far away upstream of bottleneck B2. 1-min average traffic data measured by road detectors

6.1 Features of Empirical Induced Traffic Breakdown at Bottleneck

synchronized flow, the flow rate within the wide moving jam is usually as low as zero (Fig. 6.2 (b, right)).

> Independent of the cause of synchronized flow, the flow rate within the synchronized flow can be as large as in free flow (Figs. 5.1 (c, right) and 6.2 (b, right)).

> Spatiotemporal features of synchronized flow at a bottleneck do not depend on whether this synchronized flow has emerged due to the empirical spontaneous traffic breakdown or due to the empirical induced traffic breakdown at the bottleneck.

As explained in Sect. 5.4, the origin of the localization of the local speed decrease at a bottleneck before traffic breakdown has occurred is the same as the origin of the localization of the downstream front of synchronized flow at the bottleneck after traffic breakdown has occurred. This is valid for both the empirical spontaneous traffic breakdown and the empirical induced traffic breakdown at the bottleneck.

At an on-ramp bottleneck, slow vehicles merging from the on-ramp onto the main road maintain the localization of the downstream front of synchronized flow at the bottleneck. In other words, the self-maintaining of synchronized flow resulting from induced traffic breakdown at the on-ramp bottleneck is realized due to slow vehicles merging from the on-ramp onto the main road (some of the slow merging vehicles are labeled by "on-1", "on-2", and "on-3" in Fig. 6.1b).

Studies of probe vehicle data show that the empirical induced traffic breakdown is also observed at off-ramp bottlenecks. As in the case of the empirical spontaneous traffic breakdown studied in Sect. 5.4, after the induced traffic breakdown has already occurred at an off-ramp bottleneck, the self-maintaining of synchronized flow at the bottleneck is realized due to slow vehicles going to the off-ramp.

> The cause of the empirical induced traffic breakdown is as follows: After a wide moving jam that has induced synchronized flow at the bottleneck does not maintain this synchronized flow any more, the synchronized flow is self-maintained at the bottleneck.

6.2 Explanation of Nucleation Nature of Traffic Breakdown at Bottleneck

The observation of the empirical induced traffic breakdown (F→S transition) at a bottleneck (Sect. 6.1) proves that traffic breakdown at the bottleneck exhibits the empirical nucleation nature. A consequence of this empirical result is as follows.

> The empirical nucleation nature of traffic breakdown (F→S transition) at a bottleneck contradicts basically all standard traffic and transportation theories.

6.2.1 Nucleation of Traffic Breakdown and Metastability of Free Flow with Respect to F→S Transition at Bottleneck

Results of the study of real field traffic data made Sect. 6.1 lead to the conclusion that there is a range of the flow rate in free flow within which there can be either free flow or synchronized flow at the bottleneck. This empirical result is qualitatively presented in Figs. 6.3 and 6.4.[4]

In qualitative Figs. 6.3 and 6.4, we take into account the common empirical result (Figs. 5.1c, left and 5.2b, c) that the average speed in synchronized flow is considerably lower than the average speed in free flow (Figs. 6.3 and 6.4a, d). Therefore, at a given flow rate (see formula (5.1)) the vehicle density in synchronized flow is larger than in free flow (Fig. 6.3b). For a chosen flow rate, free flow is labeled by green circle F and synchronized flow by yellow circle S (Fig. 6.3a, b).[5]

The empirical nucleation nature of traffic breakdown means that there is a flow-rate range within which free flow is *metastable with respect to an F→S transition* at the bottleneck. The origin of this metastability of free flow at the bottleneck is the

---

[4] For a qualitative explanation of the nucleation nature of traffic breakdown made with the use of Figs. 6.3 and 6.4 we still do not need detailed knowledge of features of synchronized flow that will be explained in Chap. 8 and Appendix A.

[5] It must be emphasized that arrows in Fig. 6.3 can be considered only as illustrations of traffic breakdown in the speed–flow plane and the fundamental diagram. This is because traffic is a process that occurs in space and time. Basic features of the induced traffic breakdown at the bottleneck that is a spatiotemporal phenomenon cannot be understood, if empirical traffic data are measured at one road location only as made in the fundamental diagram. Indeed, many spatiotemporal features of traffic breakdown are lost in the fundamental diagram as well as in the data presented in the speed–flow plane (Sect. 5.2). Therefore, the induced traffic breakdown and its features should be studied with the use of *spatiotemporal* traffic data (Figs. 6.1 and 6.2). Only after features of the induced traffic breakdown have been understood, the illustration of this traffic breakdown can be made on the fundamental diagram and the speed–flow plane (arrows in Fig. 6.3).

6.2 Explanation of Nucleation Nature of Traffic Breakdown at Bottleneck

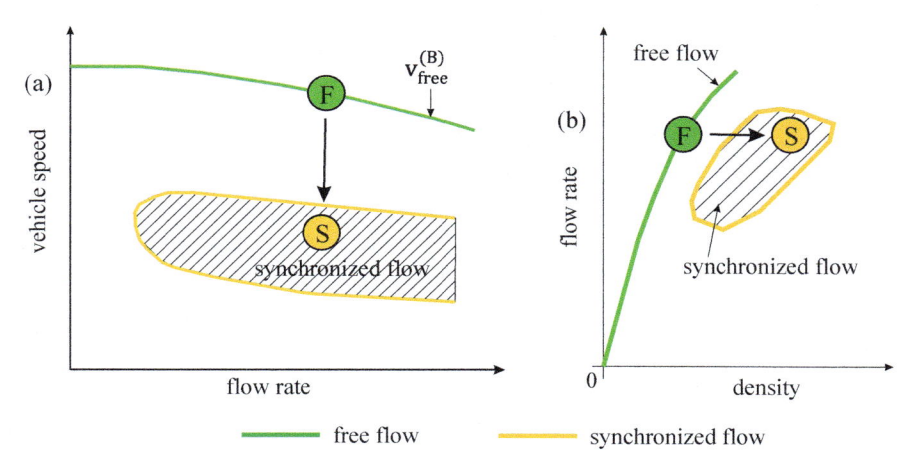

Fig. 6.3 Qualitative explanation of the empirical nucleation nature of traffic breakdown at a bottleneck with the three-phase traffic theory: F→S transition labeled by arrows from a free flow state (labeled by circle F) to a synchronized flow state (labeled by circle S) in the speed–flow (**a**) and flow–density (**b**) planes. Curves for free flow labeled by $v_{\text{free}}^{(B)}$ in (**a**) and by "free flow" in (**b**) as well as two-dimensional (2D) states of synchronized flow labeled by "synchronized flow" are qualitatively related to empirical states of free flow and synchronized flow shown in Fig. 5.2a, c for the flow–density and speed–flow planes, respectively (see explanations of 2D-states of synchronized flow in Appendix A). In (**a**), a qualitative flow-rate dependence of the minimum speed $v_{\text{free}}^{(B)}$ within the average local speed decrease in free flow at the bottleneck is taken from Fig. 5.3b

empirical nucleation nature of traffic breakdown. The metastability of free flow with respect to an F→S transition at the bottleneck can be defined as follows:

- There is a flow-rate range within which either states of free flow or states of synchronized flow can exist at the bottleneck (Fig. 6.3). Within this range of the flow rate, free flow is metastable with respect to an F→S transition at the bottleneck. This means that a small enough time-limited local speed decrease in free flow at the bottleneck does not lead to the F→S transition: Free flow persists at the bottleneck. However, a large enough time-limited local speed decrease in free flow at the bottleneck is a nucleus for traffic breakdown. The occurrence of the nucleus does lead to traffic breakdown (F→S transition) at the bottleneck (symbolically this transition is shown by arrows in Fig. 6.3 from a state of free flow F to a state of synchronized flow S).

> The empirical nucleation nature of the F→S transition at the bottleneck means that traffic breakdown does occur when free flow is in a metastable state with respect to the F→S transition at the bottleneck *and* a nucleus for traffic breakdown appears at the bottleneck.

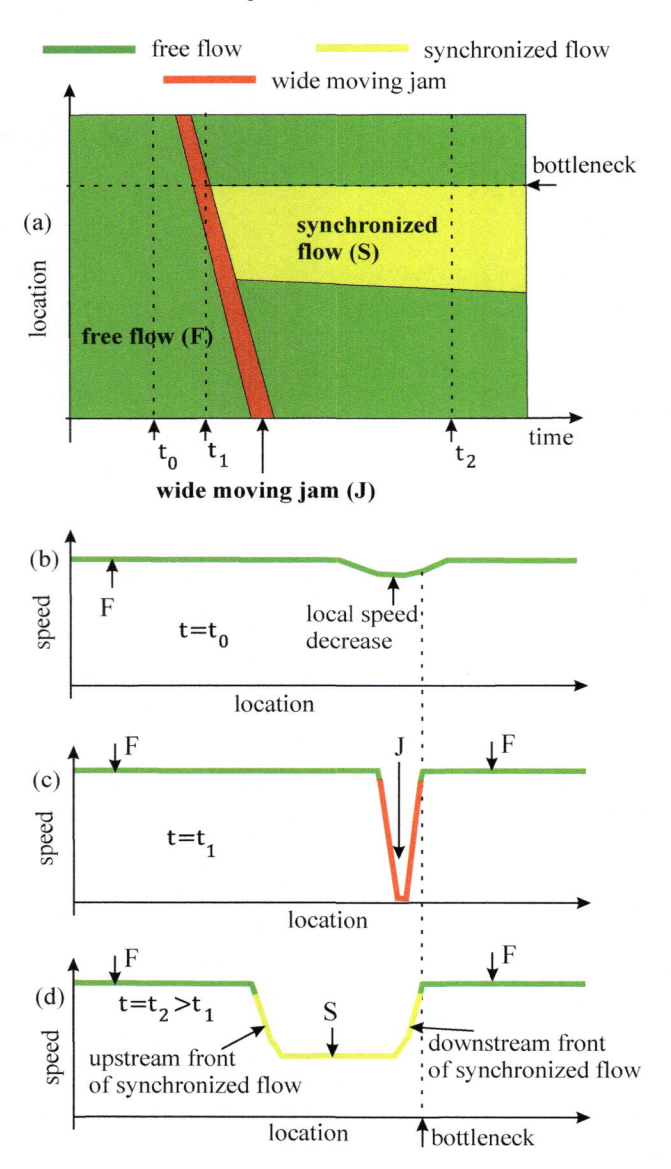

Fig. 6.4 Qualitative explanations of empirical induced traffic breakdown at bottleneck: **a** Free flow (green), synchronized flow (yellow), and wide moving jam (red) in the road location–time plane; at time instant $t = t_1$ the wide moving jam induces traffic breakdown at the bottleneck resulting in the occurrence of synchronized flow (yellow) propagating upstream. **b**–**d** Qualitative space dependencies of speed at time instant $t = t_0$ (**b**) at which the wide moving jam is downstream of the bottleneck, at time instant $t = t_1$ at which the wide moving jam reaches the bottleneck, and at time instant $t_2 > t_1$ (**d**) at which the wide moving jam is far way upstream of the bottleneck

6.2.2 Effect of Empirical Nucleation Nature of Traffic Breakdown at Bottleneck on Definition of Synchronized Flow

The evidence of empirical induced F→S transition (induced traffic breakdown) at a bottleneck (Sect. 6.1) allows us to define the synchronized flow traffic phase (S) of congested traffic as follows:

- The synchronized flow traffic phase ensures the nucleation nature of the F→S transition at a highway bottleneck. The downstream front of the resulting synchronized flow can be localized at the bottleneck.[6]

> The basic feature of the synchronized flow traffic phase formulated in the three-phase traffic theory leads to the nucleation nature of the F→S transition. In this sense, the conception of the synchronized flow traffic phase, which ensures the nucleation nature of the F→S transition at a highway bottleneck, and the three-phase traffic theory can be considered synonymous.

Driver behavioral assumptions that explain the basic feature of synchronized flow will be considered in Chap. 8.

6.3 Empirical Induced Traffic Breakdown at Bottleneck: A Summary

From the analysis of empirical induced traffic breakdown at a bottleneck made by the consideration of two different empirical examples of congested traffic patterns shown in Figs. 6.1b and 6.2a, we can make the conclusion that a wide moving jam can cause traffic breakdown in metastable free flow with respect to an F→S transition at the bottleneck when the jam propagates through the bottleneck location. However, this traffic breakdown can be considered as an empirical induced traffic breakdown at the bottleneck only in the case, when synchronized flow is further self-maintained at the bottleneck after the wide moving jam is far away upstream of the bottleneck and, therefore, traffic congestion within the jam does not affect the bottleneck any more. Thus, traffic breakdown at a bottleneck caused by the propagation of the wide moving jam (Figs. 6.1b and 6.2a) to the bottleneck location can be considered *empirical induced traffic breakdown (F→S transition)* only when conditions are satisfied:

---

[6] Recall that in contrast with this basic feature of the synchronized flow traffic phase, the wide moving jam traffic phase *cannot* be localized at the bottleneck (Sect. 2.1).

- Before the wide moving jam reaches the bottleneck, free flow should be downstream of the bottleneck as well as at the bottleneck.
- Free flow at the bottleneck should be in a metastable state with respect to an F→S transition at the bottleneck.
- When the wide moving jam reaches the bottleneck, the jam affects the bottleneck during a finite time interval only: Due to the subsequent propagation of the wide moving jam upstream from the bottleneck, the wide moving jam does not maintain congested traffic at the bottleneck any more.
- After this finite time interval, free flow should be realized downstream from the bottleneck *and* synchronized flow initially induced by the jam is further self-maintained at the bottleneck.

Under these conditions, the wide moving jam propagating through the bottleneck acts as a nucleus for traffic breakdown (F→S transition) at the bottleneck. These conditions are realized for both empirical examples of empirical induced traffic breakdown shown in Figs. 6.1b and 6.2a.

6.4 Empirical Proof of Nucleation Nature of Traffic Breakdown Using Opposite Assumption

To prove the empirical nucleation nature of traffic breakdown, we make an opposite assumption that traffic breakdown at bottleneck B3 (Fig. 6.1b) *does not exhibit* the empirical nucleation nature and, therefore, there is also *no* free flow metastability with respect to the F→S transition. In this case, at $t > 17:18$ when the wide moving jam shown in Fig. 6.1b does not maintain congested traffic at bottleneck B3 any more, free flow should return at this bottleneck. Indeed, at $t > 17:18$ there is no downstream congestion that could affect on bottleneck B3 (trajectories 4, 5, and 6 in Fig. 6.1c). However, rather than free flow recovers at bottleneck B3, synchronized flow does persist at bottleneck B3 during a long time interval $17:18 < t < 17:35$. This self-maintaining of synchronized flow at bottleneck B3 is realized although the wide moving jam does not maintain synchronized flow at bottleneck B3 any more *and* there is no traffic congestion downstream of bottleneck B3 (Figs. 2.2 and 6.1b). Thus, the above-mentioned assumption that the empirical traffic breakdown at the bottleneck does not exhibit the empirical nucleation nature is *incorrect*.

The same conclusion can be made from a consideration of empirical induced traffic breakdown shown in Fig. 6.2a. The advantage of this empirical example is as follows. The empirical induced traffic breakdown at on-ramp bottleneck B3 in Fig. 6.1b is only one of many traffic phenomena that can be found in Fig. 2.2. Therefore, the features of the empirical induced traffic breakdown discussed above might be not easy to understand. In contrast with Fig. 2.2, in Fig. 6.2a a congested pattern is much simpler one: While propagating through on-ramp bottleneck B2, a single wide moving jam induces traffic breakdown at bottleneck B2 (labeled by "induced traffic breakdown" in Fig. 6.2). Due to the induced traffic breakdown, synchronized flow emerges at bottleneck B2. The synchronized flow is self-maintained during a long time interval at bottleneck B2.

We make an opposite assumption that the empirical traffic breakdown at bottleneck B2 (Fig. 6.2) *does not exhibit* the empirical nucleation nature. In this case, when the wide moving jam shown in Fig. 6.2 is far away upstream of bottleneck B2, the jam does not maintain traffic congestion at bottleneck B2 any more (see location $x = 7.9$ km in Fig. 6.2a, c at which the jam propagates within free flow that is far away upstream of bottleneck B2). Therefore, under assumption that the empirical traffic breakdown at bottleneck B2 does not exhibit the nucleation nature, free flow should return at this bottleneck. Indeed, within a long time interval about $07\!:\!15 < t < 08\!:\!30$ there is no downstream congestion that could affect on bottleneck B2 (Fig. 6.2a). However, rather than free flow recovers at bottleneck B2, synchronized flow is self-maintained at bottleneck B2 during the long time interval $07\!:\!15 < t < 08\!:\!15$, when the wide moving jam does not maintain traffic congestion at bottleneck B2 (Fig. 6.2a). Thus, as in the case of the empirical induced traffic breakdown shown in Fig. 6.1b, the above-mentioned assumption that the empirical traffic breakdown at bottleneck B2 (labeled by "induced traffic breakdown" in Fig. 6.2a) does not exhibit the empirical nucleation nature is *incorrect*.

The empirical induced F→S transition at a bottleneck (at bottleneck B3 in Fig. 6.1b or at bottleneck B2 in Fig. 6.2a) is only possible when within some flow-rate range there can exist either the free flow traffic phase (F) or the synchronized flow traffic phase (S). Qualitatively, these two different traffic phases existing at the same flow rate at the bottleneck have been illustrated in Figs. 6.3 and 6.4.

As defined above, a time-limited local speed decrease at a bottleneck resulting in traffic breakdown in metastable free flow at the bottleneck is a nucleus for traffic breakdown (F→S transition). It is useful to introduce the term a *critical speed* and the associated term *critical nucleus* required for traffic breakdown at a bottleneck.

> The critical nucleus is the "smallest" nucleus required for the occurrence of traffic breakdown (F→S transition) at the bottleneck. The speed within the critical nucleus is called the critical speed: Any time-limited local speed decrease in free flow at the bottleneck within which the speed is *less* than the critical speed is a nucleus for traffic breakdown in metastable free flow. Contrarily, when the speed within a time-limited local speed decrease at the bottleneck is *larger* than the critical speed, then *no* traffic breakdown occurs in the metastable free flow, i.e., the local speed decrease is not a nucleus for traffic breakdown.

Empirical studies of traffic breakdown show that the critical speed is usually not lower than the speed of synchronized flow resulting from the F→S transition at the bottleneck.[7] The average speed within a wide moving jam that is less than the

---

[7] For example, this conclusion is consistent with empirical studies of nuclei for traffic breakdown presented in Sects. 9.1 and 9.2 as well as with the empirical traffic breakdown induced by a moving synchronized flow pattern (Fig. 11.2 of Sect. 11.1.1).

speed in synchronized flow is usually as low as zero: The wide moving jam can be considered the "largest" nucleus for traffic breakdown in free flow at the bottleneck. Therefore, the average speed within the jam is considerably lower than the critical speed within the critical nucleus: While propagating through metastable free flow with respect to an F→S transition at the bottleneck, the wide moving jam does act as a nucleus for traffic breakdown inducing the F→S transition at the bottleneck.

> While propagating through metastable free flow with respect to an F→S transition at the bottleneck, the wide moving jam does act as a nucleus for traffic breakdown at the bottleneck. After the wide moving jam that has induced synchronized flow at the bottleneck does not maintain this synchronized flow any more, the synchronized flow is self-maintained at the bottleneck.

It should be emphasized that the wide moving jams in Figs. 6.1b and 6.2a are only some empirical examples of congested traffic patterns whose propagation to the bottleneck location induce traffic breakdown (F→S transition) at the bottleneck. In other words, the empirical induced traffic breakdown is a *general traffic phenomenon* observed when a congested traffic pattern reaches a highway bottleneck at which free flow is in a metastable state with respect to an F→S transition (see other examples of empirical induced traffic breakdown at highway bottlenecks in Sects. 7.1 and 11.1.1).

> The speed within a congested pattern is less than the critical speed for an F→S transition in metastable free flow at a bottleneck: While reaching the bottleneck, any congested pattern induces traffic breakdown at the bottleneck.

> The empirical induced traffic breakdown at a bottleneck is the definitive proof of the empirical nucleation nature of traffic breakdown at the bottleneck.

> The basic assumption of the three-phase traffic theory is that empirical traffic breakdown (F→S transition) at a bottleneck exhibits the nucleation nature.

References

1. B.S. Kerner, Phys. Rev. Lett. **81**, 3797–3800 (1998)
2. B.S. Kerner, in *Proceedings of the 3rd Symposium on Highway Capacity and Level of Service*, ed. by R. Rysgaard (Road Directorate, Copenhagen, Ministry of Transport – Denmark, 1998), pp. 621–642
3. B.S. Kerner, Transp. Res. Rec. **1678**, 160–167 (1999)
4. B.S. Kerner, in *Transportation and Traffic Theory*, ed. by A. Ceder (Elsevier Science, Amsterdam, 1999), pp. 147–171
5. B.S. Kerner, Phys. World **12**, 25–30 (August 1999)
6. B.S. Kerner, J. Phys. A: Math. Gen. **33**, L221–L228 (2000)
7. B.S. Kerner, Transp. Res. Rec. **1710**, 136–144 (2000)
8. B.S. Kerner, Netw. Spat. Econ. **1**, 35–76 (2001)
9. B.S. Kerner, Transp. Res. Rec. **1802**, 145–154 (2002)
10. B.S. Kerner, Math. Comp. Mod. **35**, 481–508 (2002)
11. B.S. Kerner, in *Traffic and Transportation Theory in the 21st Century*, ed. by M.A.P. Taylor (Elsevier Science, Amsterdam, 2002), pp. 417–439
12. B.S. Kerner, Phys. Rev. E **65**, 046138 (2002)
13. B.S. Kerner, *The Physics of Traffic* (Springer, Berlin, Heidelberg, New York, 2004)
14. B.S. Kerner, *Introduction to Modern Traffic Flow Theory and Control* (Springer, Heidelberg, Dordrecht, London, New York, 2009)
15. B.S. Kerner, Physica A **392**, 5261–5282 (2013)
16. B.S. Kerner, Elektrotech. Inf. **132**, 417–433 (2015)
17. B.S. Kerner, Phys. Rev. E **92**, 062827 (2015)
18. B.S. Kerner, Physica A **450**, 700–747 (2016)
19. B.S. Kerner, *Breakdown in Traffic Networks* (Springer, Berlin, New York, 2017)
20. B.S. Kerner, in *Complex Dynamics of Traffic Management*, ed. by B.S. Kerner, Encyclopedia of Complexity and Systems Science Series (Springer, New York, NY, 2019), pp. 21–77
21. B.S. Kerner, M. Koller, S.L. Klenov, H. Rehborn, M. Leibel, Physica A **438**, 365–397 (2015)

Chapter 7
Empirical Induced Traffic Breakdown—Understanding Stochastic Highway Capacity

In this chapter,[1] we show that the empirical nucleation nature of traffic breakdown (Chap. 6) leads to the ultimate conclusion that *at any time instant* there is a *range* of stochastic highway capacities of free flow at a bottleneck.

7.1 Empirical Induced Traffic Breakdown as the Usual Reason for Traffic Congestion on Long Highway Sections

The usual empirical scenario of the emergence and development of real traffic congestion on a long highway section is shown in Fig. 7.1. First, spontaneous traffic breakdown (F→S transition) occurs randomly at a downstream bottleneck (off-ramp bottleneck in Fig. 7.1). Then, a spatiotemporal congested traffic pattern that has emerged due to this traffic breakdown begins to propagate upstream. When the pattern reaches an upstream on-ramp bottleneck (on-ramp bottleneck 1 in Fig. 7.1), the spatiotemporal congested pattern *induces* traffic breakdown at on-ramp bottleneck 1 (labeled by "induced traffic breakdown 1" in Fig. 7.1). Indeed, this traffic breakdown satisfies the definition of induced traffic breakdown of Sects. 2.3 and 6.3: The congested pattern affects on-ramp bottleneck 1 during a time-limited interval. During the further upstream pattern propagation, the pattern is between the locations of on-ramp bottleneck 1 and on-ramp bottleneck 2 (Fig. 7.1). This means that the pattern does not maintain traffic congestion at on-ramp bottleneck 1 any more. After the pattern has been upstream of on-ramp bottleneck 1, there is a time interval within which there is no downstream congestion that could affect on on-ramp

---

[1] A consideration of stochastic highway capacity in the framework of the three-phase traffic theory presented in this chapter is based on Refs. [1–5].

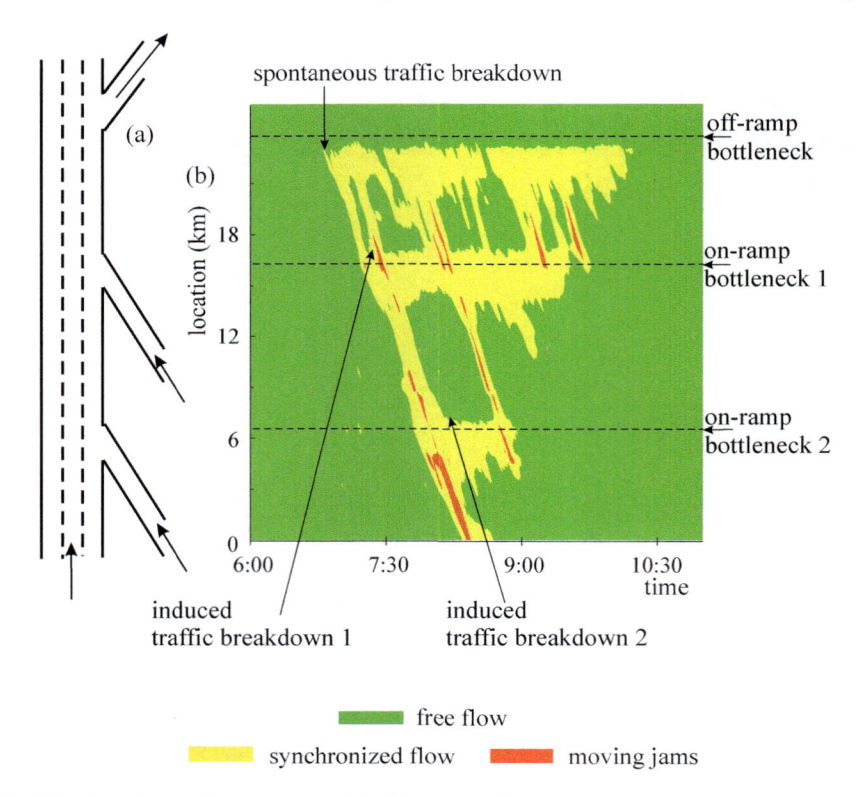

Fig. 7.1 Overview of features of empirical induced traffic breakdown (induced F→S transition). Real field traffic data measured by road detectors on three-lane freeway A5-South in Germany on September 03, 1998 (1 min averaged field data): **a** Sketch of section of three-lane highway in Germany with off- and on-ramps bottlenecks. **b** Speed data measured with road detectors installed along road section in (**a**); data are presented in space and time with the use of averaging method described in Sect. C.2 of [7]. Off-ramp bottleneck, on-ramp bottleneck 1, and on-ramp bottleneck 2 marked by dashed lines in (**b**) are, respectively, bottlenecks explained in Fig. 2.1 of the book [2]

bottleneck 1. Nevertheless, free flow does not return at on-ramp bottleneck 1. This means that while propagating through on-ramp bottleneck 1, the pattern does induce traffic breakdown at this bottleneck (labeled by "induced traffic breakdown 1" in Fig. 7.1).

Later, the spatiotemporal congested pattern propagates upstream of on-ramp bottleneck 1. While reaching the next upstream bottleneck on highway section (on-ramp bottleneck 2 in Fig. 7.1), the spatiotemporal congested pattern induces traffic breakdown at on-ramp bottleneck 2 (labeled by "induced traffic breakdown 2" in Fig. 7.1). Indeed, as in the case of on-ramp bottleneck 1, the spatiotemporal congested pattern affects on on-ramp bottleneck 2 during a time-limited interval only: During the further upstream pattern propagation, the pattern is upstream of on-ramp bottleneck 2 (Fig. 7.1). This means that the pattern does not maintain congested traffic at on-ramp

7.1 Empirical Induced Traffic Breakdown as the Usual ... 87

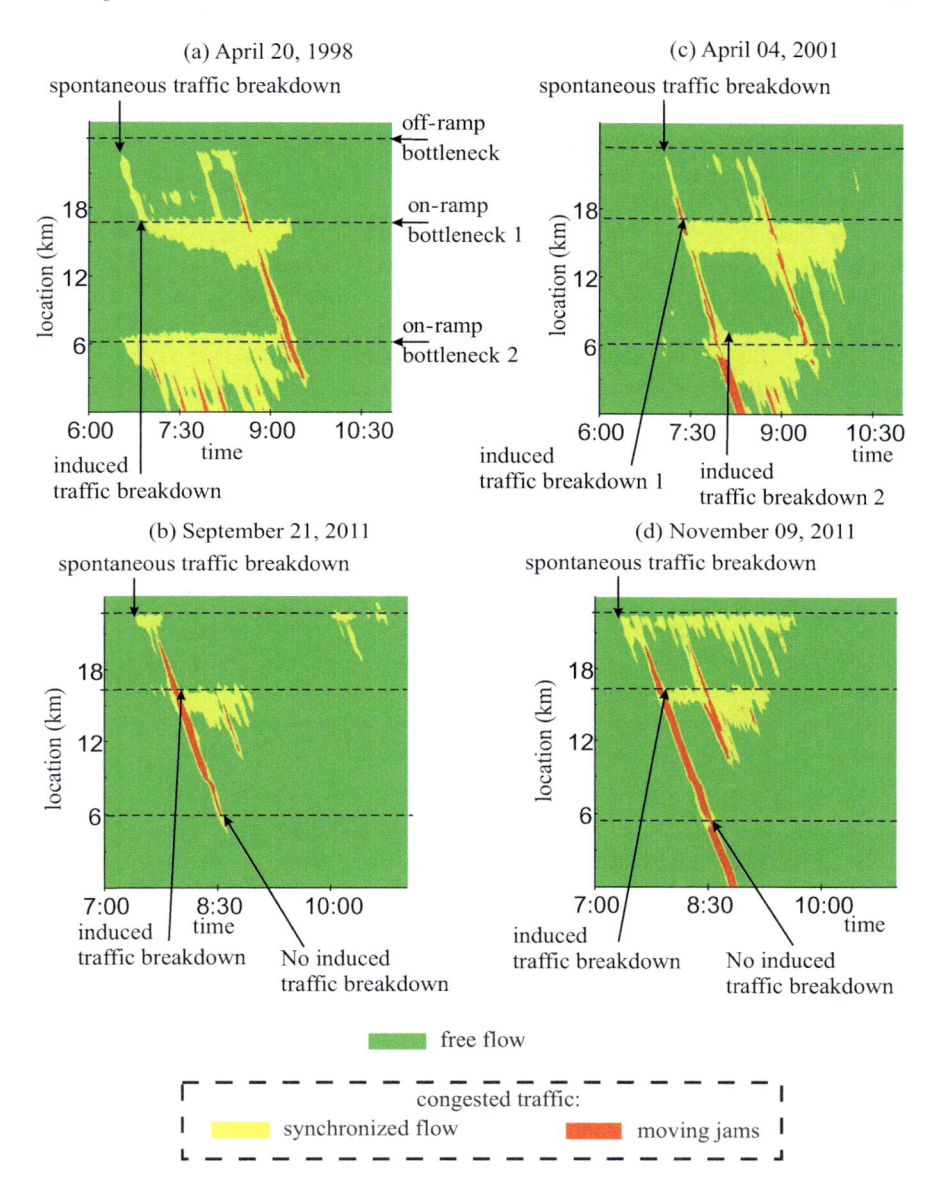

Fig. 7.2 Examples of empirical induced traffic breakdown at highway bottlenecks measured on four different days on freeway A5-South in Germany (1 min averaged field data): Speed data measured with road detectors installed along road section; data are presented in space and time with the use of averaging method described in Sect. C.2 of [7]. Off-ramp bottleneck, on-ramp bottleneck 1, and on-ramp bottleneck 2 marked by dashed lines are, respectively, the same ones shown in Fig. 7.1 and explained in Fig. 2.1 of the book [2]

bottleneck 2 any more. After the pattern has been upstream of on-ramp bottleneck 2, there is a time interval within which there is no downstream congestion that could affect on on-ramp bottleneck 2. Nevertheless, free flow does not return at on-ramp bottleneck 2. This means that while propagating through on-ramp bottleneck 2, the pattern does induce traffic breakdown at this bottleneck (labeled by "induced traffic breakdown 2" in Fig. 7.1).

As a result of the spontaneous traffic breakdown at the off-ramp bottleneck and subsequent induced traffic breakdowns at on-ramp bottlenecks 1 and 2, a complex congested traffic pattern has been built (Fig. 7.1). At each of the bottlenecks the downstream front of congested traffic is localized at the related bottleneck. Therefore, this congested traffic belongs to the synchronized flow traffic phase.

Qualitatively the same scenario of empirical induced traffic breakdown is observed on many other days (Fig. 7.2). On April 20, 1998 spontaneous traffic breakdown (F→S transition) (labeled by "spontaneous traffic breakdown" in Fig. 7.2a) occurs randomly at the downstream off-ramp bottleneck leading to a spatiotemporal congested pattern. The subsequent propagation of the pattern to upstream on-ramp bottleneck 1 induces traffic breakdown at the bottleneck (labeled by "induced traffic breakdown" in Fig. 7.2a). A scenario of congested pattern formation that is very similar shown in Fig. 7.1 is observed on April 04, 2001 (Fig. 7.2c). In this case, as in Fig. 7.1, a spatiotemporal congested pattern that has emerged after spontaneous traffic breakdown at the downstream off-ramp bottleneck causes subsequent induced traffic breakdown at upstream on-ramp bottlenecks 1 and 2.

In all scenarios of empirical induced traffic breakdown at a bottleneck discussed above (Figs. 7.1 and 7.2), we observe the following stages:

1. There is a spatiotemporal congested traffic pattern (congested pattern for short) that has initially emerged downstream of the bottleneck.[2]
2. The congested pattern propagates upstream.
3. Before the congested pattern reaches the bottleneck, free flow conditions are at the bottleneck.
4. When the congested pattern reaches the bottleneck, the pattern causes traffic congestion at the bottleneck only during a finite time interval.
5. After this finite time interval, the congested pattern does not maintain traffic congestion at the bottleneck.
6. Free flow recovers downstream of the bottleneck.
7. Although free flow recovers downstream of the bottleneck, traffic congestion remains at the bottleneck.

---

[2] In Figs. 7.1 and 7.2, a spatiotemporal congested pattern that induces traffic breakdown at upstream on-ramp bottleneck 1 propagates upstream from the downstream off-ramp bottleneck at which the spatiotemporal congested pattern has initially emerged. However, it can occur that a spatiotemporal congested pattern propagates *downstream*. Then, while reaching a downstream bottleneck, the pattern can induce traffic breakdown at the downstream bottleneck. This specific case occurs, for example, when the spatiotemporal congested pattern is a moving synchronized flow pattern (MSP) occurring initially at a moving bottleneck (MB) (see simulations of the occurrence of such an MSP that propagates downstream in Fig. 2b of Ref. [6]); qualitative explanations of the MSP will be given in Fig. 11.4b of Sect. 11.1.4 in this book.

8. This self-maintaining of traffic congestion at the bottleneck means that while propagating through the bottleneck the congested pattern has induced traffic breakdown at the bottleneck.
9. In accordance with the definition of synchronized flow (Sect. 6.2.2), traffic congestion, which is self-maintained at the bottleneck after the empirical induced traffic breakdown has occurred at the bottleneck, belongs to the synchronized flow traffic phase. Indeed, the downstream front of synchronized flow resulting from the induced traffic breakdown is localized at the bottleneck. This synchronized flow ensures the nucleation nature of the F→S transition at the bottleneck. Therefore, as already explained in Sects. 6.1 and 6.2, any empirical induced traffic breakdown can be called empirical induced F→S transition at the bottleneck.

7.2 Range of Highway Capacities at Any Time Instant

7.2.1 Minimum and Maximum Highway Capacities

In Sect. 6.2.1, we have explained that the empirical nucleation nature of traffic breakdown (F→S transition) at a bottleneck means that free flow is in a metastable state with respect to the F→S transition at the bottleneck: If a nucleus for traffic breakdown occurs in the metastable state of free flow at the bottleneck, traffic breakdown does occur. In contrast, as long as no nucleus appears, no breakdown occurs in the metastable state of free flow at the bottleneck. Thus, when free flow is metastable with respect to the F→S transition at the bottleneck, then a spatiotemporal congested pattern propagating through the bottleneck acts as a nucleus for induced traffic breakdown at the bottleneck. Empirical examples of the empirical induced traffic breakdown have been shown in Figs. 6.1, 6.2, 7.1, and 7.2.

There should be a *limited range* of the flow rate in free flow at a bottleneck within which empirical free flow is in a metastable state (Figs. 7.3 and 7.4).[3] Indeed, in real field traffic data at small enough values of the flow rate in free flow no traffic breakdown (F→S transition) is observed at the bottleneck. In this case, free flow

---

[3] It must be emphasized that the range of highway capacities presented in Figs. 7.3 and 7.4 can be considered only as illustrations of the metastability of free flow with respect to traffic breakdown at a bottleneck. This is because traffic is a process that occurs in space and time. The metastability of free flow with respect to traffic breakdown at the bottleneck is a spatiotemporal phenomenon. Basic empirical features of the spatiotemporal traffic phenomenon "metastability of free flow with respect to traffic breakdown (F→S transition)" cannot be understood, if empirical traffic data are measured at one road location only as made in the fundamental diagram and in the speed–flow plane (Figs. 7.3 and 7.4). Indeed, many spatiotemporal features of traffic breakdown are lost in the fundamental diagram as well as in the data presented in the speed–flow plane (Sect. 5.2). Therefore, the range of highway capacities and its features should be studied with the use of *spatiotemporal* traffic data (Figs. 7.1 and 7.2). Only after such a study of spatiotemporal traffic data has been made, the illustration of the range of highway capacities can be made in the fundamental diagram and in the speed–flow plane (Figs. 7.3 and 7.4).

Fig. 7.3 Qualitative explanation of range of highway capacities at a bottleneck in speed–flow plane. States of free and synchronized flow are taken from Fig. 6.3a

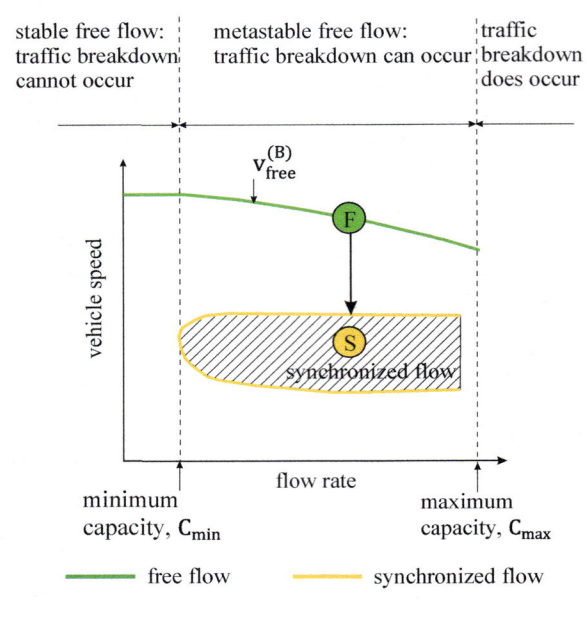

Fig. 7.4 Qualitative explanation of range of highway capacities of free flow at a bottleneck in flow–density plane. States of free and synchronized flow are taken from Fig. 6.3b

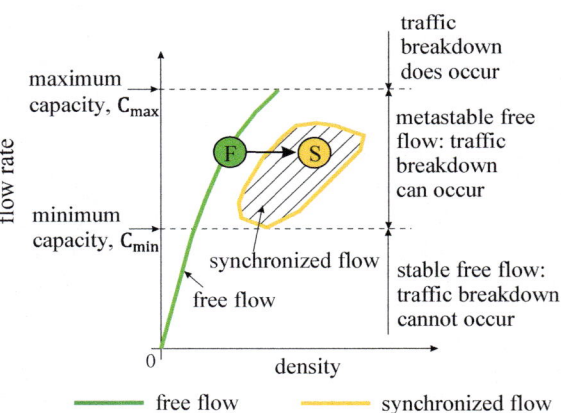

is *stable* with respect to traffic breakdown (F→S transition) at the bottleneck. We denote the minimum flow rate in free flow at the bottleneck that separates the stable and metastable states of free flow by $q = C_{\min}$. This minimum flow rate defines a minimum highway capacity $C_{\min}$ of free flow at the bottleneck (Figs. 7.3 and 7.4). Therefore, at

$$q < C_{\min}, \qquad (7.1)$$

free flow is stable at the bottleneck, i.e., no nuclei can appear in the free flow. Consequently, no traffic breakdown can be induced in stable free flow at the bottleneck.

7.2 Range of Highway Capacities at Any Time Instant 91

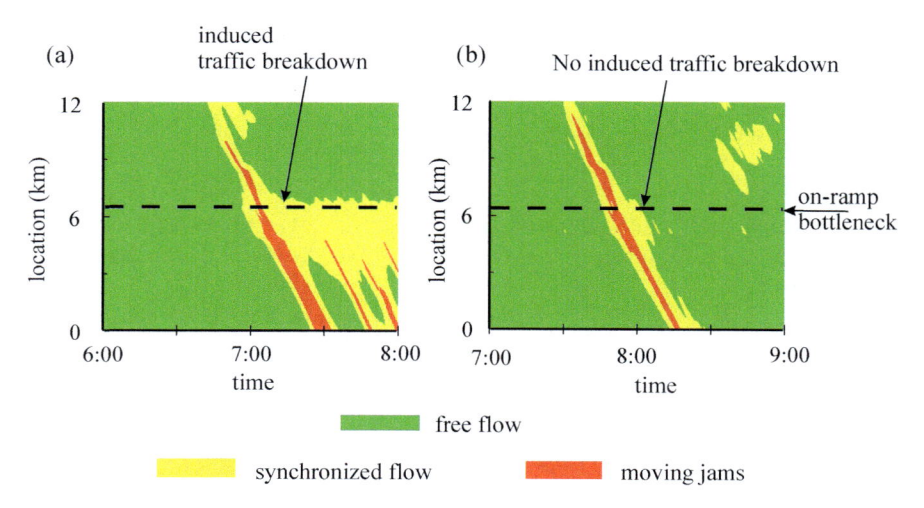

Fig. 7.5 Empirical jam propagation through on-ramp bottleneck: **a** Empirical induced traffic breakdown. **b** Moving jam propagation through the bottleneck without induced traffic breakdown. Real field traffic data measured by road detectors on three-lane freeway A5-South in Germany on March 22, 2001 (**a**) and June 23, 1998 (**b**). On-ramp bottleneck marked by dashed lines in (**a, b**) is the same one as that labeled by "on-ramp bottleneck 2" in Figs. 7.1 and 7.2

To explain the minimum highway capacity, we consider the empirical propagation of wide moving jams through the same on-ramp bottleneck occurring on two different days. The first case shown in Fig. 7.5a is related to an empirical induced traffic breakdown at the bottleneck. This case is qualitatively the same one as those shown Figs. 6.1b, 6.2, 7.1, and 7.2 in which a spatiotemporal congested pattern induces traffic breakdown at a bottleneck when the congested pattern has reached the bottleneck. The empirical induced traffic breakdown has been labeled by "induced traffic breakdown" in Figs. 6.1b, 6.2, 7.1, 7.2, and 7.5a.

Contrary to the case of empirical induced traffic breakdown shown in Fig. 7.5a, in Fig. 7.5b no induced traffic breakdown occurs when a wide moving jam propagates through the same on-ramp bottleneck on another day. The same effect has been also observed on other days (labeled by "No induced traffic breakdown" in Figs. 7.2b, d).

The vehicle speed within a wide moving jam is usually as low as zero. For this reason, as we have already discussed in Sect. 6.4, the wide moving jam can be considered the "largest" nucleus for traffic breakdown in free flow at the bottleneck: The speed within this largest nucleus is considerably lower than the speed within the critical nucleus required for traffic breakdown. Nevertheless, no induced traffic breakdown occurs when the jam propagates through the bottleneck in Fig. 7.5b. However, if free flow is in a metastable state at the bottleneck, the wide moving jam should be a nucleus for traffic breakdown at the bottleneck. Therefore, the case shown in Fig. 7.5b, in which no traffic breakdown has been induced, should be related to the opposite condition: This free flow is in a stable state with respect to traffic breakdown (F→S transition). In other words, the flow rate q in free flow at the bottleneck should satisfy condition (7.1). Thus, under condition (7.1) there can be no nuclei in free flow that can cause traffic breakdown at the bottleneck and, therefore, traffic breakdown cannot occur.

Contrary to condition (7.1), at

$$q \geq C_{\min} \qquad (7.2)$$

free flow is in a metastable state with respect to traffic breakdown at the bottleneck (Figs. 7.3 and 7.4). Therefore, under condition (7.2) traffic breakdown can occur at the bottleneck.

We can assume that the more the flow rate q in free flow at the bottleneck exceeds the minimum highway capacity $C_{\min}$, the larger should be the speed within the critical nucleus required for traffic breakdown at the bottleneck. There should be a large enough flow rate $q > C_{\min}$ at which any small local speed decrease in free flow at the bottleneck is a nucleus for traffic breakdown (F→S transition) at the bottleneck. At this flow rate

$$q = C_{\max} \qquad (7.3)$$

that we denote by a maximum highway capacity $C_{\max}$ of free flow at the bottleneck traffic breakdown (F→S transition) does occur at the bottleneck (Figs. 7.3 and 7.4). This means that under condition

$$q \geq C_{\max} \qquad (7.4)$$

free flow cannot exist at the bottleneck: During some time interval traffic breakdown does occur in an initial free flow. From this analysis it follows that the maximum highway capacity should be larger than the minimum highway capacity:

$$C_{\max} > C_{\min}. \qquad (7.5)$$

Thus, from formulas (7.2) and (7.5) it follows that within the flow-rate range

$$C_{\min} \leq q < C_{\max} \qquad (7.6)$$

free flow is in a metastable state with respect to the F→S transition at the bottleneck (Figs. 7.3 and 7.4).

7.2.2 Stochastic Highway Capacity in Three-Phase Traffic Theory

Any flow rate q in free flow at a highway bottleneck that satisfies (7.6) is highway capacity C. This is because at each of the flow rates (7.6) traffic breakdown can occur at the bottleneck. At any time instant, there are the infinite number of the flow rates (7.6) at which traffic breakdown can occur in free flow at the bottleneck: There are the infinite number of highway capacities C at any time instant that satisfy conditions

7.2 Range of Highway Capacities at Any Time Instant

Fig. 7.6 Qualitative explanation of range of stochastic highway capacities at a bottleneck at each time instant: Hypothetical fragment of qualitative time dependencies of minimum capacity $C_{\min}(t)$ and maximum capacity $C_{\max}(t)$

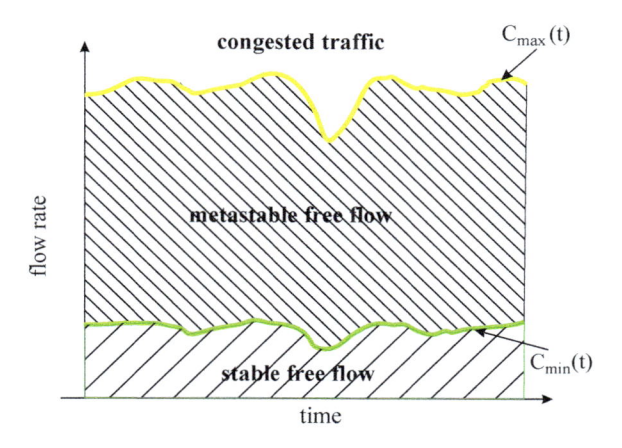

$$C_{\min} \leq C \leq C_{\max}. \qquad (7.7)$$

Conditions (7.7) mean that at any time instant there is a *range* of highway capacities between the minimum capacity $C_{\min}$ and maximum capacity $C_{\max}$: At any flow rate within the range (7.6), traffic breakdown can *randomly* occur at the bottleneck.

> The existence of an infinite number of highway capacities at any time instant (7.7) means that highway capacity is *stochastic*.

> The range of stochastic highway capacities is defined as follows: *At any time instant t*, there are the infinite number of stochastic highway capacities between the minimum highway capacity $C_{\min}$ and maximum highway capacity $C_{\max}$.

A consideration of the minimum highway capacity $C_{\min}$ and maximum highway capacity $C_{\max}$ as some time-independent values made above is a rough simplification of real traffic. Indeed, as already emphasized in Sect. 5.1, in real traffic the share of trucks as well as the share of vehicles with different vehicle and driver characteristics are usually stochastic time functions. Therefore, values $C_{\min}$ and $C_{\max}$ that depend on vehicle and driver characteristics should be also stochastic time functions. In accordance with conditions (7.5) that are valid for any time instant, stochastic values of the minimum and maximum highway capacities designated, respectively, by $C_{\min}(t)$ and $C_{\max}(t)$ satisfy the following condition (Fig. 7.6):

$$C_{\max}(t) > C_{\min}(t). \tag{7.8}$$

Thus, in a general case of heterogeneous traffic flow, the range of stochastic highway capacities can be formulated as follows. *At any time instant t*, there are the infinite number of highway capacity values between the minimum highway capacity $C_{\min}(t)$ and maximum highway capacity $C_{\max}(t)$ (Fig. 7.6). The existence of the infinite number of highway capacity values at any time instant means that highway capacity is stochastic. The definition of the range of stochastic highway capacities made in the three-phase traffic theory changes also the meaning of the term *stochastic highway capacity* as follows.

> When at some time instant t the flow rate $q(t)$ in free flow at a bottleneck is equal to or larger than the minimum highway capacity $C_{\min}(t)$ at this time instant but it is less than the maximum highway capacity $C_{\max}(t)$ at the same time instant, i.e., conditions
>
> $$C_{\min}(t) \leq q(t) < C_{\max}(t) \tag{7.9}$$
>
> are satisfied, then traffic breakdown can occur at the bottleneck with some probability but traffic breakdown should not necessarily occur (Figs. 7.3 and 7.4).

> The metastability of free flow with respect to the F→S transition (Sect. 6.2.1) is the origin of the range of stochastic highway capacities at any time instant.

As explained in Chap. 6, to understand the metastability of free flow with respect to traffic breakdown (F→S transition) at a bottleneck, empirical traffic data should be studied in space and time. Therefore, the range of stochastic highway capacities at any time instant can only be understood, if a spatiotemporal analysis of empirical traffic data is performed.

7.3 Empirical Induced Traffic Breakdown at Bottleneck as Empirical Proof for Range of Highway Capacities

To explain that empirical induced traffic breakdown is empirical proof for *a range* of stochastic highway capacities *at any given time instant*, we consider the opposite assumption made in standard traffic and transportation science (Sect. 5.1): Rather than the capacity range of the three-phase traffic theory, in standard traffic and transportation science it is assumed that *at any given time instant* there is *a particular*

7.3 Empirical Induced Traffic Breakdown at Bottleneck ...

value of stochastic highway capacity. This means that when the flow rate $q(t)$ in free flow at a bottleneck at a time instant exceeds a value of stochastic highway capacity $C(t)$ related to this time instant, i.e., $q(t) > C(t)$, traffic breakdown must occur. Otherwise, when the flow rate in free flow at the bottleneck at another time instant is less than a value of stochastic highway capacity related to this time instant, i.e., $q(t) < C(t)$, traffic breakdown at the bottleneck cannot be possible.

If the assumption made in standard traffic and transportation science about stochastic highway capacity would be correct, then at a time instant for any flow rate $q(t)$ that is less than the value of stochastic highway capacity $C(t)$ related to this time instant *no* traffic breakdown can occur at a bottleneck. This contradicts empirical observations in which traffic breakdown is induced at the bottleneck through the propagation of a congested pattern to the bottleneck location (labeled by "induced traffic breakdown" in Figs. 6.1b, 6.2, 7.1, 7.2, and 7.5a):

- Stochastic highway capacity of free flow at a bottleneck *cannot* depend on whether there is a congested pattern, which has initially occurred outside the bottleneck, or not.

Thus, the assumption made in standard traffic and transportation science about a particular value of stochastic highway capacity at any time instant (Sect. 5.1) contradicts basically the evidence of the empirical induced traffic breakdown at the bottleneck. Because this incorrect understanding of stochastic highway capacity is one of the most important fundamentals of standard traffic engineering, in Sect. C.1 we consider this subject in more details.

> The understanding of stochastic highway capacity made in standard traffic and transportation science is incorrect for real traffic.

> In accordance with Sect. 6.4, the empirical induced traffic breakdown is the empirical proof for the metastability of free flow with respect to an F→S transition at a bottleneck. The free flow metastability is the origin of the existence of the range of stochastic highway capacities at any time instant. Therefore, the empirical induced traffic breakdown is indeed the empirical proof for the range of stochastic highway capacities at any time instant.

7.4 Empirical Induced Traffic Breakdown as One of Consequences of Spill-Over Effect

In standard traffic science, upstream propagation of a traffic congested pattern occurring at a downstream bottleneck is often called the *spill-back effect*. This congested traffic pattern can reach an adjacent upstream bottleneck resulting in traffic congestion at the bottleneck; this case of the occurrence of congestion at the upstream bottleneck is often called the *spill-over effect*. In other words, empirical induced traffic breakdown at the bottleneck could also be considered the spill-over effect at the bottleneck. However, we do not use the term *spill-over effect* in the book. This is because *not* any spill-over effect can be considered the phenomenon "empirical induced traffic breakdown" at the bottleneck.

An empirical example of the spill-over effect at bottleneck B2 that *cannot* be considered the phenomenon "empirical induced traffic breakdown" at bottleneck B2 is shown in Fig. 7.7: During the existence of traffic congestion at bottleneck B2, there is congested traffic downstream of this bottleneck caused by the propagation of traffic congestion from downstream bottleneck B1 to the location of bottleneck B2. This can be seen on vehicle speed distributions on trajectories 3 and 4 in Fig. 7.7c: There is congested traffic downstream of the location of bottleneck B2.

To explain why we cannot consider the empirical example (Fig. 7.7b, c) as the empirical induced traffic breakdown at bottleneck B2, we consider the traffic data shown in Fig. 7.7b at time $t < 16{:}30$. At time $t < 16{:}30$, free flow is at bottleneck B2. This is because congested traffic propagating upstream from bottleneck B1 has not still reached the location of bottleneck B2. However, we do *not* know whether this free flow at bottleneck B2 is in a metastable state with respect to an F→S transition, i.e., conditions (7.6) are satisfied or, in contrast, the free flow is stable, i.e., condition (7.1) is satisfied. Later, at about 16:35 congested traffic propagating upstream from bottleneck B1 has reached the location of bottleneck B2. As a result, within the time interval $16{:}35 < t \leq 17{:}15$ (Fig. 7.7b) congested traffic is both downstream and upstream of bottleneck B2. For this reason, we cannot state whether congested traffic propagating upstream from bottleneck B1 induces traffic breakdown at bottleneck B2, or, in contrast, traffic congestion at bottleneck B2 is caused by the existence of congested traffic between bottleneck B1 and bottleneck B2.

We can make the following conclusions. While propagating upstream, a congested traffic pattern can reach an upstream bottleneck resulting in traffic congestion at the bottleneck (spill-over effect at the bottleneck). There can be two qualitative different consequences of the spill-over effect at the upstream bottleneck:

1. After traffic congestion has occurred at the upstream bottleneck due to the spill-over effect, there is traffic congestion *downstream* from the bottleneck. In this case, the occurrence of traffic congestion at the bottleneck due to the spill-over effect *cannot* be considered the empirical induced traffic breakdown at the bottleneck (Fig. 7.7).

7.4 Empirical Induced Traffic Breakdown as One of Consequences ... 97

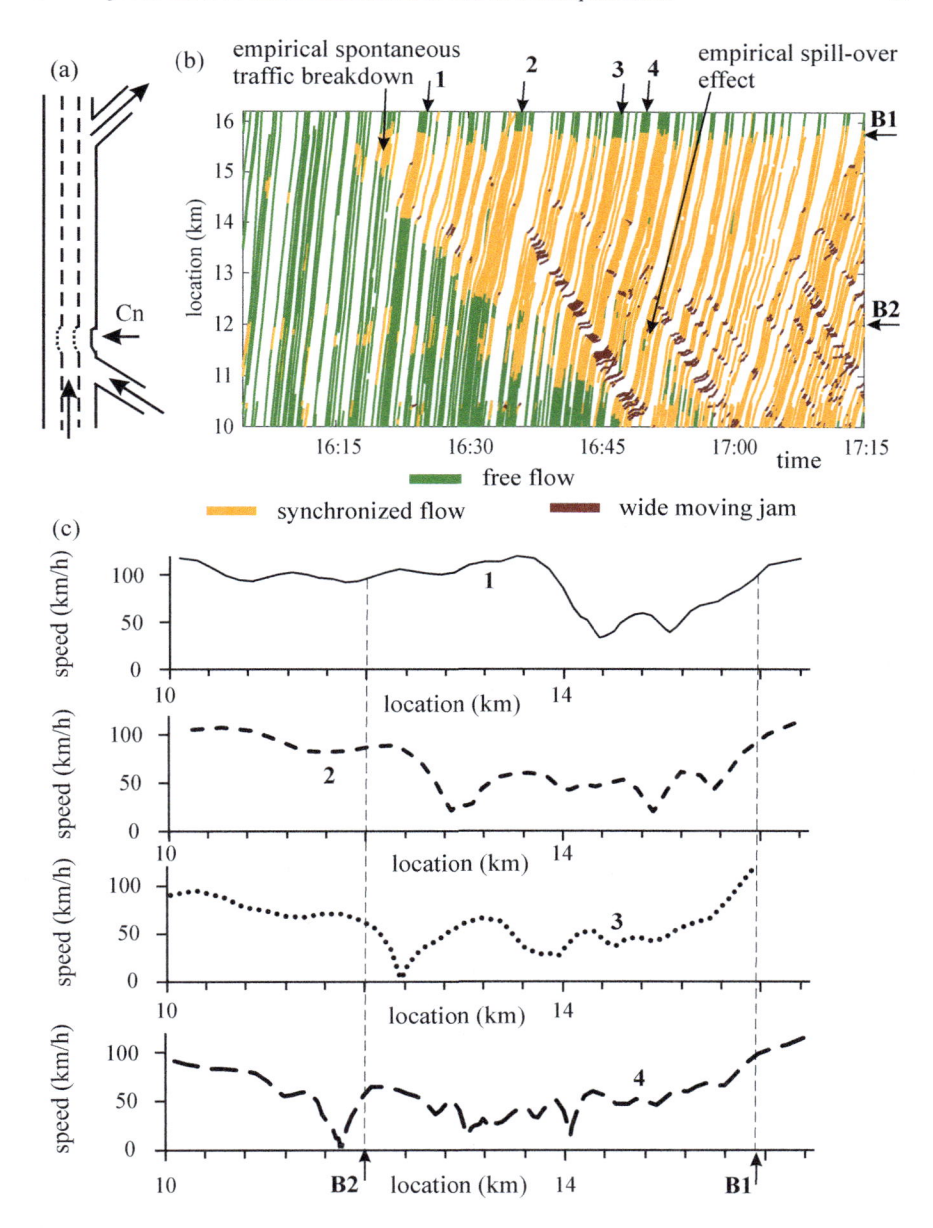

Fig. 7.7 Overview of features of empirical spill-over effect: **a** Fragment of highway section of Fig. 2.1a. **b** Fragment of vehicular trajectories taken from Fig. 2.2. **c** Microscopic speed along vehicle trajectories 3 and 4 whose numbers are the same as those in (**b**); dashed vertical lines show locations of bottlenecks B1 and B2, respectively. Off-ramp bottleneck B1 and bottleneck B2 have been explained in caption to Fig. 2.1

2. After synchronized flow has occurred at the upstream bottleneck due to the spill-over effect, free flow is realized downstream of the bottleneck *and* the synchronized flow is further self-maintained at the bottleneck. In this case, the occurrence of synchronized flow at the bottleneck due to the spill-over effect *can* be considered the empirical induced traffic breakdown at the bottleneck (Figs. 6.1b, 6.2, 7.1, 7.2, and 7.5a).

In more details, conditions for the empirical induced traffic breakdown at the bottleneck have been considered in Sect. 6.3.

7.5 Perception of Highway Capacity Resulting from Empirical Induced Traffic Breakdown at Bottleneck

The necessity of vehicular traffic control and management as well as other ITS applications is associated with the occurrence of traffic breakdown in a traffic network at a large enough traffic demand. Traffic breakdown results in a considerable increase in travel time, fuel consumption, and other travel costs as well as in a decrease in traffic safety and comfort. Therefore, through ITS applications the negative effects of traffic congestion should be decreased.

Accordingly the results of this chapter, the perception of highway capacity resulting from empirical induced traffic breakdown at a bottleneck is as follows:

- Traffic breakdown (F→S transition) at a highway bottleneck is a nucleation phenomenon. This means that at any time instant there should be a range of highway capacities at the bottleneck. The existence of the range of highway capacities at any time instant means that highway capacity is stochastic. When at a time instant the flow rate in free flow at the bottleneck is within the capacity range related to this time instant, then traffic breakdown occurs only if a nucleus for traffic breakdown appears at the bottleneck.

This perception of vehicular traffic results from the empirical nucleation nature of traffic breakdown at a highway bottleneck. For this reason, standard traffic and transportation theories, which cannot explain the nucleation nature of traffic breakdown (F→S transition) at highway bottlenecks, cannot also be applied for the development of reliable traffic control, dynamic traffic assignment as well as other reliable ITS applications in traffic and transportation networks.

> The empirical nucleation nature of traffic breakdown at a bottleneck can be considered an empirical fundamental of traffic and transportation science.

> Perception of vehicular traffic in the three-phase traffic theory is associated with the understanding of traffic breakdown (F→S transition) at a highway bottleneck as a *nucleation phenomenon*: Traffic breakdown occurs when free flow is in a metastable state with respect to the F→S transition at the bottleneck *and* a nucleus for traffic breakdown appears at the bottleneck.

References

1. B.S. Kerner, Phys. A **333**, 379–440 (2004)
2. B.S. Kerner, *The Physics of Traffic* (Springer, Berlin, 2004)
3. B.S. Kerner, *Introduction to Modern Traffic Flow Theory and Control* (Springer, Heidelberg, 2009)
4. B.S. Kerner, *Breakdown in Traffic Networks* (Springer, Berlin, 2017)
5. B.S. Kerner, in *Complex Dynamics of Traffic Management*, ed. by B.S. Kerner. Encyclopedia of Complexity and Systems Science Series (Springer, New York, 2019), pp. 21–77
6. B.S. Kerner, S.L. Klenov, J. Phys. A: Math. Theor. **43**, 425101 (2010)
7. B.S. Kerner, H. Rehborn, R.-P. Schäfer, S.L. Klenov, J. Palmer, S. Lorkowski, N. Witte, Phys. A **392**, 221–251 (2013)

Chapter 8
Empirical Nucleation Nature of Traffic Breakdown—Emergence of Three-Phase Traffic Theory

In Chaps. 6 and 7, we have shown that traffic breakdown (F→S transition) at a highway bottleneck exhibits the empirical nucleation nature.

> The fundamental requirement for a traffic flow theory that claims to explain real traffic is as follows: The theory should show the empirical nucleation nature of traffic breakdown (F→S transition) at a highway bottleneck.

The empirical nucleation nature of traffic breakdown (F→S transition) at the bottleneck has *no* sense for standard traffic and transportation theories (see explanations in Appendix C). This explains the failure of the standard traffic and transportation theories in the real world.

> The three-phase traffic theory has been introduced to explain the empirical nucleation nature of traffic breakdown (F→S transition) at a highway bottleneck.

As we will show in Chaps. 9–11, the three-phase traffic theory explains the entire *empirical complexity* of real traffic phenomena in space and time. In this chapter, we try to answer on the following question:

- What are the driver behaviors that can be responsible for the empirical nucleation nature of traffic breakdown (F→S transition) at a highway bottleneck?

We will show that the empirical nucleation nature of traffic breakdown (F→S transition) at the bottleneck can be explained by a discontinuous character of over-acceleration together with a spatiotemporal competition between driver speed adaptation and driver over-acceleration occurring within a local speed decrease in

© The Author(s), under exclusive license to Springer Nature Switzerland AG 2021
B. S. Kerner, *Understanding Real Traffic*,
https://doi.org/10.1007/978-3-030-79602-0_8

metastable free flow with respect to an F→S transition at the bottleneck. The crucial role in this mechanism of the empirical nucleation nature of traffic breakdown plays a critical speed for traffic breakdown: The occurrence of a nucleus for traffic breakdown in metastable free flow at the bottleneck is associated with a case when the speed within the local speed decrease at the bottleneck becomes less than the critical speed. In this case, traffic breakdown (F→S transition) does occur. Otherwise, when the speed within the local speed decrease at the bottleneck is larger than the critical speed, no nucleus appears and, therefore, no traffic breakdown occurs in metastable free flow at the bottleneck.[1]

From the author's experience, the understanding of the mechanism of the empirical nucleation nature of traffic breakdown introduced in the three-phase traffic theory[2] might be difficult for many readers. For this reason, we consider in this chapter a very simplified theoretical explanation of the driver speed adaptation and driver over-acceleration as well as their spatiotemporal competition leading to the occurrence of a nucleus for an F→S transition at a highway bottleneck.[3] An empirical study of the occurrence of a nucleus for spontaneous traffic breakdown (F→S transition) at the bottleneck will be considered in Sects. 9.1 and 9.2 of Chap. 9.

8.1 Discontinuous Character of Over-Acceleration

8.1.1 Driver Speed Adaptation and Over-Acceleration

To explain the driver behaviors that are responsible for the empirical nucleation nature of traffic breakdown (F→S transition) at a highway bottleneck, we should mention that when a driver approaches a slower moving preceding vehicle and the driver cannot immediately pass the slow vehicle, the driver must decelerate to the speed of the slow moving preceding vehicle. This well-known effect can be called *driver speed adaptation* or *speed adaptation effect* (Fig. 8.1a).

To escape from this car-following of the slow moving preceding vehicle, the driver searches for the opportunity to accelerate. We call vehicle acceleration from the car-following of the slow moving preceding vehicle as *over-acceleration* or *over-acceleration effect*. The term *car-following* means the dynamic behavior of a vehicle that follows the preceding vehicle.

---

[1] There are a number of applications and further developments of the three-phase traffic theory that are out of scope of this book. In particular, readers can find developments of the three-phase traffic theory for city traffic in [26, 30, 31, 44, 53, 58], applications of this theory for intelligent transportation systems (ITS) in [17, 19–21, 34], and applications of the three-phase traffic theory for traffic dynamic assignment problems in traffic and transportation networks in [25, 26, 35, 37].

[2] A theoretical consideration of the occurrence of a nucleus for traffic breakdown (F→S transition) at a bottleneck made in this chapter is based on results of Ref. [1–16, 24, 29, 32, 33, 36, 38, 39, 42, 43, 45–47, 49–51, 60–62].

[3] Readers that are interested in a more detailed explanation of the origin of the empirical nucleation nature of traffic breakdown (F→S transition) at the bottleneck can find it in Appendix A.

8.1 Discontinuous Character of Over-Acceleration

(a) **speed adaptation effect**: deceleration to the speed of the preceding vehicle

(b) **over-acceleration effect** due to lane changing to a faster lane

▨→v   vehicle under consideration

▬→v_ℓ slower moving preceding vehicle

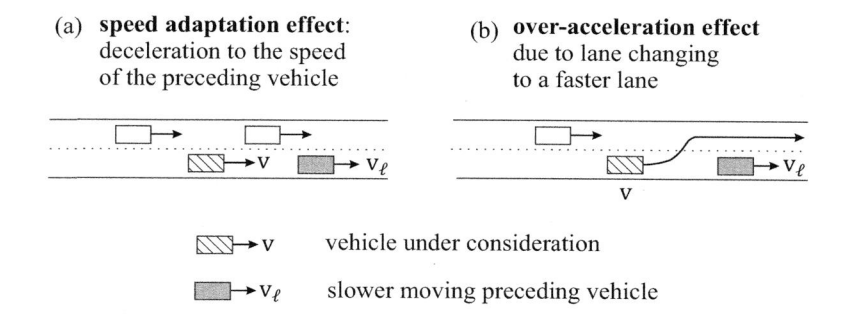

Fig. 8.1 Qualitative explanation of driver speed adaptation (**a**) and driver over-acceleration through lane changing (**b**)

In general, over-acceleration is a vehicle maneuver leading to a higher speed from an initial car-following of a slow moving preceding vehicle. We should mention that a driver accelerates in the car-following when the vehicle speed is less than the speed of the preceding vehicle or when the preceding vehicle accelerates. Contrary to this "usual" driver acceleration, the term *over-acceleration* should emphasize that to escape from the initial car-following of a slow moving preceding vehicle, a driver can also accelerate when the vehicle speed is equal to or it is even higher than the speed of the preceding vehicle. In more details, the cause of the use of the term *over-acceleration* will be explained in Sect. 8.1.5.

8.1.2 Time Delay in Over-Acceleration

The over-acceleration is possible both on single-lane and multi-lane roads.[4] For a simple qualitative explanation of the discontinuous character of over-acceleration that we intent to do in this chapter a mechanism of over-acceleration on a multi-lane road is more convenient than that for a single-lane road. For this reason, we consider a vehicle approaching a slow preceding vehicle moving in the right line of a multi-lane road. We assume that the vehicle must decelerate to a lower speed of the preceding vehicle because it cannot immediately pass the slow vehicle ("speed adaptation effect" in Fig. 8.1a).

On the multi-lane road, over-acceleration leading to the vehicle escaping from the car-following is often possible through lane changing to a faster lane with the subsequent passing of the slow moving vehicle (Fig. 8.1b). However, after the driver has begun the speed adaptation to the speed of the slow moving preceding vehicle

---

[4] Models of over-acceleration on single-lane and multi-lane roads can be found in Sect. 5.10 of the book [19].

Fig. 8.2 Hypothesis of three-phase traffic theory about the discontinuous character of over-acceleration: Qualitative dependence of the mean time delay in over-acceleration on the flow rate in traffic flow. F is a state of free flow, S is a state of synchronized flow

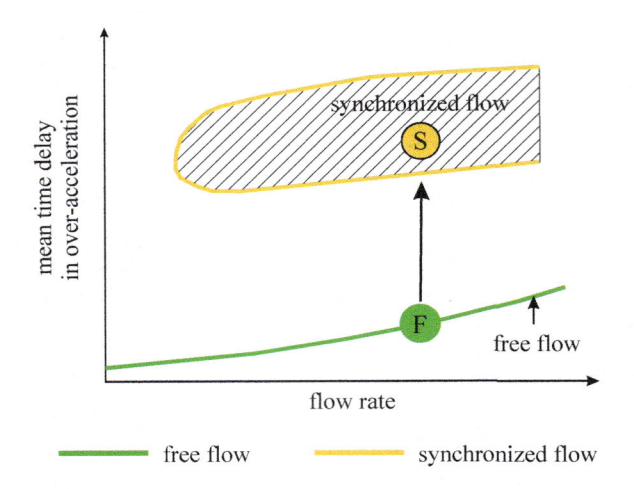

(Fig. 8.1a) there can be a waiting time before the lane changing maneuver leading to vehicle acceleration with the subsequent passing of the slow moving vehicle is successful. We call this waiting time as a *time delay in over-acceleration*.[5]

We have above-mentioned that the vehicle density in synchronized flow is larger than the density is in free flow at the same flow rate (Fig. 6.3b). We can assume that the larger the vehicle density, the more difficult to escape from the car-following of the slow moving preceding vehicle through over-acceleration. Respectively, the larger the vehicle density, the longer should be the *mean* time delay in over-acceleration.

8.1.3 Discontinuity of Mean Time Delay in Over-Acceleration

Through a large density in synchronized flow, vehicles prevent each other to accelerate to a higher speed. This prevention of vehicle acceleration in synchronized flow distinguishes synchronized flow from free flow: In free flow, the vehicle density is considerably smaller than in synchronized flow at the same flow rate and, therefore, vehicles in free flow can more easily escape from the car-following of the slow moving preceding vehicle. This has led to the assumption made in the three-phase traffic theory about the *discontinuous character* of over-acceleration (Fig. 8.2).

> The term *discontinuous character of over-acceleration* is defined as follows: In synchronized flow, the mean time delay in over-acceleration should be considerably longer than it is in free flow.

---

[5] It should be emphasized that the time delay in over-acceleration should not be confounded with a *driver reaction time* that will be considered in Sect. 10.2. We will consider a crucial difference between the time delay in over-acceleration and the driver reaction time in Sect. 10.3.

Therefore, during traffic breakdown (F→S transition) at a bottleneck there should be a jump from a short mean time delay in over-acceleration in free flow to a considerably longer mean time delay in over-acceleration in synchronized flow (up arrow in Fig. 8.2).[6]

- It must be emphasized that the discontinuous character of over-acceleration is basically a *macroscopic* effect associated with the collective behavior of many vehicles in traffic flow: The discontinuity in the mean time delay in over-acceleration (Fig. 8.2) results from a considerable difference in the vehicle density in synchronized flow and free flow at the same flow rate.

8.1.4 Driver Behaviors Explaining the Range of Highway Capacities at Bottleneck

In Sect. 6.2.1, we have shown that the evidence of the empirical induced traffic breakdown at a highway bottleneck leads to the conclusion that at a given flow rate there can be either synchronized flow or free flow at the bottleneck: There is a speed gap between the states of free flow and synchronized flow (Figs. 6.3a and 7.3).

The discontinuous character of over-acceleration (Fig. 8.2) can explain both the empirical metastability of free flow with respect to traffic breakdown (F→S transition) at a highway bottleneck (see Fig. 6.3 of Sect. 6.2.1) and the empirical range of highway capacities (see Fig. 7.3 of Sect. 7.2). Indeed, the gap in the mean time of over-acceleration between states of synchronized flow and free flow (Fig. 8.3a) results in the speed gap between the states of synchronized flow and free flow (Fig. 8.3b). The discontinuous character of over-acceleration remains in a range of the flow rate (Fig. 8.3a). Therefore, this flow-rate range determines conditions (7.6) of the empirical metastability of free flow with respect to traffic breakdown (F→S transition) at a bottleneck as well as the associated range of highway capacities discussed in Sect. 7.2.

---

[6] The discontinuous character of over-acceleration was introduced by the author in 1999 [3–5]. In [3–5], the discontinuous character of over-acceleration was explained and presented through the discontinuous character of the probability that over-acceleration is realized during a given time interval (probability of over-acceleration for short). In [3–5] the probability of over-acceleration was called the probability of passing for the case of over-acceleration due to lane changing (see a discussion in Sect. 5.2.5 of the book [16]). It must be emphasized that the assumption about the discontinuous character of the mean time delay in driver over-acceleration (Fig. 8.2) is equivalent to the assumption about the discontinuous character of the probability of over-acceleration. Indeed, the probability of over-acceleration is the larger, the shorter the mean time delay in over-acceleration (Fig. 8.2). Therefore, traffic breakdown (F→S transition) at the bottleneck that leads to a jump in the mean time delay in over-acceleration (up arrow in Fig. 8.2) is also accompanied by a drop in the probability of over-acceleration. The discontinuous character of over-acceleration [3–5] was incorporated in 2002 by Kerner and Klenov in a mathematical stochastic microscopic traffic flow model [45] and in 2006 in a mathematical deterministic microscopic traffic flow model [48].

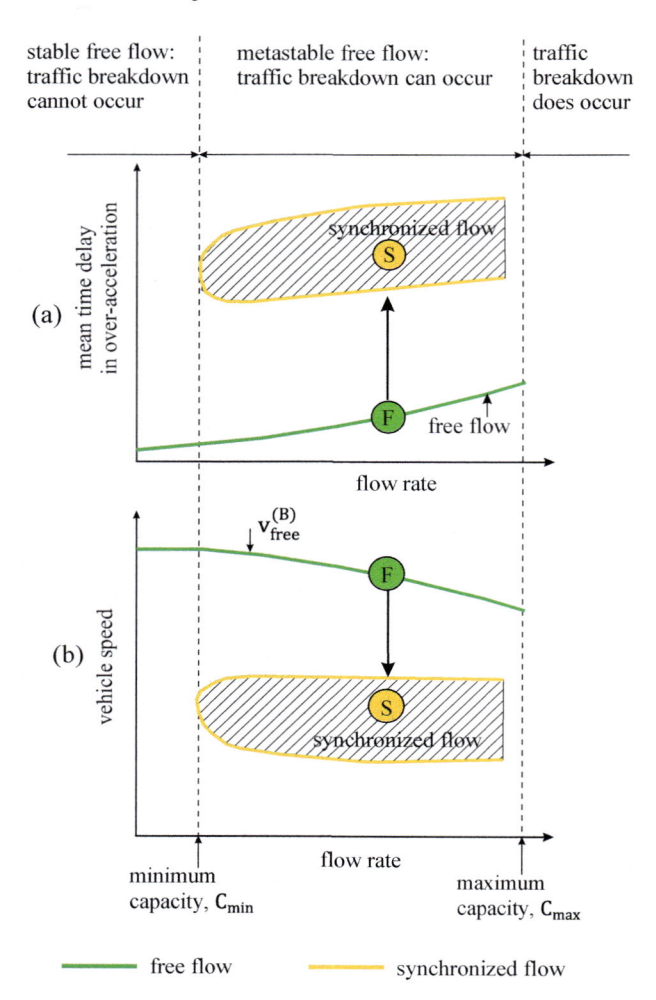

Fig. 8.3 Qualitative explanation of range of highway capacities and the metastability of free flow with respect to traffic breakdown (F→S transition) at a bottleneck based on the discontinuous character of over-acceleration: **a** Qualitative flow-rate dependence of the mean time in over-acceleration taken from Fig. 8.2. **b** Qualitative flow-rate dependencies of states of free flow labeled by $v^{(B)}_{\text{free}}$ and states of synchronized flow in the speed–flow-rate plane taken from Fig. 7.3. $v^{(B)}_{\text{free}}$ is the minimum speed within the average local speed decrease in free flow at the bottleneck

> The discontinuous character of over-acceleration (Fig. 8.3a) is the cause for both the free flow metastability with respect to an F→S transition at a bottleneck and the range of highway capacities between the minimum highway capacity $C_{\rm min}$ and maximum highway capacity $C_{\rm max}$ (Fig. 8.3b).

8.1.5 Explanation of the Choice of the Term "Over-Acceleration"

We should mention that in the car-following a driver accelerates when the vehicle speed is less than the speed of the preceding vehicle and/or the preceding vehicle accelerates. This "usual" driver acceleration does *not* lead to the free flow metastability with respect to the F→S transition at a bottleneck. Contrary to the "usual" driver acceleration, the driver acceleration called *over-acceleration* results in this free flow metastability. This is the most important reason for the necessity of the distinguishing between "usual" driver acceleration and over-acceleration.

It should be emphasized that the prefix *over-* in the term *over-acceleration* can lead to confusion. Indeed, the prefix *over-* in the term *over-acceleration* literally implies too much acceleration. To explain the application of the term *over-acceleration*, we should recall that over-acceleration is a vehicle maneuver leading to a higher speed from initial car-following of a slow moving preceding vehicle; contrary to the "usual" driver acceleration, over-acceleration occurs even when the vehicle speed is equal to or it is even higher than the speed of the preceding vehicle.

In this book, we discuss only over-acceleration due to lane changing to a faster lane (Fig. 8.1b). However, in the three-phase traffic theory, driver acceleration that causes the free flow metastability with respect to the F→S transition at the bottleneck is also realized, when a driver tries to accelerate from the car-following at a low speed of the preceding vehicle that can occur on a single-lane road when no passing is possible.[7] Therefore, we should find a generic term for the driver acceleration that causes the free flow metastability. The choice of the term *over-acceleration* seems to be suitable for this:

- Contrary to "usual" driver acceleration, over-acceleration occurs even when the vehicle speed is equal to or it is even higher than the speed of the preceding vehicle.
- The term *over-acceleration* distinguishes driver acceleration behaviors that do result in the free flow metastability with respect to the F→S transition at the bottleneck from usual driver acceleration behaviors that do not cause the free flow metastability.

---

[7] Explanations of over-acceleration on a single-lane road can be found in Sect. 5.10.2 of the book [38].

8.2 Nucleus Occurrence for Spontaneous Traffic Breakdown in Free Flow at Bottleneck

We have shown that the discontinuous character of over-acceleration introduced in the three-phase traffic theory explains both the range of highway capacities and the empirical metastability of free flow with respect to traffic breakdown (F→S transition) at a bottleneck (Fig. 8.3). As explained in Sect. 6.1, the empirical nucleation nature of traffic breakdown means that there should be a nucleus for traffic breakdown in metastable free flow with respect to the F→S transition at the bottleneck: While a *small enough* local speed decrease at the bottleneck does *not* initiate an F→S transition, the nucleus is a *large enough* local speed decrease that does initiate the F→S transition at the bottleneck. Therefore, the following question arises: How is a transition from a metastable state of free flow to a state of synchronized flow (arrows F→S in Fig. 8.3) initiated? This question can also be formulated as follows:

- What are the driver behaviors that determine features of a nucleus initiating the F→S transition (traffic breakdown) in free flow at the bottleneck?

8.2.1 Competition Between Speed Adaptation and Over-Acceleration Within Local Speed Decrease at Bottleneck

As we will explain below, features of a local speed decrease in free flow at a bottleneck result from a competition between driver speed adaptation and over-acceleration in the vicinity of the bottleneck (Fig. 8.4).

Fig. 8.4 Qualitative explanation of the competition of speed adaptation and over-acceleration within a local speed decrease in free flow at road bottleneck: **a** On-ramp bottleneck. **b** Off-ramp bottleneck. Vehicles labeled by colored light-green move at a lower speed than vehicles labeled by colored green

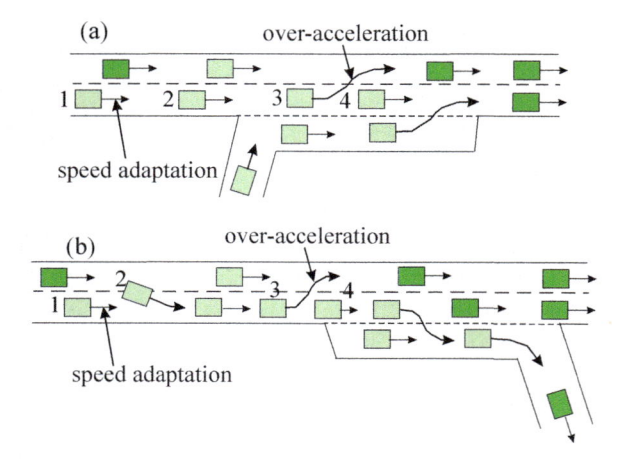

8.2 Nucleus Occurrence for Spontaneous Traffic Breakdown ... 109

For an on-ramp bottleneck (Fig. 8.4a), vehicles moving on the main road should decelerate while adapting their speeds to a lower speed of vehicles merging from the on-ramp onto the main road. For an off-ramp bottleneck (Fig. 8.4b), some of the vehicles going to the off-ramp can decelerate considerably while changing from the left lane to the right lane; consequently, following vehicles in the right lane should decelerate while adapting their speeds to a lower speed of vehicles going to the off-ramp. This speed adaptation effect occurring at the bottlenecks is labeled by "speed adaptation" for vehicle 1 adapting its speed to the speed of the preceding vehicle 2 in Figs. 8.4a, b, respectively, for the on- and off-ramp bottlenecks. An empirical example of such speed adaptation can be seen in Fig. 4.3 for off-ramp bottleneck B1: Vehicles that remain on the main road must decelerate while adapting their speeds to the low speed of vehicles "off-c" and "off-d" going to the off-ramp.

In the vicinity of the bottlenecks (Figs. 8.4a, b), additionally to the speed adaptation effect there is also the opposite effect of over-acceleration through lane changing from

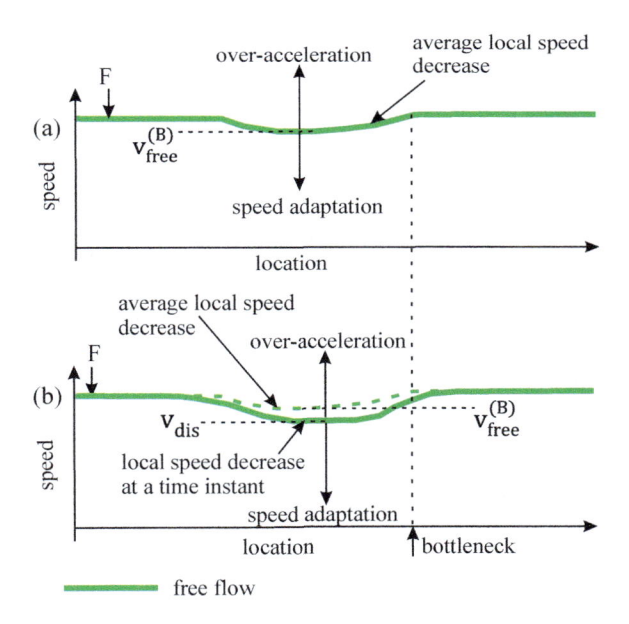

Fig. 8.5 Qualitative explanation of the competition between speed adaptation and over-acceleration at a bottleneck leading to a time-dependent local speed decrease: **a** The space distribution of the speed within the average local speed decrease in free flow at the bottleneck taken from Fig. 5.3a that is related to a given flow rate q satisfying conditions (7.6); the average local speed decrease corresponds to the averaging of the speed over a long enough time interval. **b** The space distribution of the speed within a local speed decrease at the bottleneck related to a time instant (solid curve) for the same value of the flow rate as that in (**a**); dashed curve shows the average local speed decrease taken from (**a**). Arrows show symbolically over-acceleration (up arrow) that tends to increase the speed and speed adaptation (down arrow) that tends to decrease the speed within the local speed decrease. v_{dis} is the minimum speed within the local speed decrease at a time instant; $v_{free}^{(B)}$ is the minimum speed within the average local speed decrease at the bottleneck

the right lane to the left faster lane (Fig. 8.1b). Qualitatively, this over-acceleration due to lane changing to a faster lane is labeled by "over-acceleration" for vehicle 3 that passes a slower moving preceding vehicle 4 in Figs. 8.4a, b. The speed increase caused by over-acceleration prevents a continuous speed decrease due to speed adaptation within a local speed decrease in free flow at the bottleneck. Symbolically, the competition between the speed adaptation and over-acceleration is shown by arrows labeled by "over-acceleration" and "speed adaptation" in Fig. 8.5a.

> The competition between driver speed adaptation and over-acceleration determines features of a local speed decrease at a bottleneck.

8.2.2 Critical Speed Within Local Speed Decrease at Bottleneck

In a real local speed decrease in free flow at a bottleneck, speeds of different vehicles change over time considerably. Therefore, there is a *time-dependence* of the speed distribution within the local speed decrease at the bottleneck. For this reason, we should distinguish between the *average* local speed decrease discussed in Sect. 5.2 (Fig. 5.3a) that is labeled by "average local speed decrease" in Fig. 8.5a and the local speed decreases related to different time instants. One of the possible spatial distributions of the speed within the local speed decrease at a time instant is shown by a solid curve in Fig. 8.5b. Respectively, the time-dependent minimum speed within the local speed decrease will be designated by $v_{\rm dis}$ (Fig. 8.5b), to distinguish it from the time-independent minimum speed $v_{\rm free}^{(B)}$ within the average local speed decrease at the bottleneck. Over time, $v_{\rm dis}$ can increase or decrease in comparison with $v_{\rm free}^{(B)}$. We consider below only those speed changes in the local speed decrease at the bottleneck that cause the *decrease* in $v_{\rm dis}$ in comparison with $v_{\rm free}^{(B)}$, i.e., $v_{\rm dis} < v_{\rm free}^{(B)}$ (Figs. 8.5b and 8.6).

Here we apply the discontinuous character of over-acceleration (Fig. 8.2) together with the competition between speed adaptation and over-acceleration (Fig. 8.5) for the explanation of the driver behaviors that are responsible for the nucleation nature of traffic breakdown (F→S transition) at the bottleneck (Figs. 8.6 and 8.7). We assume that the flow rate q in free flow at the bottleneck is within the flow-rate range $C_{\min} \leq q < C_{\max}$ (7.6). From Fig. 8.3, we can see that in this case free flow is in a metastable state with respect to the F→S transition (traffic breakdown) at the bottleneck.

At two different time instants, possible speed distributions within a local speed decrease at a bottleneck are qualitatively shown in Figs. 8.6a, b. At the first time instant, the minimum speed $v_{\rm dis}$ within the local speed decrease (Fig. 8.6a) that is taken from Fig. 8.5b is only slightly lower than the minimum speed $v_{\rm free}^{(B)}$ within the

8.2 Nucleus Occurrence for Spontaneous Traffic Breakdown ...   111

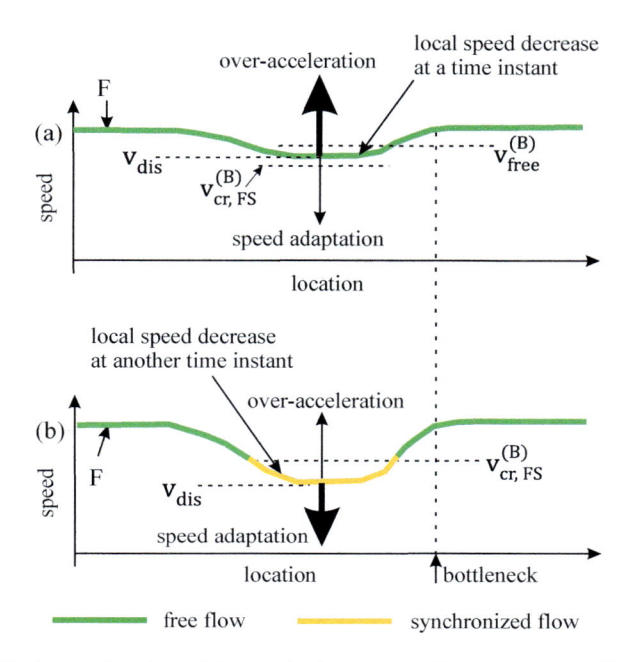

Fig. 8.6 Qualitative explanation of the nucleation nature of spontaneous traffic breakdown in metastable free flow with respect to an F→S transition at a bottleneck: **a** At a time instant, the minimum speed $v_{\rm dis}$ within the local speed decrease at a bottleneck is larger than the critical speed $v_{\rm cr, FS}^{\rm (B)}$ for the F→S transition: No traffic breakdown occurs because the over-acceleration (thick up arrow) overcomes on average speed adaptation (thin down arrow). **b** At another time instant, the minimum speed $v_{\rm dis}$ within the local speed decrease at the bottleneck is less than the critical speed $v_{\rm cr, FS}^{\rm (B)}$: Traffic breakdown does occur because the speed adaptation (thick down arrow) overcomes on average over-acceleration (thin up arrow). In (**a**, **b**), the space distributions of the speed are related to the same flow rate q as that in Fig. 8.5; this flow rate satisfies conditions of the free flow metastability (7.6). In (**a, b**), $v_{\rm dis}$ is less than the minimum speed $v_{\rm free}^{\rm (B)}$ within the average local speed decrease at the bottleneck

average local speed decrease. We assume that the speed $v_{\rm dis}$ is related to a free flow speed (Fig. 8.6a).

In accordance with the discontinuous character of over-acceleration shown in Fig. 8.2, the mean time delay of over-acceleration in free flow is short. Therefore, when the minimum speed $v_{\rm dis}$ within the local speed decrease is large enough (Figs. 8.6a and 8.7b), over-acceleration overcomes on average speed adaptation within the local speed decrease at the bottleneck. In this case, the occurrence of the local speed decrease at the bottleneck does not lead to traffic breakdown. This means that the local speed decrease in free flow at the bottleneck shown in Figs. 8.6a and 8.7b is *not* a nucleus for traffic breakdown in metastable free flow at the bottleneck. This case is related to time $t < t_0$ in Fig. 8.7a, when free flow persists at the bottleneck.

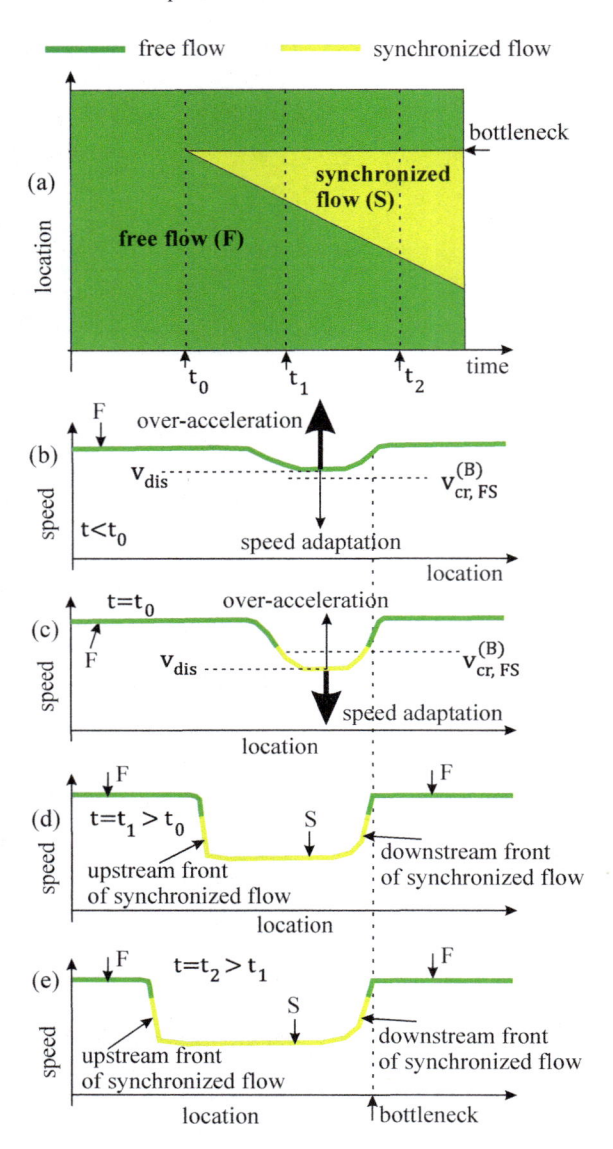

Fig. 8.7 Qualitative explanation of spontaneous traffic breakdown shown in Fig. 5.7 through the nucleation nature of an F→S transition at a bottleneck: **a** Free flow (green) and synchronized flow (yellow) in the space–time plane taken from Fig. 5.7a. **b** Local speed decrease at the bottleneck at $t < t_0$ that does not lead to traffic breakdown; the local speed decrease is taken from Fig. 8.6a. **c** Local speed decrease at the bottleneck at $t = t_0$ causing traffic breakdown; the local speed decrease is taken from Fig. 8.6b. **d, e** Development of synchronized flow at two time instants $t = t_1 > t_0$ (**d**) and $t = t_2 > t_1$ (**e**) after traffic breakdown has occurred related to Figs. 5.7c, d, respectively. The flow rate q is the same one as that in Fig. 8.6 that satisfies conditions of the free flow metastability (7.6); v_{dis} is the minimum speed within the local speed decrease at the bottleneck at a time instant; $v_{\text{cr, FS}}^{(B)}$ is the critical speed for the F→S transition at the bottleneck

8.2 Nucleus Occurrence for Spontaneous Traffic Breakdown ...

In contrast to the case shown in Figs. 8.6a and 8.7b, we assume now that at the second time instant, the minimum speed $v_{\rm dis}$ within the local speed decrease (Figs. 8.6b and 8.7c) is related to a synchronized flow speed. In accordance with the discontinuous character of over-acceleration shown in Fig. 8.2, the mean time delay of over-acceleration in synchronized flow is long. Therefore, when the minimum speed $v_{\rm dis}$ within the local speed decrease is low enough, speed adaptation overcomes on average over-acceleration within this local speed decrease. This case is related to time instant $t = t_0$ in Fig. 8.7a, c, when traffic breakdown does occur at the bottleneck: Synchronized flow occurring within the local speed decrease propagates upstream of the bottleneck (Figs. 8.7a, c–e). Thus, we can conclude that there are the following two opposite cases:

1. When at a time instant the minimum speed $v_{\rm dis}$ within a local speed decrease at the bottleneck is large enough and it is related to a free flow speed, then, in accordance with Fig. 8.2, the mean time delay in over-acceleration is short. Therefore, over-acceleration overcomes on average speed adaptation within the local speed decrease in free flow at the bottleneck: No traffic breakdown occurs.
2. In contrast, when at another time instant the minimum speed $v_{\rm dis}$ within a local speed decrease at the bottleneck is low enough and it is related to a synchronized flow speed, then, in accordance with Fig. 8.2, the mean time delay in over-acceleration is long. Therefore, speed adaptation overcomes on average over-acceleration within the local speed decrease at the bottleneck. In this case, the local speed decrease grows and traffic breakdown does occur at the bottleneck: Synchronized flow begins to propagate upstream of the bottleneck over time as this is qualitatively shown in Figs. 8.7d, e.

These two opposite cases exist for any metastable free flow state at a bottleneck. This means that for any given flow rate q at the bottleneck within the flow-rate range $C_{\min} \leq q < C_{\max}$ (7.6), there should be a *critical speed* within a local speed decrease for the occurrence of spontaneous traffic breakdown (F→S transition) at the bottleneck (Figs. 8.6 and 8.7b, c). We denote the critical speed for the F→S transition at the bottleneck by $v^{(B)}_{\rm cr,\,FS}$ (Figs. 8.6 and 8.7).[8] The critical speed has already been introduced in Sect. 6.4, to explain the evidence of the empirical induced traffic breakdown at a bottleneck. The presented study of the driver behaviors that

---

[8] It should be emphasized that many random local speed decreases can occur in free flow outside bottlenecks. The three-phase traffic theory states [16] that there is also another critical speed for an F→S transition in free flow *outside bottlenecks*. However, the F→S transition is observed very seldom outside bottlenecks (see, e.g., [4]). This is because a bottleneck introduces a local speed decrease in free flow (Chap. 4); this increases the probability of traffic breakdown (F→S transition) at the bottleneck considerably in comparison with the F→S transition outside the bottleneck. Nevertheless, it might be assumed that in real traffic there can be cases when a long enough road without bottlenecks is realized. This is possible, for example, due to a road work made at a long section of a highway where such a road is dedicated to passenger vehicles only (sometimes, the length of the road is longer than 10 km). However, we have no real field traffic data for such cases. Because in this book we consider mostly the traffic phenomena that have been observed in real vehicular traffic, we do not consider the F→S transition outside bottlenecks.

are responsible for the existence of the critical speed allows us to make the following conclusion.

> The cause of the existence of the critical speed within the critical nucleus required for traffic breakdown (F→S transition) at a bottleneck is the discontinuous character of over-acceleration together with the competition between speed adaptation with over-acceleration occurring within the local speed decrease at the bottleneck.

Based on a study of the critical speed made in this section we can define the critical speed as follows[9]:

(i) If at a time instant the minimum speed $v_{\rm dis}$ within a local speed decrease at a bottleneck is larger than the critical speed $v^{(B)}_{\rm cr,\, FS}$, no traffic breakdown occurs in metastable free flow at the bottleneck (Figs. 8.6a and 8.7b).

(ii) However, if at another time instant the minimum speed $v_{\rm dis}$ within another local speed decrease at the bottleneck is equal to or less than the critical speed $v^{(B)}_{\rm cr,\, FS}$, then the local speed decrease is a *nucleus* for traffic breakdown (F→S transition) in the metastable free flow at the bottleneck (Figs. 8.6b and 8.7c): Spontaneous traffic breakdown does occur at the bottleneck with resulting upstream propagation of synchronized flow (Fig. 8.7d, e).[10]

> A local speed decrease at a bottleneck within which the speed is equal to or less than the critical speed is a nucleus for traffic breakdown (F→S transition) in metastable free flow with respect to the F→S transition at the bottleneck.

---

[9] A dependence of the critical speed $v^{(B)}_{\rm cr,\, FS}$ on the flow rate q in free flow at a bottleneck will be considered in Sect. A.8.1 of Appendix A.

[10] In a metastable state of free flow with respect to traffic breakdown (F→S transition) at a bottleneck, traffic breakdown can occur independent of traffic control at the bottleneck. This has been the reason for a *congested pattern control approach* introduced by the author in [17–23, 25–28, 35, 37, 52, 54–57, 59]. In the congested pattern control approach, *no control* of traffic flow at the bottleneck is applied as long as free flow is realized at the bottleneck. This means that the occurrence of random traffic breakdown is permitted to occur at the bottleneck. Only after traffic breakdown has occurred, traffic control starts. A detailed consideration of the congested pattern control approach can be found in [17, 19, 24, 40, 41].

8.3 Driver Behaviors Resulting in Nucleation Nature of Traffic Breakdown (F→S Transition) at Bottleneck: A Summary

We can summarize predictions of the three-phase traffic theory as follows:

1. The origin of the empirical nucleation nature of traffic breakdown (F→S transition) at a bottleneck is the discontinuous character of over-acceleration.
2. The discontinuous character of over-acceleration together with the competition between speed adaptation and over-acceleration within a local speed decrease at the bottleneck determine the following empirical features of vehicular traffic:
   - The range of highway capacities of free flow at the bottleneck.
   - The metastability of free flow with respect to traffic breakdown (F→S transition) at the bottleneck.
   - The existence of the critical speed for traffic breakdown at the bottleneck.
3. In a metastable state of free flow, any local speed decrease at the bottleneck within which the speed is equal to or less than the critical speed is a *nucleus* for traffic breakdown (F→S transition) at the bottleneck.[11]
4. The nucleus for traffic breakdown can *randomly* appear in the metastable state of free flow at the bottleneck. When the nucleus appears, *spontaneous* traffic breakdown (spontaneous F→S transition) does occur at the bottleneck and synchronized flow begins to propagate upstream of the bottleneck.
5. When free flow at the bottleneck is in the metastable state with respect to traffic breakdown (F→S transition), then *any congested pattern* that propagates through the bottleneck acts as a nucleus initiating *induced* traffic breakdown (induced F→S transition) at the bottleneck.[12]

> The cause of the empirical nucleation nature of traffic breakdown (F→S transition) at a bottleneck is the discontinuous character of over-acceleration together

---

[11] It should be emphasized that a nucleus for traffic breakdown (F→S transition) at a bottleneck can be observed in real field traffic data (see Sect. 9.1). Contrary to the nucleus for traffic breakdown, it is probably impossible to measure in empirical traffic data the exact value of the critical speed for traffic breakdown (F→S transition) at the bottleneck. Indeed, the critical speed separates two cases: (i) A local speed decrease at the bottleneck, within which the speed is higher than the critical speed, does not lead to traffic breakdown. (ii) Another local speed decrease at the bottleneck, within which the speed is equal to or less than the critical speed is a nucleus for traffic breakdown (F→S transition) at the bottleneck. In empirical data, we can only find that either case (i) or case (ii) is realized, whereas the critical speed separating cases (i) and (ii) is very difficult to measure.

[12] It should be noted that a congested pattern can be considered the nucleus for the empirical induced traffic breakdown at a bottleneck only if conditions for the empirical induced traffic breakdown at the bottleneck discussed in Sects. 6.3 and 7.4 are satisfied.

with the competition between speed adaptation and over-acceleration within the local speed decrease in free flow at the bottleneck.

The occurrence of a nucleus for traffic breakdown in metastable free flow with respect to an F→S transition at a bottleneck explains the empirical spontaneous traffic breakdown at the bottleneck.

The empirical *spontaneous* traffic breakdown at a bottleneck occurs through the random occurrence of a nucleus for traffic breakdown in metastable free flow at the bottleneck. A congested pattern propagating through the bottleneck acts as a nucleus for the empirical *induced* traffic breakdown in metastable free flow at the bottleneck. Spontaneous and induced traffic breakdowns are distinguished only through the origin of the nucleus for traffic breakdown (F→S transition) at the bottleneck (for more details, see Sect. 9.3).

References

1. B.S. Kerner, Phys. Rev. Lett. **81**, 3797–3800 (1998)
2. B.S. Kerner, in *Proceedings of the 3rd Symposium on Highway Capacity and Level of Service*, ed. by R. Rysgaard (Road Directorate, Copenhagen, Ministry of Transport – Denmark 1998), pp. 621–642
3. B.S. Kerner, Transp. Res. Rec. **1678**, 160–167 (1999)
4. B.S. Kerner, in *Transportation and Traffic Theory*, ed. by A. Ceder (Elsevier Science, Amsterdam, 1999), pp. 147–171
5. B.S. Kerner, Phys. World **12**, 25–30 (August 1999)
6. B.S. Kerner, J. Phys. A: Math. Gen. **33**, L221–L228 (2000)
7. B.S. Kerner, in *Traffic and Granular Flow '99: Social, Traffic and Granular Dynamics*, ed. by D. Helbing, H.J. Herrmann, M. Schreckenberg, D.E. Wolf (Springer, Heidelberg, 2000), pp. 253–284
8. B.S. Kerner, Transp. Res. Rec. **1710**, 136–144 (2000)
9. B.S. Kerner, Netw. Spat. Econ. **1**, 35–76 (2001)
10. B.S. Kerner, Transp. Res. Rec. **1802**, 145–154 (2002)
11. B.S. Kerner, Math. Comput. Model. **35**, 481–508 (2002)
12. B.S. Kerner, in *Traffic and Transportation Theory in the 21st Century*, ed. by M.A.P. Taylor (Elsevier Science, Amsterdam, 2002), pp. 417–439
13. B.S. Kerner, Phys. Rev. E **65**, 046138 (2002)
14. B. S. Kerner, in *Traffic and Granular Flow' 01*, ed. by M. Fukui, Y. Sugiyama, M. Schreckenberg, D.E. Wolf (Springer, Berlin, 2003), pp.13–50
15. B.S. Kerner, Phys. A **333**, 379–440 (2004)
16. B.S. Kerner, *The Physics of Traffic* (Springer, Berlin, 2004)

References

17. B.S. Kerner, Phys. A **355**, 565–601 (2005)
18. B.S. Kerner, in *Traffic and Transportation Theory*, ed. by H. Mahmassani (Elsevier Science, Amsterdam, 2005), pp. 181–203
19. B.S. Kerner, IEEE Trans. ITS **8**, 308–320 (2007)
20. B.S. Kerner, Transp. Res. Rec. **1999**, 30–39 (2007)
21. B.S. Kerner, Transp. Res. Rec. **2088**, 80–89 (2008)
22. B.S. Kerner, J. Phys. A: Math. Theor. **41**, 215101 (2008)
23. B.S. Kerner, in *Transportation Research Trends*, ed. by P.O. Inweldi (Nova Science Publishers Inc., New York, 2008), pp. 1–92
24. B.S. Kerner, *Introduction to Modern Traffic Flow Theory and Control* (Springer, Heidelberg, 2009)
25. B.S. Kerner, J. Phys. A: Math. Theor. **44**, 092001 (2011)
26. B.S. Kerner, Phys. Rev. E **84**, 045102(R) (2011)
27. B.S. Kerner, Transp. Res. Circ. **E-C149**, 22–44 (2011)
28. B.S. Kerner, Phys. Rev. E **85**, 036110 (2012)
29. B.S. Kerner, Phys. A **392**, 5261–5282 (2013)
30. B.S. Kerner, Europhys. Lett. **102**, 28010 (2013)
31. B.S. Kerner, Phys. A **397**, 76–110 (2014)
32. B.S. Kerner, Elektrotech. Inf. **132**, 417–433 (2015)
33. B.S. Kerner, Phys. Rev. E **92**, 062827 (2015)
34. B.S. Kerner, in *Vehicular Communications and Networks*, ed. by W. Chen (Woodhead Publishings, Cambridge, 2015), pp. 223–254
35. B.S. Kerner, Eur. Phys. B J. **89**, 199 (2016)
36. B.S. Kerner, Phys. A **450**, 700–747 (2016)
37. B.S. Kerner, Phys. A **466**, 626–662 (2017)
38. B.S. Kerner, *Breakdown in Traffic Networks* (Springer, Berlin, 2017)
39. B.S. Kerner, Phys. Rev. E **97**, 042303 (2018)
40. B.S. Kerner, in *Complex Dynamics of Traffic Management*, ed. by B.S. Kerner. Encyclopedia of Complexity and Systems Science Series (Springer, New York, 2019), pp. 21–77
41. B.S. Kerner, in *Complex Dynamics of Traffic Management*, ed. by B.S. Kerner. Encyclopedia of Complexity and Systems Science Series (Springer, New York, 2019), pp. 195–283
42. B.S. Kerner, Phys. Rev. E **100**, 012303 (2019)
43. B.S. Kerner, Phys. A **562**, 125315 (2021)
44. B.S. Kerner, P. Hemmerle, M. Koller, G. Hermanns, S.L. Klenov, H. Rehborn, M. Schreckenberg, Phys. Rev. E **90**, 032810 (2014)
45. B.S. Kerner, S.L. Klenov, J. Phys. A: Math. Gen. **35**, L31–L43 (2002)
46. B.S. Kerner, S.L. Klenov, Phys. Rev. E **68**, 036130 (2003)
47. B.S. Kerner, S.L. Klenov, J. Phys. A: Math. Gen. **37**, 8753–8788 (2004)
48. B.S. Kerner, S.L. Klenov, J. Phys. A: Math. Gen. **39**, 1775–1809 (2006)
49. B.S. Kerner, S.L. Klenov, Transp. Res. Rec. **1965**, 70–78 (2006)
50. B.S. Kerner, S.L. Klenov, Phys. Rev. E **80**, 056101 (2009)
51. B.S. Kerner, S.L. Klenov, J. Phys. A: Math. Theor. **43**, 425101 (2010)
52. B.S. Kerner, S.L. Klenov, G. Hermanns, M. Schreckenberg, Phys. A **392**, 4083–4105 (2013)
53. B.S. Kerner, S.L. Klenov, G. Hermanns, P. Hemmerle, H. Rehborn, M. Schreckenberg, Phys. Rev. E **88**, 054801 (2013)
54. B.S. Kerner, S.L. Klenov, A. Hiller, J. Phys. A: Math. Gen. **39**, 2001–2020 (2006)
55. B.S. Kerner, S.L. Klenov, A. Hiller, Nonlinear Dyn. **49**, 525–553 (2007)
56. B.S. Kerner, S.L. Klenov, A. Hiller, H. Rehborn, Phys. Rev. E **73**, 046107 (2006)
57. B.S. Kerner, S.L. Klenov, M. Schreckenberg, Phys. Rev. E **84**, 046110 (2011)
58. B.S. Kerner, S.L. Klenov, M. Schreckenberg, J. Stat. Mech. **P03001** (2014)
59. B.S. Kerner, S.L. Klenov, M. Schreckenberg, Phys. Rev. E **89**, 052807 (2014)
60. B.S. Kerner, S.L. Klenov, D.E. Wolf, J. Phys. A: Math. Gen. **35**, 9971–10013 (2002)
61. B.S. Kerner, M. Koller, S.L. Klenov, H. Rehborn, M. Leibel, Phys. A **438**, 365–397 (2015)
62. B.S. Kerner, H. Rehborn, R.-P. Schäfer, S.L. Klenov, J. Palmer, S. Lorkowski, N. Witte, Phys. A **392**, 221–251 (2013)

Chapter 9
Understanding Empirical Nuclei for Traffic Breakdown (F→S Transition) at Bottleneck

The three-phase traffic theory (Chap. 8) predicts that even if there is no traffic congestion downstream and upstream of a bottleneck at which metastable free flow with respect to an F→S transition is realized, a nucleus for traffic breakdown can *randomly* appear at the bottleneck resulting in spontaneous traffic breakdown at the bottleneck. However, the following question can arise:

- Can a nucleus for spontaneous traffic breakdown (spontaneous F→S transition) be observed in real traffic?

The main objective of this chapter is to show that nuclei for spontaneous traffic breakdown can indeed be observed in real vehicular traffic (Sects. 9.1–9.3). In the remainder of this chapter, we discuss empirical proof of the existence of the finite value of the mean time delay in over-acceleration (Sect. 9.4).[1]

9.1 Nucleus for Empirical Spontaneous Traffic Breakdown

One of the main results of the three-phase traffic theory presented in Chap. 8 is that there should be a critical speed within a local speed decrease at a bottleneck: When the speed within the local speed decrease is less than the critical speed, then this local speed decrease should be a nucleus for traffic breakdown in free flow that is in a metastable state with respect to an F→S transition at the bottleneck.

To prove empirically this theoretical prediction of the three-phase traffic theory, we should note that real vehicular traffic is usually a very heterogeneous traffic: Besides passenger vehicles, there can be different trucks and other heavy long vehicles in the

---

[1] The methodological basis of this chapter is related to Refs. [2–6, 9, 11–13]. Empirical results discussed in this chapter have been derived in [1–5, 12–15]. For more details, see the books [7, 8, 10, 16].

© The Author(s), under exclusive license to Springer Nature Switzerland AG 2021
B. S. Kerner, *Understanding Real Traffic*,
https://doi.org/10.1007/978-3-030-79602-0_9

heterogeneous traffic. Some of the trucks have a lower maximum speed in free flow in comparison with passenger vehicles. Moreover, in some countries (for example, in Germany) trucks and other heavy vehicles have a limitation of the maximum speed. Thus, we can expect that in real free heterogeneous flow, there can be trucks that are slow vehicles in comparison with faster passenger vehicles; therefore, a truck in heterogeneous free flow can act as a moving bottleneck (MB) discussed in Sect. 4.3. In this section, we will show that this generally well-known fact allows us to observe empirical nuclei for spontaneous traffic breakdown (F→S transition) at a bottleneck.

9.1.1 Waves in Heterogeneous Free Flow: Qualitative Consideration

As explained in Sect. 4.3, an MB causes the wave of the flow-rate increase (blue wave in Fig. 4.4f) and the wave of the local decrease in the average speed of the passenger vehicles (gray wave in Fig. 4.4g). However, a single slow vehicle that acts as the MB in free flow shown in Fig. 4.4a, e is a rough simplification of heterogeneous free flow: Many slow vehicles can exist in heterogeneous free flow. Sometimes slow vehicles are distributed very non-homogeneously in traffic flow. For example, there can be local regions of the increase in the share of slow vehicles within which the share of slow vehicles is considerably larger than it is in other regions of heterogeneous traffic flow.

To understand empirical results presented in Sect. 9.1.2 below, we consider firstly a hypothetical case, when each of the regions of the increase in the share of slow vehicles can be considered a group of several slow vehicles moving at the same speed v_{MB} in the right lane of the road. The group of the slow vehicles can be considered a wave of the increase in the share of the slow vehicles. Each group of the slow vehicles acts as an MB with the speed v_{MB}.

In real heterogeneous traffic flow, there can be a random sequence of such groups of slow moving vehicles resulting in the random sequence of MBs that move at the speed v_{MB} (green waves in Fig. 9.1a). In accordance with explanations of the effect of an MB on free flow made in Sect. 4.3 (Fig. 4.4), any MB related to a group of the slow vehicles should also cause a wave of the flow-rate increase and a wave of the local decrease in the average speed of the passenger vehicles. Therefore, the random sequence of MBs (labeled by "waves of increase in share of slow vehicles (sequence of MBs)" in Fig. 9.1a) causes the associated sequences of the waves of the flow-rate increase (blue waves in Fig. 9.1b) and the waves of the local speed decrease (gray waves in Fig. 9.1c) localized in the vicinity of the associated MBs (green waves in Fig. 9.1a); all waves move at the same speed v_{MB}.

9.1 Nucleus for Empirical Spontaneous Traffic Breakdown

Fig. 9.1 Qualitative explanation of waves in free flow caused by a random sequence of moving bottlenecks (MBs): The sequence of MBs (**a**) results in the waves of the increase in the flow rate (**b**) and in the waves of the decrease in the average speed (**c**). Each of the waves of the decrease in the average speed (**c**) can be considered a local speed decrease localized at the MB that moves at the speed v_{MB}. The cause of each of the waves in (**a**–**c**) is the same as that for the waves shown in Figs. 4.4e–g, respectively

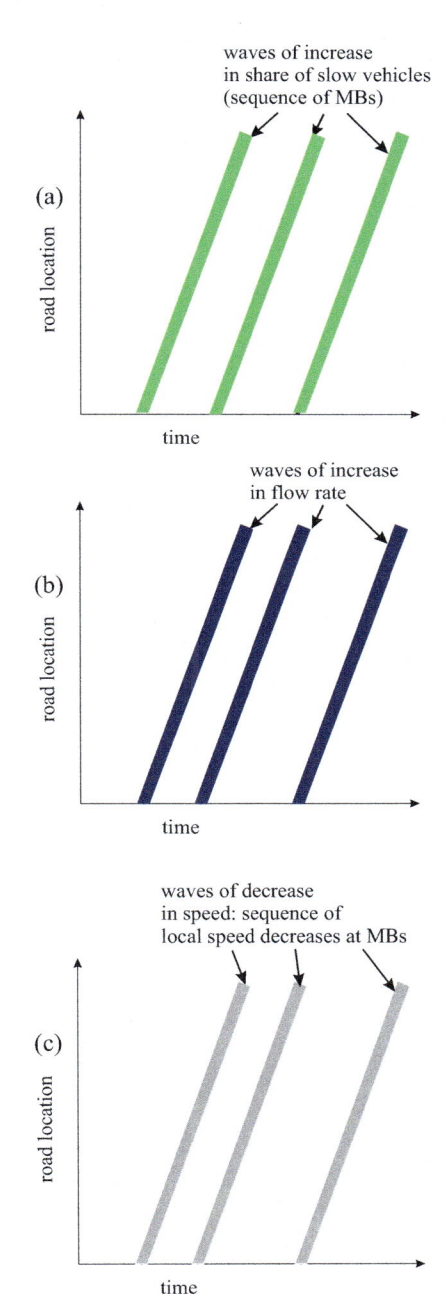

9.1.2 Empirical Speed Waves in Heterogeneous Free Flow: Local Speed Decreases at Sequence of Moving Bottlenecks

We consider empirical sequences of MBs caused by trucks and other slow vehicles in free flow found in real field traffic data measured on German highways (Fig. 9.2). It should noted that in Germany trucks have a speed limit 80 km/h (in reality, trucks usually move at the speed within a speed range 80–90 km/h). In accordance with the theoretical waves of the share of slow vehicles shown in Fig. 9.1a, we have found qualitatively similar empirical waves of the share of slow vehicles in free flow (green waves in Fig. 9.2a).

In accordance with theoretical explanations made in Sect. 9.1.1 (Fig. 9.1b, c), the empirical waves of the increase in the share of slow vehicles in free flow (green waves in Fig. 9.2a) act as a sequence of MBs while resulting in empirical waves of the increase in the flow rate (blue waves in Fig. 9.2b) and empirical waves of the decrease in the average vehicle speed (gray waves in Fig. 9.2c).

As predicted in Sect. 9.1.1, each of the waves propagates downstream with the mean wave velocity that is approximately equal to the mean speed of slow vehicles that changes within range 85–88 km/h (Fig. 9.2). The empirical waves can propagate through the whole road section. Because vehicles enter the main road from on-ramps and leave the main road to off-ramps, some of the waves can appear at on-ramps or disappear at off-ramps.

> In empirical heterogeneous free flow, there is a random sequence of waves of the increase in the share of slow vehicles. Each of the waves can act as an MB propagating in the flow direction on average with the speed of slow vehicles: At the MB, the share of slow vehicles and the flow rate are larger, whereas the average vehicle speed is lower than outside the MB.

> In accordance with the empirical results (Fig. 9.2), we define the term *moving bottleneck* (MB) in free flow as follows: A single vehicle or a group of vehicles that move slower than other vehicles can be considered an MB in heterogeneous free flow if there is a local speed decrease localized at the MB; the local speed decrease moves together with the MB at the MB speed.

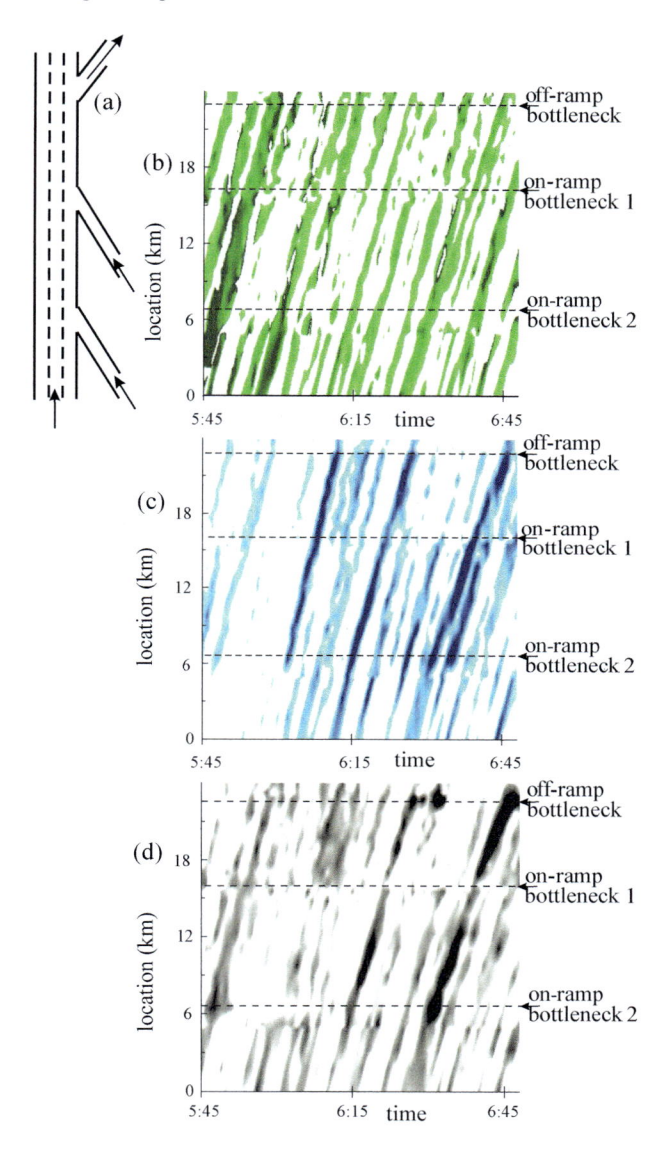

9.1 Nucleus for Empirical Spontaneous Traffic Breakdown 123

Fig. 9.2 Empirical waves in free flow averaged across highway. Real field traffic data measured by road detectors on three-lane freeway A5-South in Germany on September 03, 1998: **a** Sketch of section of three-lane highway in Germany with off- and on-ramps bottlenecks. **b** Waves of dimensionless increase in the share of slow vehicles (MBs) are presented by regions with variable shades of green color (darker color green in the waves indicates a larger increase in the share of slow vehicles). **c** Waves of increase in the flow rate are presented by regions with variable shades of blue color (darker color blue in the waves indicates a larger increase in the flow rate). **d** Waves of decrease in the average speed at MBs are presented by regions with variable shades of gray (darker color gray in the waves indicates a larger decrease in the average speed). Adapted from [12]

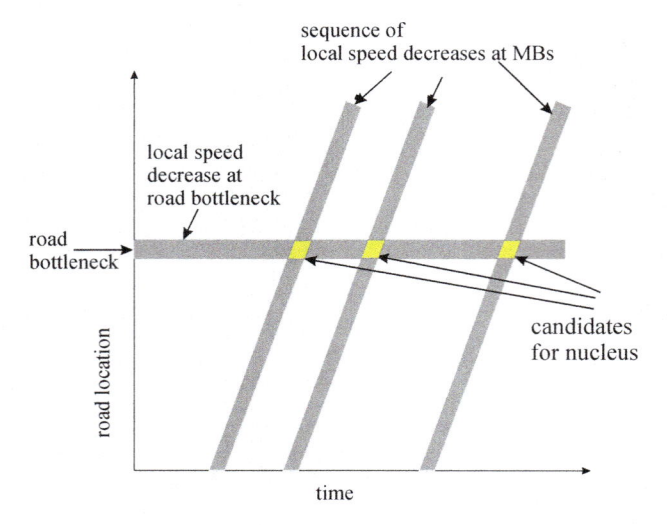

Fig. 9.3 Qualitative explanation of the occurrence of a nucleus for traffic breakdown at a road bottleneck during the propagation of an MB through the bottleneck: There is a local speed decrease localized at the road bottleneck (labeled by "local speed decrease at road bottleneck"). Additionally, there is a sequence of local speed decreases at MBs taken from Fig. 9.1c propagating downstream in free flow (labeled by "sequence of local speed decreases at MBs"). While propagating through the location of the road bottleneck, each of the local speed decreases at the MBs causes some additional speed decrease within the local speed decrease localized at the road bottleneck. Each of the additional speed decreases at the road bottleneck caused by the MB propagation through the road bottleneck can be considered a candidate for a nucleus for traffic breakdown at the road bottleneck (additional speed decreases are labeled by yellow regions "candidates for nucleus")

9.1.3 A Mechanism of Nucleus Occurrence in Heterogeneous Free Flow at Road Bottleneck

To understand the mechanism of the nucleus occurrence in heterogeneous flow at a road bottleneck, we recall that there is a local speed decrease in free flow at the road bottleneck (Sect. 4.1.1). The speed within the local speed decrease is less than outside the local speed decrease. The local speed decrease in free flow at the bottleneck is qualitatively shown by a gray region localized in the bottleneck vicinity (labeled by "local speed decrease at road bottleneck" in Fig. 9.3).

The sequence of MBs causes the associated sequence of local speed decreases moving together with the MBs at the MB speed (Figs. 9.1c and 9.2d). The local speed decreases at the MBs taken from Fig. 9.1c are labeled by "sequence of local speed decreases at MBs" in Fig. 9.3. It can be expected that when a local speed decrease at an MB reaches the road bottleneck location, the local speed decrease localized at the MB leads to an additional decrease in the speed within the local speed decrease localized at the road bottleneck.

9.1 Nucleus for Empirical Spontaneous Traffic Breakdown   125

Due to this additional decrease in the speed caused by the MB propagation through the local speed decrease localized at the road bottleneck, a critical speed for traffic breakdown (F→S transition) at the road bottleneck (see Sect. 8.2.2) can more easily be reached. For this reason, a nucleus for traffic breakdown can occur with a considerably larger probability during a time interval, when one of the MBs propagates through the road bottleneck. For this reason, each of the additional speed decreases within the local speed decrease at the road bottleneck caused by the MB propagation through the road bottleneck can be considered a candidate for a nucleus for traffic breakdown at the road bottleneck (regions "candidates of nucleus" labeled by colored yellow in Fig. 9.3).

9.1.4 Random Occurrence of Nucleus for Empirical Spontaneous Traffic Breakdown

As we explain below, the qualitative consideration of the mechanism of the nucleus occurrence in free flow of Sect. 9.1.3 is confirmed by empirical data presented in Figs. 9.4 and 9.5. It should be noted that the empirical data shown in Figs. 9.2, 9.4, and 9.5 are the same as those considered already in Fig. 7.1: Traffic breakdown that occurs at the off-ramp bottleneck leads to the emergence of a complex spatiotemporal congested pattern upstream of the off-ramp bottleneck (Fig. 7.1). Thus, with the use of empirical Figs. 9.4 and 9.5 we discuss below the occurrence of the empirical nucleus that has led to the empirical spontaneous traffic breakdown at the off-ramp bottleneck labeled by "spontaneous traffic breakdown" in Fig. 7.1.

During a long time interval, waves of the decrease in speed caused by local speed decreases at MBs in heterogeneous free flow do not lead to traffic breakdown while the waves propagate through the off-ramp bottleneck (Fig. 9.2). However, when we consider a longer time interval as that shown in Fig. 9.2, we find that randomly one of the waves of the decrease in speed, while propagating through the off-ramp bottleneck, initiates traffic breakdown at the off-ramp bottleneck: An empirical nucleus for an F→S transition (traffic breakdown) at the off-ramp bottleneck appears when this wave reaches the off-ramp bottleneck (Fig. 9.4). Thus, the propagation of the wave through the off-ramp bottleneck leads to the occurrence of the empirical nucleus for traffic breakdown at the off-ramp bottleneck (Fig. 9.4c).

- The propagation of the MB through the road bottleneck can lead to the occurrence of a nucleus for empirical sponataneous traffic breakdown at the road bottleneck.
- The physics of the occurrence of a nucleus for empirical spontaneous traffic breakdown at a road bottleneck can be explained by the interaction of the local speed decrease localized at an MB with the local speed decrease localized at the road bottleneck.

To see that in the empirical data (Fig. 9.4) a local speed decrease localized at an MB initiates a nucleus for traffic breakdown while the MB propagates through another local speed decrease localized at the road bottleneck, we consider empirical

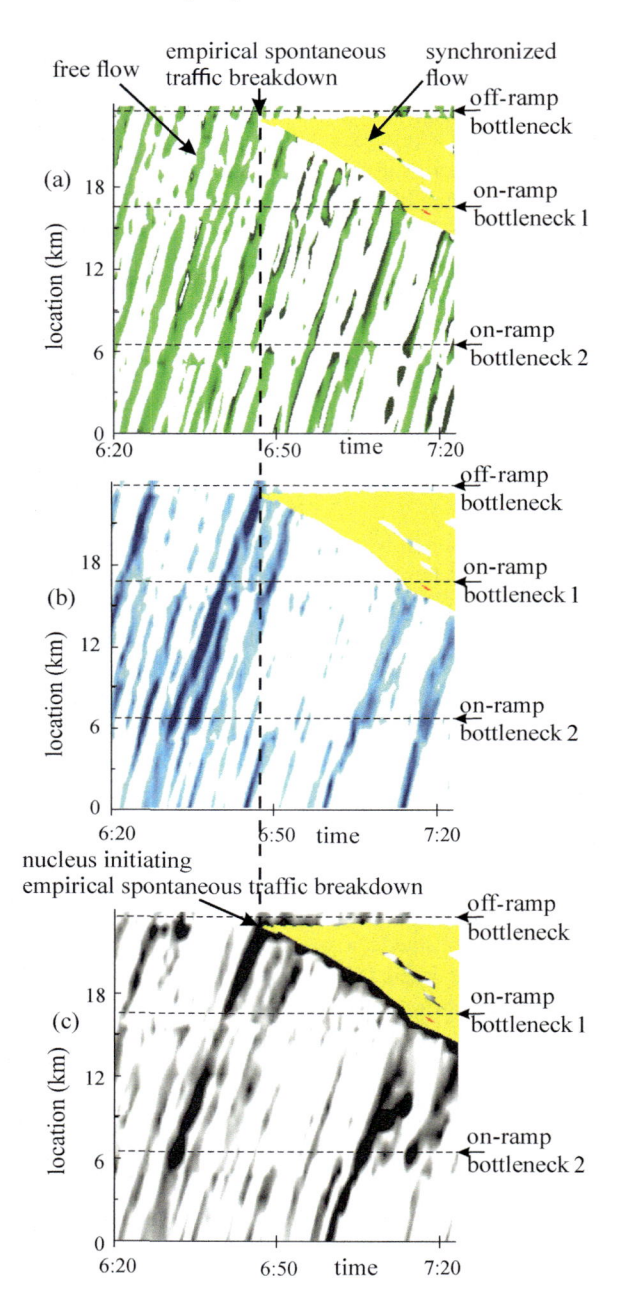

Fig. 9.4 Nucleus for empirical spontaneous traffic breakdown shown in Fig. 7.1. Empirical waves of dimensionless increase in the share of slow vehicles (**a**), of increase in the flow rate (**b**) and of decrease in the average speed (**c**) for a longer time interval than that shown in Fig. 9.2. In **a**–**c**, region labeled by "synchronized flow" shows symbolically synchronized flow occurring after traffic breakdown. Adapted from [12]

9.1 Nucleus for Empirical Spontaneous Traffic Breakdown

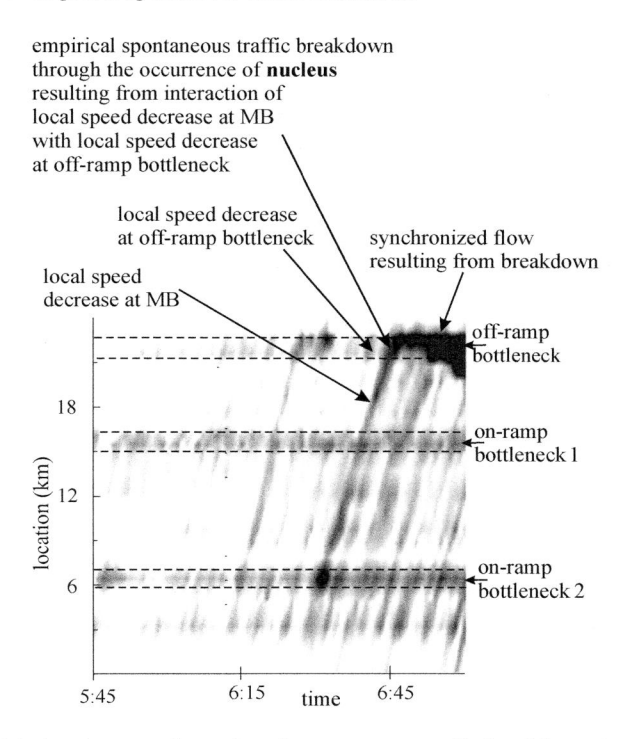

Fig. 9.5 Empirical emergence of a nucleus for spontaneous traffic breakdown shown in Figs. 7.1 and 9.4: A local speed decrease at an MB (gray wave labeled by "local speed decrease at MB") causes the nucleus when the MB propagates through a local speed decrease localized at the off-ramp bottleneck (double dashed lines labeled by "local speed decrease at off-ramp bottleneck"). Empirical data for the speed presented in space and time by regions with variable shades of gray; in white regions $v \geq 115$ km/h, in black regions $v \leq 80$ km/h. Off-ramp bottleneck, on-ramp bottleneck 1, and on-ramp bottleneck 2 are the same as those in Figs. 7.1 and 9.4. Adapted from [12]

waves of the speed averaged across the road (Fig. 9.5). It can be seen in Fig. 9.5 that additionally to local speed decreases localized at MBs propagating downstream, there are three narrow road regions, which are localized in neighborhoods of the locations of the off-ramp bottleneck, on-ramp bottleneck 1, and on-ramp bottleneck 2, respectively. Within these narrow regions, the speed is less than outside them (Fig. 9.5). The narrow regions are related to empirical local speed decreases in free flow at these road bottlenecks, respectively.

> An empirical local speed decrease localized at an MB in free flow can initiate a nucleus for empirical spontaneous traffic breakdown at a road bottleneck, when the MB propagates through the local speed decrease at the road bottleneck.

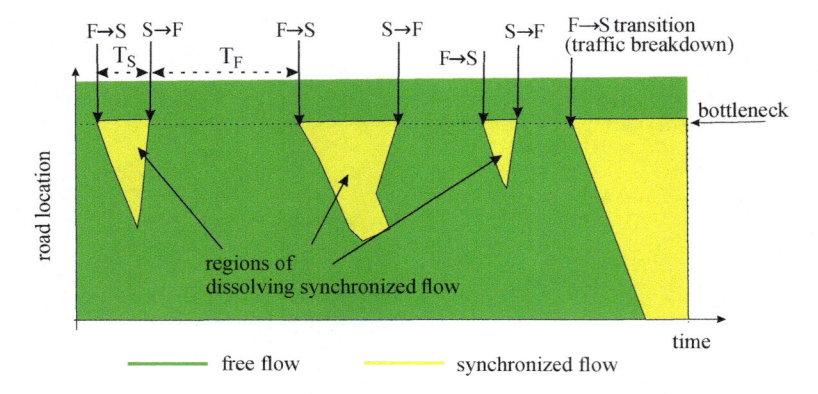

Fig. 9.6 Qualitative explanation of sequences of F→S→F transitions at a bottleneck in the road location–time plane. T_S and T_F are, respectively, a duration of dissolving synchronized flow caused by F→S→F transitions and a duration of free flow between two subsequent F→S→F transitions at the bottleneck; T_S and T_F are random values. Green—free flow, yellow—synchronized flow

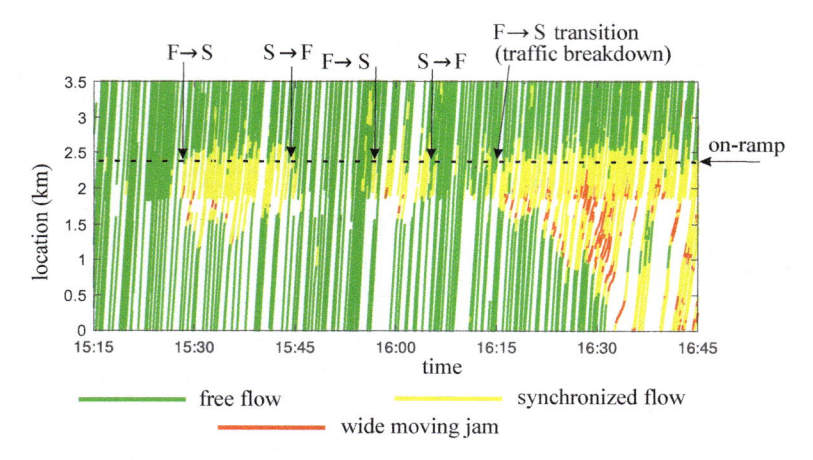

Fig. 9.7 Empirical sequences of F→S→F transitions at on-ramp bottleneck: Empirical trajectories of probe vehicles measured on October 25, 2016 on highway A81-North in Germany. Green—free flow, yellow—synchronized flow, red—wide moving jam. Adapted from [1]

9.2 Empirical Transitions from Free Flow to Synchronized Flow and Backwards Before Traffic Breakdown (F→S→F Transitions)

The objective of this section is a consideration of some empirical spatiotemporal features and characteristics of empirical nuclei for traffic breakdown.

We recall that a nucleus for traffic breakdown (F→S transition) at the bottleneck should result from a competition between driver speed adaptation and over-

9.2 Empirical Transitions from Free Flow to Synchronized ...

acceleration occurring within the local speed decrease at the bottleneck (Sect. 8.2.1): Due to speed adaptation, there is a tendency to the emergence of synchronized flow at the bottleneck. However, due to over-acceleration, there is an opposite tendency to free flow at the bottleneck. Over-acceleration occurs with a time delay. Therefore, we can assume that over-acceleration can start considerably later than speed adaptation. In this case, first the development of speed adaptation leads to an F→S transition at the bottleneck (labeled by the first arrow "F→S" in Fig. 9.6). The F→S transition results in synchronized flow that begins to propagate upstream. However, later the over-acceleration effect that is realized within the emergent synchronized flow at the bottleneck can lead to a return S→F transition (labeled by the first arrow "S→F" in Fig. 9.6).[2] Due to this S→F transition, free flow recovers at the bottleneck.

The occurrence of the emergent synchronized flow (F→S transition) following by the recovering of free flow at the bottleneck (S→F transition) can be considered a sequence of an F→S transition and a return S→F transition at the bottleneck (called as the F→S→F transitions) (Fig. 9.6). Due to the F→S→F transitions, a region of dissolving synchronized flow occurs (the first of the synchronized flow regions labeled by "regions of dissolving synchronized flow" in Fig. 9.6): No traffic breakdown leading to a long-living congested pattern is realized at the bottleneck.

After free flow has been recovered at the bottleneck, a new F→S transition can randomly occur at the bottleneck (labeled by the second arrow "F→S" in Fig. 9.6). As explained above, after some random time interval, due to the over-acceleration effect within the emergent synchronized flow, a return S→F transition can randomly occur (labeled by the second arrow "S→F" in Fig. 9.6). Therefore, new F→S→F transitions appear resulting in a new region of dissolving synchronized flow at the bottleneck (the second of the synchronized flow regions labeled by "regions of dissolving synchronized flow" in Fig. 9.6), and so on. In other words, sequences of the F→S→F transitions can randomly occur at the bottleneck (Fig. 9.6).

Recall that the occurrence of a nucleus for empirical spontaneous traffic breakdown is a *random event* (Sect. 9.1.4). Therefore, we can expect that the duration of synchronized flow at the bottleneck caused by the F→S→F transitions (T_S in Fig. 9.6), the duration of free flow between two subsequent F→S→F transitions (T_F in Fig. 9.6), and characteristics of associated different regions of dissolving synchronized flow should be random values.

An empirical example of sequences of the F→S→F transitions at an on-ramp bottleneck is presented in Fig. 9.7. In Fig. 9.7, traffic breakdown with the formation of a long-living congested pattern is observed at about 16:14. However, considerably earlier two subsequent sequences of the F→S→F transitions are observed. At about 15:28, due to an F→S transition at the bottleneck, synchronized flow appears and propagates upstream (labeled by first arrow "F→S" in Fig. 9.7). Then, an S→F transition occurs in the emergent synchronized flow: Synchronized flow dissolves

---

[2] The return S→F transition is realized through the development of an S→F instability in synchronized flow. For simplicity, we do not consider in the main text of the book the S→F instability that is the origin of the S→F transition in the synchronized flow. The S→F instability is briefly discussed in Sect. A.5 of Appendix A.

and free flow recovers at the bottleneck (labeled by first arrow "S→F" in Fig. 9.7). For this reason, the synchronized flow is called dissolving synchronized flow. Later, the subsequent F→S→F transitions are realized (labeled by second arrows "F→S" and "S→F" in Fig. 9.7).

There are many other empirical examples of sequences of the F→S→F transitions observed on different days at different bottlenecks in empirical traffic data.[3] It has been found that either the F→S transition or the S→F transition occurs at random time instants. Therefore, in accordance with the three-phase traffic theory (Fig. 9.6) random values of the duration of dissolving synchronized flow as well as random values of the duration of free flow that recovers between different F→S→F transitions have been observed in real field traffic data.

> In real traffic, there can be a sequence of the emergence of synchronized flow (F→S transition) following by the subsequent dissolution of synchronized flow (S→F transition) at a bottleneck. Such empirical F→S→F transitions can be explained by the discontinuous character of over-acceleration together with the competition between speed adaptation and over-acceleration within the emergent synchronized flow at the bottleneck.

The competition between speed adaptation and over-acceleration that causes the F→S→F transitions at the bottleneck is as follows. At a time instant, firstly, speed adaptation overcomes on average over-acceleration within the local speed decrease at the bottleneck. As a result, the nucleation of synchronized flow has occurred (F→S transition). At a later time instant, during upstream propagation of this synchronized flow, in contrast, over-acceleration overcomes on average speed adaptation within the emergent synchronized flow at the bottleneck. As a result, the nucleation of free flow within the synchronized flow (S→F transition) has occurred and, therefore, synchronized flow dissolves. This nucleation of free flow within the synchronized flow can also be considered as the *interruption* of the development of the synchronized flow at the bottleneck.[4]

The occurrence of empirical F→S→F transitions leads to the conclusion that a nucleus for traffic breakdown (F→S transition) at a bottleneck should be characterized by *spatiotemporal distributions* of traffic variables within the local speed decrease at the bottleneck. Therefore, the theoretical consideration of a critical speed within the local speed decrease at the bottleneck as the sole parameter of a nucleus required for traffic breakdown (F→S transition) at the bottleneck, as made in Sects. 6.4 and 8.2.2, is a rough simplification of real spatiotemporal characteristics of the nucleus.[5]

---

[3] See Refs. [1, 15].

[4] Therefore, we can consider the F→S→F transitions as a *nucleation-interruption* phenomenon in free flow at the bottleneck [8].

[5] A deeper insight in the spatiotemporal behavior of the nucleation of traffic breakdown can be found in Appendix A.

9.3 Is There a Difference Between Empirical Spontaneous and Induced Traffic Breakdowns?

As emphasized in Sect. 6.1.3, there can be found no qualitative difference between features of synchronized flow that has emerged at a bottleneck due to the empirical *spontaneous* traffic breakdown and the features of synchronized flow that has been induced at a bottleneck due to the empirical *induced* traffic breakdown. All empirical studies of empirical spontaneous and induced traffic breakdowns at highway bottlenecks confirm the following common result.

> Microscopic features of synchronized flow do not qualitatively depend on whether the empirical induced traffic breakdown or empirical spontaneous traffic breakdown is the cause for the synchronized flow.

As found in Sects. 6.2, 7.1, 7.2, and 8.2.2, when a nucleus for traffic breakdown appears at a bottleneck, then either the empirical induced traffic breakdown or empirical spontaneous traffic breakdown occurs in metastable free flow with respect to an F→S transition at the bottleneck. It must be emphasize that the *sole* difference between the empirical induced traffic breakdown and empirical spontaneous traffic breakdown is *the source* of the nucleus for traffic breakdown:

1. The nucleus for the empirical induced traffic breakdown at a bottleneck is a spatiotemporal congested pattern that reaches the bottleneck location due to the congested pattern propagation on the road. The congested pattern that induces traffic breakdown at the bottleneck has initially emerged outside the bottleneck. This congested pattern is relatively easily to be identified in traffic flow, as we have shown in Sects. 6.2 and 7.1.
2. In contrast with the empirical induced traffic breakdown, the identification of a nucleus for the empirical spontaneous traffic breakdown at a bottleneck can be very difficult. Indeed, additionally to the mechanism for the empirical nucleus occurrence considered in Sect. 9.1, there can be other mechanisms for the occurrence of the nucleus for empirical spontaneous traffic breakdown at the bottleneck.[6]
3. Contrary to empirical spontaneous traffic breakdown, the empirical induced traffic breakdown does not depend on the inhomogeneity of vehicular traffic as well as other traffic characteristics: If free flow is in a metastable state with respect to traffic breakdown (F→S transition) at a bottleneck, then, while reaching the bottleneck, a congested pattern does induce traffic breakdown at the bottleneck.

Thus, the sole difference between the empirical induced traffic breakdown and empirical spontaneous traffic breakdown is associated with the different source of the nucleus occurrence initiating traffic breakdown. In contrast with a congested

---

[6] Some mechanisms for the empirical nucleus occurrence have been discussed in the books [7, 8].

pattern that is the nucleus for the empirical induced traffic breakdown, there can be a considerable difficulty for the identification of a nucleus of empirical spontaneous traffic breakdown. It seems that the difficulty for the identification of a nucleus of empirical spontaneous traffic breakdown is one of the main reasons for the difficulty in *understanding real traffic*. This is the reason for the conclusion made above that the definitive empirical proof for the empirical nucleation nature of traffic breakdown is the evidence of the empirical induced traffic breakdown, not the evidence of the empirical spontaneous traffic breakdown.

> Rather than the nature of traffic breakdown, the terms *empirical spontaneous* and *empirical induced* traffic breakdowns at a bottleneck distinguish different *sources* of a nucleus that occurrence leads to traffic breakdown (F→S transition) at the bottleneck.

> The empirical induced traffic breakdown at a highway bottleneck is the definitive empirical proof of the nucleation nature of empirical traffic breakdown (F→S transition) at the bottleneck.

> Qualitative features and characteristics of empirical synchronized flow that has emerged at a bottleneck due to traffic breakdown (F→S transition) do not depend on whether empirical spontaneous traffic breakdown or empirical induced traffic breakdown is the reason for synchronized flow at the bottleneck.

9.4 Empirical Proof of Time Delay in Over-Acceleration Using Opposite Assumption

The existence of a time delay in over-acceleration is of the crucial importance for the three-phase traffic theory (Sect. 8.1). Empirical studies of synchronized flow allows us to perform the empirical proof of the existence of the time delay in over-acceleration using the opposite assumption.

We assume firstly that the time delay in over-acceleration is zero. This means that after a vehicle begins to adapt its speed to the speed of a slower moving preceding vehicle, the vehicle accelerates to a higher speed without any delay. In this case, no synchronized flow can exist during a finite time interval. This contradicts empirical observations in which synchronized flow is observed. Thus, the assumption that the time delay in over-acceleration were equal to zero is *incorrect*.

We assume now that the time delay in over-acceleration is infinitely long. This means that after a vehicle has decelerated while adapting its speed to the speed of a slower moving preceding vehicle, the vehicle cannot accelerate to free flow any more if the preceding vehicle moves further at the low speed. In this case, when synchronized flow has occurred, no return S→F transition to free flow is possible. This contradicts empirical observations in which the S→F transition is observed. Thus, the above assumption that the time delay in over-acceleration is infinitely long is also *incorrect*.

> The empirical observation of free flow and synchronized flow as well as of empirical phase transitions between these traffic phases is empirical proof of the assumption of the three-phase traffic theory that the time delay in over-acceleration is a finite value.

References

1. Y. Dülgar, S.-E. Molzahn, H. Rehborn, M. Koller, B.S. Kerner, D. Wegerle, M. Schreckenberg, M. Menth, S.L. Klenov, J. Intell. Transp. Syst. **24**, 539–555 (2020)
2. B.S. Kerner, Transp. Res. Rec. **1678**, 160–167 (1999)
3. B.S. Kerner, in *Transportation and Traffic Theory*, ed. by A. Ceder (Elsevier Science, Amsterdam, 1999), pp. 147–171
4. B.S. Kerner, J. Phys. A: Math. Gen. **33**, L221–L228 (2000)
5. B.S. Kerner, Phys. Rev. E **65**, 046138 (2002)
6. B.S. Kerner, *The Physics of Traffic* (Springer, Berlin, 2004)
7. B.S. Kerner, *Introduction to Modern Traffic Flow Theory and Control* (Springer, Berlin, 2009)
8. B.S. Kerner, *Breakdown in Traffic Networks* (Springer, Berlin, 2017)
9. B.S. Kerner, Phys. Rev. E **92**, 062827 (2015)
10. B.S. Kerner (ed.), *Complex Dynamics of Traffic Management*, Encyclopedia of Complexity and Systems Science Series (Springer, New York, 2019)
11. B.S. Kerner, S.L. Klenov, J. Phys. A: Math. Theor. **43**, 425101 (2010)
12. B.S. Kerner, M. Koller, S.L. Klenov, H. Rehborn, M. Leibel, Phys. A **438**, 365–397 (2015)
13. B.S. Kerner, H. Rehborn, R.-P. Schäfer, S.L. Klenov, J. Palmer, S. Lorkowski, N. Witte, Phys. A **392**, 221–251 (2013)
14. S.-E. Molzahn, B.S. Kerner, H. Rehborn, J. Intell. Transp. Syst. **24**, 569–584 (2020)
15. S.-E. Molzahn, B.S. Kerner, H. Rehborn, S.L. Klenov, M. Koller, IET Intell. Transp. Syst. **11**(9), 604–612 (2017)
16. H. Rehborn, M. Koller, S. Kaufmann, *Data-Driven Traffic Engineering: Understanding of Traffic and Applications Based on Three-Phase Traffic Theory* (Elsevier, Amsterdam, 2021)

Chapter 10
Origin of Emergence of Empirical Moving Traffic Jams: F→S→J Transitions

10.1 Empirical Moving Jam Emergence in Synchronized Flow (S→J Transition)

It should be noted that empirical moving jams were studied already in 1950s–1960s.[1] Sequences of moving jams are often observed in real congested traffic. In the literature, sequences of moving jams are often also called traffic oscillations of congested traffic or "stop-and-go" traffic.

In Sect. 2.4, we have already mentioned that empirical moving jams do not emerge spontaneously in free flow.[2] On contrast, empirical moving jams emerge spontaneously only in synchronized flow. Thus, in real traffic, the spontaneous emergence of moving jams is realized through a sequence of an F→S transition and an S→J transition (called as the F→S→J transitions).[3] The sequence of the F→S→J transitions is as follows. Firstly, an F→S transition (traffic breakdown) occurs at a bottleneck. Due to traffic breakdown, synchronized flow emerges at the bottleneck. Later, growing moving jams can spontaneously emerge in synchronized flow. While propagating upstream, a growing moving jam can transform into a wide moving jam (S→J transition).

> Empirical moving jams can spontaneously emerge in synchronized flow *only*. In empirical observations, *no* moving jams emerge spontaneously in free flow.

---

[1] Empirical moving jams have been studied by many authors, in particular, in classic empirical works by Edie et al. [3–6], Treiterer et al. [29–31], and Koshi et al. [21]. Other references can be found, e.g., in reviews and books [7, 10, 24].

[2] The result of empirical observations of real traffic that moving jams do not emerge spontaneously in real free flow will be explained in Sect. A.10 of Appendix A.

[3] Empirical moving jams emergence through the F→S→J transitions was found and explained in [13].

© The Author(s), under exclusive license to Springer Nature Switzerland AG 2021
B. S. Kerner, *Understanding Real Traffic*,
https://doi.org/10.1007/978-3-030-79602-0_10

An empirical example of the emergence of a moving jam in synchronized flow is presented in Fig. 10.1. The synchronized flow in Fig. 10.1 has occurred earlier at the off-ramp bottleneck (bottleneck B1 in Fig. 10.1a, b), as explained in Sect. 5.4. Firstly, a local speed decrease occurs in synchronized flow[4] (labeled by "local speed decrease in synchronized flow" on vehicle trajectory 1 in Fig. 10.1c). Then, the local speed decrease grows while transforming into a growing moving jam; the jam propagates upstream in synchronized flow (labeled by "moving jam" on vehicle trajectory 2 in Fig. 10.1c). Finally, the moving jam transforms into a wide moving jam (labeled by "wide moving jam" on trajectories 3 and 4 in Fig. 10.1c).

10.2 Qualitative Explanation of Moving Jam Emergence in Synchronized Flow (S→J Instability)

10.2.1 Driver Reaction Time and Classical Traffic Flow Instability

Any human driver exhibits a finite reaction time. The driver reaction time leads to a time delay in the driver reaction on some unexpected change in a current driving situation (usually the driver reaction time for different humans is approximately within a range 0.7–2 sec). In 1958–1961, it was discovered that due to the existence of the driver reaction time, a traffic flow instability in an initially homogeneous traffic flow can occur.[5] In general, the traffic flow instability called as the classical traffic flow instability is a growing wave of a local speed decrease in traffic flow; the growing wave propagates upstream.

The classical traffic flow instability caused by the driver reaction time is explained as follows. We assume that in an initially homogeneous traffic flow a vehicle decelerates. If this vehicle deceleration is strong enough and it begins unexpectedly for the following driver, then due to the finite driver reaction time the following driver starts the deceleration with a delay. Such a driver deceleration can be called as *driver over-deceleration*. The prefix *over-* in the term *over-deceleration* can be explained as follows: when the driver reaction time is long enough, the driver of the following vehicle decelerates stronger than it is needed to avoid collisions. As a result of driver over-deceleration, the speed of the following vehicle becomes lower than the speed of the preceding vehicle. When each of the following drivers exhibits over-deceleration, then a growing speed wave of a local speed decrease occurs in traffic

---

[4] To avoid confusions, we should emphasize that local speed decreases in *synchronized flow* considered in this chapter should not be confounded with a local speed decrease in *free flow* at a bottleneck discussed in Chaps. 4–9.

[5] This classical traffic flow instability caused by the driver reaction time was introduced by Herman, Gazis, Rothery, Montroll, Chandler, and Potts [1, 8, 9, 11] as well as Kometani and Sasaki [17–20] (see also papers and reviews [7, 23–25]).

10.2 Qualitative Explanation of Moving Jam Emergence ... 137

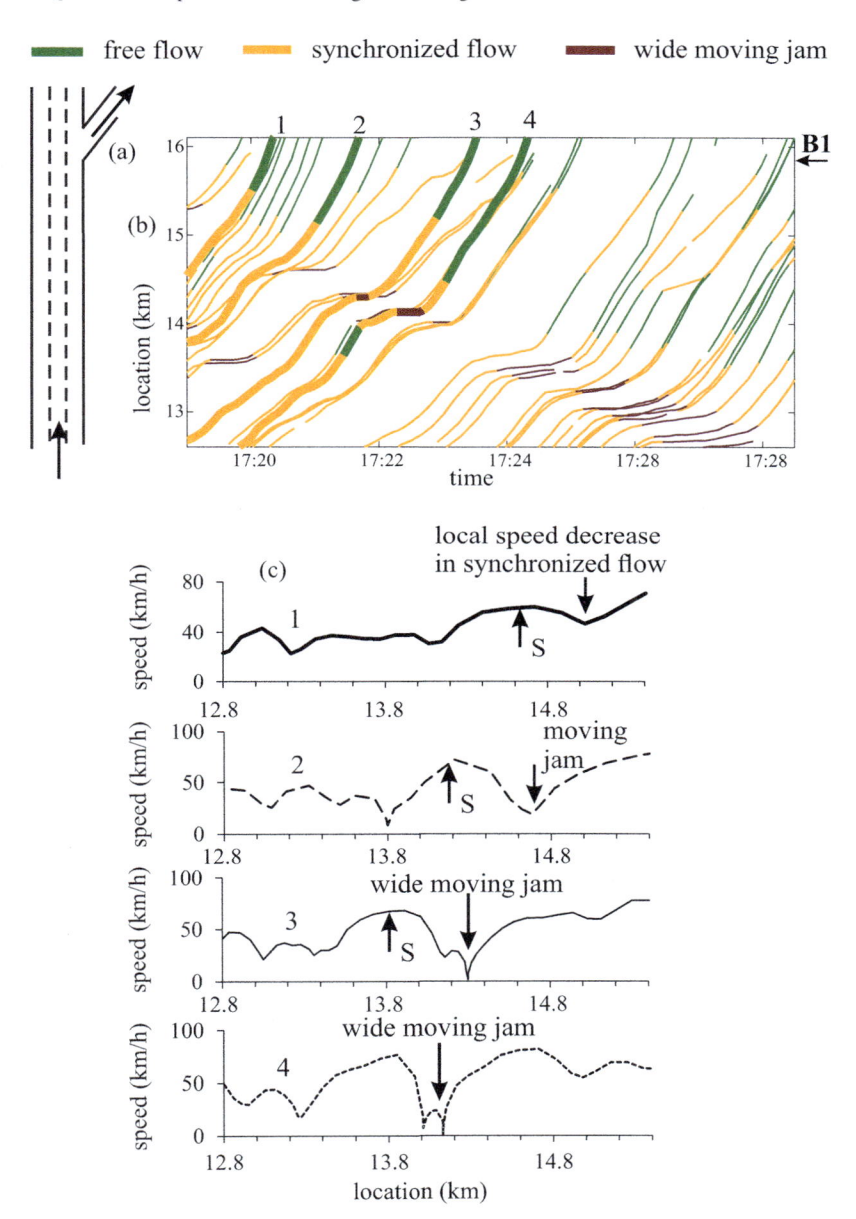

Fig. 10.1 Empirical emergence of moving jam in synchronized flow. **a** Schema of highway section with off-ramp bottleneck B1 explained in caption of Fig. 2.1. **b** A part of vehicle trajectories of Fig. 2.2 shown in the vicinity of a growing moving jam in synchronized flow; the synchronized flow has earlier occurred due to traffic breakdown at off-ramp bottleneck B1 (traffic breakdown has been studied in Fig. 5.5 of Sect. 5.4). **c** Dependencies of the microscopic vehicle speed on road location for vehicles 1–4 those trajectories are, respectively, marked by bold curves 1–4 in (**b**); S – synchronized flow

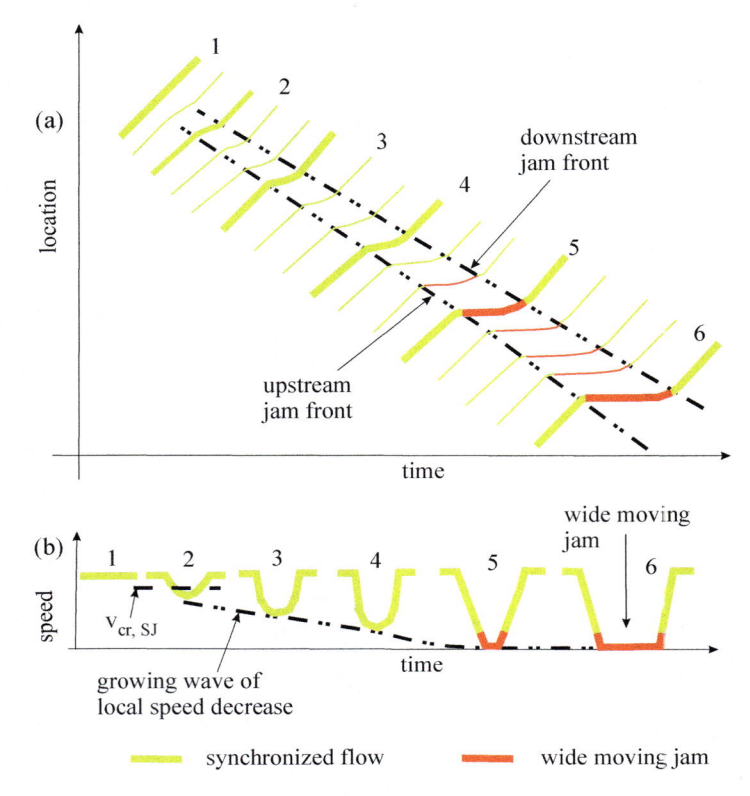

Fig. 10.2 Qualitative explanation of classical traffic flow instability in synchronized flow: **a** Vehicle trajectories in the vicinity of the growing moving jam occurring due to classical traffic flow instability in synchronized flow (S→J transition). **b** Time functions of speeds of some vehicles those trajectories 1–6 are marked by bold curves in (a). In (b), a dashed horizontal line shows qualitatively a critical speed $v_{\mathrm{cr,\,SJ}}$ for the S→J instability (see Sect. 10.2.2)

flow. The growing speed wave propagates upstream. This growing speed wave can also be considered a growing moving jam.

In the three-phase traffic theory, it has been shown that the classical traffic flow instability explains the empirical wide moving jam emergence in synchronized flow, for example, as shown in Fig. 10.1. For this reason, the development of the classical traffic flow instability in synchronized flow leading to the emergence of growing moving jams has been called an S→J instability. The S→J instability is the classical traffic flow instability occurring in synchronized flow. We assume that in initially homogeneous synchronized flow (vehicle trajectory 1 in Fig. 10.2) a local speed decrease (vehicle trajectory 2 in Fig. 10.2) occurs randomly. Within the local speed decrease, the synchronized flow speed is less than the speed is within synchronized flow downstream and upstream of the local speed decrease (vehicle trajectory 2 in Fig. 10.2). We have already mentioned that due to over-deceleration there is a

10.2 Qualitative Explanation of Moving Jam Emergence ...

tendency to the growth of the initial local speed decrease. The growing wave of the local speed decrease caused by the classical traffic flow instability in synchronized flow is qualitatively shown in Fig. 10.2.

10.2.2 Critical Speed for S→J Instability

From the above qualitatively consideration of the growth of a local speed decrease in synchronized flow due to over-deceleration, there can be the impression that over-deceleration might lead to the growth of any local speed decrease in synchronized flow. However, this impression is invalid for the reason explained below.[6]

We have mentioned that due to over-deceleration there is a tendency to the occurrence of a growing wave of a local speed decrease in synchronized flow. However, in addition to the over-deceleration effect, in synchronized flow there is also driver speed adaptation (Sect. 8.1.1). Due to driver speed adaptation, there is an opposite tendency to equalize the vehicle speed to the speed of the preceding vehicle. In the three-phase traffic theory, it has been shown that the S→J instability leading to the growing moving jams in synchronized flow results from a spatiotemporal competition between over-deceleration and driver speed adaptation. This competition occurring within local speed decreases in synchronized flow which is symbolically shown in Fig. 10.3 by down arrows for over-deceleration and by up arrows for speed adaptation.

The effect of the competition between over-deceleration and driver speed adaptation is as follows: only when the speed within a local speed decrease in synchronized flow is equal to or less than some critical speed for an S→J instability, a growing speed wave of the local speed decrease in synchronized flow (S→J instability) occurs. We denote the critical speed for the S→J instability by $v_{\text{cr, SJ}}$ (dashed horizontal lines in Figs. 10.2b and 10.3). To explain the sense of the critical speed $v_{\text{cr, SJ}}$, we consider Fig. 10.3.

A qualitative road-location dependence of the speed of vehicle 2 in Fig. 10.2 is presented in Fig. 10.3a. In this case, the minimum speed within the local speed decrease in synchronized flow is less than the critical speed $v_{\text{cr, SJ}}$ for the S→J instability. This means that the over-deceleration within the local speed decrease (thick down arrow labeled by "over-deceleration" in Fig. 10.3a) is on average stronger than speed adaptation (thin up arrow labeled by "speed adaptation"). For this reason, the local speed decrease grows over time—the S→J instability does occur (Fig. 10.2).

In contrast with the case shown in Figs. 10.2 and 10.3a, we assume now that the minimum speed within another local speed decrease in synchronized flow shown in Fig. 10.3b is larger than the critical speed $v_{\text{cr, SJ}}$ for the S→J instability.[7] This means that the speed adaptation within the local speed decrease (thick up arrow labeled

---

[6] Results presented in Sect. 10.2.2 are based on Refs. [14–16].

[7] For a simplification of the qualitative analysis of the S→J instability, we have assumed that the flow rate in the initial synchronized flow is a given value and the density of the synchronized flow is

Fig. 10.3 Qualitative explanation of critical speed $v_{cr,\ SJ}$ for S→J instability: **a**, **b** Speed as road-location function at a time instant within a local speed decrease in synchronized flow, when the minimum speed within the local speed decrease is less than the critical speed (**a**) and for the opposite case, when at another time instant the minimum speed within a local speed decrease is larger than the critical speed (**b**). Dashed horizontal lines show qualitatively the critical speed $v_{cr,\ SJ}$ for S→J instability

(a) Minimum speed within local speed decrease is less than the critical speed $v_{cr,\ SJ}$: The S→J instability does occur

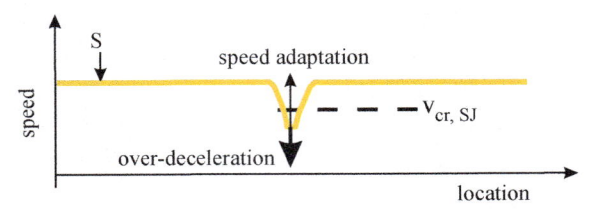

(b) Minimum speed within local speed decrease is larger than the critical speed $v_{cr,\ SJ}$: The S→J instability does not occur

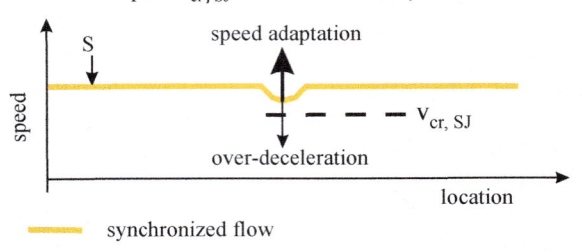

— synchronized flow

by "speed adaptation" in Fig. 10.3b) is on average stronger than over-deceleration (thin down arrow labeled by "over-deceleration"). For this reason, the local speed decrease decays over time—the S→J instability does not occur.

The existence of the critical speed $v_{cr,\ SJ}$ for the S→J instability means that the S→J instability exhibits the nucleation nature. A nucleus for the S→J instability is a local speed decrease in synchronized flow within which the speed is equal to or less than the critical speed $v_{cr,\ SJ}$ for the S→J instability. The local speed decrease shown in Fig. 10.3a is an example of the nucleus for the S→J instability. The continuous development of the S→J instability leads to the emergence of a wide moving jam in synchronized flow (S→J transition). For this reason, we can consider the critical speed for the S→J instability as the critical speed for the S→J transition.[8] This is

---

large enough for the possible occurrence of the S→J instability. A dependence of the critical speed $v_{cr,\ SJ}$ on the flow rate will be discussed in Sect. A.8.2 of Appendix A.

[8] For a simplification of the qualitative analysis, we have also assumed in Fig. 10.2 that the development of the S→J instability does lead to the emergence of a wide moving jam in synchronized flow (S→J transition). However, in real traffic, it can also occur that during the propagation of the growing wave of the local speed decrease in synchronized flow the interruption of the development of the S→J instability is realized: Rather than a wide moving jam emerges, the initially growing moving jam dissolves later [14]. Such a spontaneous emergence of the growing moving jam with the subsequent jam dissolution occurring later has been called as a nucleation-interruption effect of the moving jam emergence in synchronized flow. In this book, we do not consider the nucleation-interruption effect that can exhibit a very complex behavior in space and time.

10.2 Qualitative Explanation of Moving Jam Emergence ...

indeed observed in the empirical example of the spontaneous emergence of the wide moving jam in synchronized flow shown in Fig. 10.1.[9]

> The classical traffic flow instability (Fig. 10.2) can occur and explain wide moving jam emergence in real synchronized flow.

> The spontaneous emergence of a wide moving jam in real synchronized flow exhibits the nucleation nature.

10.3 Crucial Difference Between Driver Reaction Time and Time Delay in Over-Acceleration—A Difficulty for Understanding of Three-Phase Traffic Theory

There can be the following difficulty in the understanding of the three-phase traffic theory: Since the beginning of traffic research, the *driver reaction time* was the driver characteristic that is assumed to be very important for the understanding of traffic breakdown in free flow. Contrary to this generally accepted assumption, rather than the driver reaction time, in the three-phase traffic theory the *time delay in over-acceleration* is responsible for traffic breakdown (Sect. 8.1). Therefore, the following question arises:

- Is there a basic difference between the well-known driver reaction time and the time delay in over-acceleration of the three-phase traffic theory?

As shown in Sect. 10.1, the driver reaction time is responsible for the wide moving jam emergence in synchronized flow (S→J transition). Contrary to the driver reaction time, the time delay in over-acceleration is responsible for traffic breakdown (F→S transition) at a bottleneck (Sects. 8.1 and 8.2). Here, we emphasize that and why the

---

[9] It should be noted that Sugiyama et al. [26], Nakayama et al. [22], and Tadaki et al. [27, 28] have made driver experiments on a circular road in which the synchronized flow phase of high vehicle density has been the initial state of traffic flow. In particular, in [26], 22 vehicles have moved along a single-lane circle road with the length 230 m at the speed of about 30 km/h; the initial density in the initial traffic flow has been about 95 vehicles/km. Thus, the initial traffic flow in the experiment of Ref. [22, 26–28] has been indeed congested traffic associated with the synchronized flow phase. Due to the high vehicle density in the initial synchronized flow, over time the S→J instability has been observed on the circular road without bottlenecks. In accordance with results of this section, the development of the S→J instability in these driver experiments on the circular road has resulted in the emergence of wide moving jam(s).

time delay in over-acceleration and the driver reaction time are basically different driver characteristics.

We consider the time delay in over-acceleration due to lane changing to a faster lane (Fig. 8.1b). For a safety lane changing, a driver should wait for the occurrence of a long enough (safety) time headway between two following each other vehicles[10] in the neighborhood lane to which the driver would like to change. Time headway between vehicles in the neighborhood lane does not depend on the reaction time of the driver who wants to change to this lane. However, the waiting time for the safety lane changing depends on time headway between vehicles in the neighborhood faster lane. In turn, this waiting time is the time delay in over-acceleration through lane changing to a faster lane (Fig. 8.1b).

We can conclude that even if the driver reaction time were negligibly short, the mean time delay in over-acceleration is a finite value that exhibits the discontinuous character qualitatively shown in Fig. 8.2. In turn, the discontinuous character of the mean time delay in over-acceleration is responsible for the nucleation nature of traffic breakdown (F→S transition) at a bottleneck. Thus, traffic breakdown at the bottleneck exhibits the nucleation nature due to the discontinuous character of the mean time delay in over-acceleration; this does not depend on the existence of the driver reaction time.

> The driver reaction time is responsible for moving jam emergence in synchronized flow, *not* for traffic breakdown (F→S transition) in free flow at a bottleneck.

> Traffic breakdown (F→S transition) in free flow at the bottleneck is associated with the time delay in over-acceleration, *not* with the driver reaction time.

References

1. R.E. Chandler, R. Herman, E.W. Montroll, Oper. Res. **6**, 165–184 (1958)
2. L.C. Edie, Oper. Res. **2**, 107–138 (1954)
3. L.C. Edie, Oper. Res. **9**, 66–77 (1961)
4. L.C. Edie, R.S. Foote, in *Highway Research Board Proceedings*, (HRB, National Research Council, Washington, D.C., 1958), **37**, pp. 334–344

---

[10] A *time headway* is a net time gap between two vehicles following each other. When at a time instant the vehicle speed v is larger than zero and the net space distance between two following enough other vehicles (call as *space gap*) is equal to g, then time headway denoted by $\tau^{(\mathrm{net})}$ is equal to $\tau^{(\mathrm{net})} = g/v$. A more detailed consideration of time headway and space gap between vehicles will be done in Appendix A.

References

5. L.C. Edie, R.S. Foote, in *Highway Research Board Proceedings*, (HRB, National Research Council, Washington, D.C., 1960), **39**, pp. 492–505
6. L.C. Edie, R. Herman, T.N. Lam, Transp. Sci. **14**, 55–76 (1980)
7. D.C. Gazis, *Traffic Theory* (Springer, Berlin, 2002)
8. D.C. Gazis, R. Herman, R.B. Potts, Oper. Res. **7**, 499–505 (1959)
9. D.C. Gazis, R. Herman, R.W. Rothery, Oper. Res. **9**, 545–567 (1961)
10. D.L. Gerlough, M.J. Huber, *Traffic Flow Theory Special Report 165*, (Transp. Res. Board, Washington D.C., 1975)
11. R. Herman, E.W. Montroll, R.B. Potts, R.W. Rothery, Oper. Res. **7**, 86–106 (1959)
12. W. Huang, Y. Dülgar, H. Rehborn, B. A. Bernhardt, J. Xu, in *Proc. TRB 100th Annual Meeting*, TRB Paper TRBAM-21-01-261 (TRB, Washington DC, 2021)
13. B.S. Kerner, Phys. Rev. Lett. **81**, 3797–3800 (1998)
14. B.S. Kerner, *The Physics of Traffic* (Springer, Berlin, Heidelberg, New York, 2004)
15. B.S. Kerner, *Complex Dynamics of Traffic Management*, Encyclopedia of Complexity and Systems Science Series, ed. by B.S. Kerner (Springer, New York, NY 2019), pp. 387–500
16. B.S. Kerner, S.L. Klenov, Phys. Rev. E **68**, 036130 (2003)
17. E. Kometani, T. Sasaki, J. Oper. Res. Soc. Jap. **2**, 11–26 (1958)
18. E. Kometani, T. Sasaki, Oper. Res. **7**, 704–720 (1959)
19. E. Kometani, T. Sasaki, Oper. Res. Soc. Jap. **3**, 176–190 (1961)
20. E. Kometani, T. Sasaki, in *Theory of Traffic Flow*. ed. by R. Herman (Elsevier, Amsterdam, 1961), pp. 105–119
21. M. Koshi, M. Iwasaki, I. Ohkura, in *Proc. 8th International Symposium on Transportation and Traffic Theory*, ed. by V.F. Hurdle (University of Toronto Press, Toronto, Ontario, 1983), pp. 403
22. A. Nakayama, M. Fukui, M. Kikuchi, K. Hasebe, K. Nishinari, Y. Sugiyama, S.-I. Tadaki, S. Yukawa, New J. Phys. **11**, 083025 (2009)
23. G.F. Newell, Oper. Res. **9**, 209–229 (1961)
24. G.F. Newell, in *Proc. Second Internat. Sympos. on Traffic Road Traffic Flow*, (OECD, London, 1963), pp. 73–83
25. G.F. Newell, Trans. Res. B **36**, 195–205 (2002)
26. Y. Sugiyama, M. Fukui, M. Kikuchi, K. Hasebe, A. Nakayama, K. Nishinari, S.-I. Tadaki, S. Yukawa, New J. Phys. **10**, 033001 (2008)
27. S.-I. Tadaki, M. Kikuchi, M. Fukui, A. Nakayama, K. Nishinari, A. Shibata, Y. Sugiyama, T. Yoshida, S. Yukawa, New J. Phys. **15**, 103034 (2013)
28. S.-I. Tadaki, M. Kikuchi, A. Nakayama, A. Shibata, Y. Sugiyama, S. Yukawa, New J. Phys. **18**, 083022 (2016)
29. J. Treiterer, *Investigation of Traffic Dynamics by Aerial Photogrammetry Techniques, Ohio State University Technical Report PB 246 094* (Columbus, Ohio, 1975)
30. J. Treiterer, J.A. Myers, in *Procs. 6th International Symposium on Transportation and Traffic Theory*. ed. by D.J. Buckley (A.H. & AW Reed, London, 1974), pp. 13–38
31. J. Treiterer, J.I. Taylor, Highway Res. Rec. **142**, 1–12 (1966)

Chapter 11
Basic Types of Empirical Spatiotemporal Congested Traffic Patterns at Bottlenecks

The objective of this chapter is to answer the following question:

- What are the basic types of empirical spatiotemporal congested patterns[1] that can occur due to traffic breakdown (F→S transition) at a highway bottleneck?

11.1 Synchronized Flow Patterns (SPs)

We define a *synchronized flow pattern (SP)* as a spatiotemporal congested traffic pattern in which congested traffic consists of the synchronized flow traffic phase only.

---

[1] Empirical spatiotemporal congested patterns have studied in many works, in particular, in classical works by Edie et al. [2–6], Treiterer et al. [37–39], and Koshi et al. [32]. Other references can be found, e.g., in reviews and books [7, 10, 34]. The classical traffic flow instability introduced by Herman, Gazis, Rothery, Montroll, Chandler, and Potts [1, 8, 9, 11] as well as Kometani and Sasaki [28–31] has had a crucial importance for studies of moving jams (called also "stop-and-go" traffic or traffic oscillations) observed in congested traffic (see also papers and reviews [7, 33–35]). In 1998, it was found [13] that in real traffic the spontaneous emergence of the moving jams is realized through a sequence of the F→S→J transitions (Chap. 10). The emergence of the three-phase traffic theory has led to *understanding real spatiotemporal congested traffic patterns* presented in this chapter with the use of Refs. [14–16, 19, 21–24]. A more detailed analysis of features of empirical spatiotemporal congested traffic patterns can be found in the books [15, 17, 18, 36].

© The Author(s), under exclusive license to Springer Nature Switzerland AG 2021
B. S. Kerner, *Understanding Real Traffic*,
https://doi.org/10.1007/978-3-030-79602-0_11

11.1.1 Emergence of Moving SP (MSP) at Road Bottlenecks

In Sect. 9.2, we have already explained that in some empirical data after an F→S transition has occurred at a bottleneck, several minutes later a return S→F transition can be realized at the bottleneck. The return S→F transition has led to a dissolving synchronized flow at the bottleneck. This traffic phenomenon shown qualitatively in Fig. 9.6 has been called the F→S→F transitions.

However, in some empirical data rather than the occurrence of dissolving synchronized flow at a bottleneck (Fig. 9.6), we can observe a totally different traffic phenomenon: Due to a return S→F transition at the bottleneck (Fig. 11.1a), the region of synchronized flow departs from the bottleneck location. Then, the localized region of synchronized flow begins to propagate upstream. Such a localized region of synchronized flow is called a moving synchronized flow pattern (MSP) (labeled by "MSP" in Fig. 11.1a).

- In general, a localized pattern of synchronized flow that propagates *outside road bottlenecks* is called a moving synchronized flow pattern (MSP).

In empirical traffic data, the empirical phenomenon of the emergence of an MSP at a road bottleneck is often observed at a large enough flow rate in free flow upstream of the bottleneck (Fig. 11.2a). In the case shown in Fig. 11.2, the MSP induces traffic breakdown at the upstream on-ramp bottleneck when the MSP reaches the bottleneck (labeled by "induced traffic breakdown" in Fig. 11.2).

> The empirical phenomenon *the emergence of an MSP at a road bottleneck* can be explained by a sequence of an F→S transition and a return S→F transition at the bottleneck with the subsequent departing of synchronized flow from the bottleneck location. The synchronized flow propagates upstream as a localized synchronized pattern (MSP) outside the bottleneck.

11.1.2 Basic Types of SPs at Road Bottlenecks

There can be the following basic types of SPs at road bottlenecks:

1. Moving SP (MSP) (Fig. 11.1a). An empirical example of the MSP has been shown in Fig. 11.2. In some cases, rather than a single MSP (Fig. 11.1a), a sequence of different MSPs is realized at a bottleneck.
2. Widening SP (WSP) (Figs. 8.7a and 11.1b). The WSP is an SP whose downstream front is fixed at a bottleneck while the upstream front of the WSP does

11.1 Synchronized Flow Patterns (SPs)

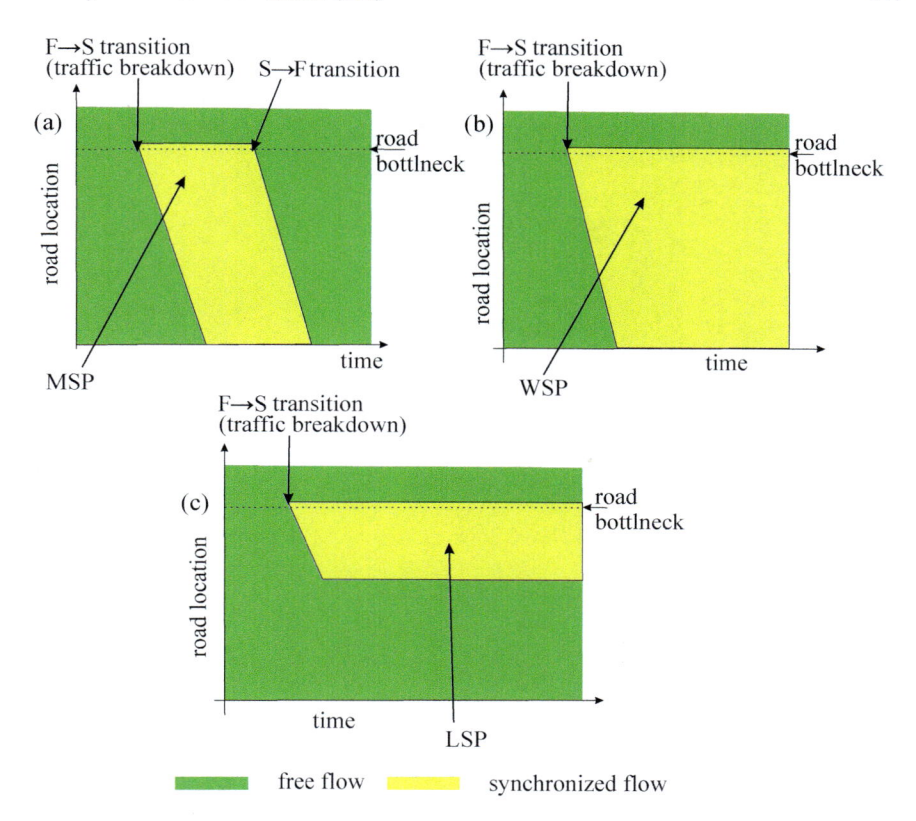

Fig. 11.1 Qualitative explanation of basic types of synchronized flow patterns (SPs) at road bottlenecks. **a** Moving SP (MSP). **b** Widening SP (WSP). **c** Localized SP (LSP). Green—free flow, yellow—synchronized flow

continuously propagate upstream of the bottleneck. Empirical WSP are often observed at road bottlenecks.[2]

3. Localized SP (LSP) (Fig. 11.1c). The LSP is an SP whose downstream front is fixed at a bottleneck while the upstream front of the LSP does not continuously propagate upstream of the bottleneck. The width of the LSP (in the traffic flow direction) can exhibit oscillations over time. An empirical example of the LSP at an on-ramp bottleneck is shown in Fig. 11.2a. In this case, the LSP has resulted from empirical induced traffic breakdown at the bottleneck (labeled by "LSP" in Fig. 11.2a); this induced traffic breakdown has been caused by the MSP propagation to the location of the on-ramp bottleneck.

---

[2] Empirical examples of WSPs can be found in the book [15]. In many empirical WSPs, moving jams emerge often over time; some of the moving jams can transform into wide moving jams (see Sect. 10.2).

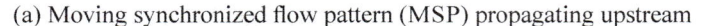

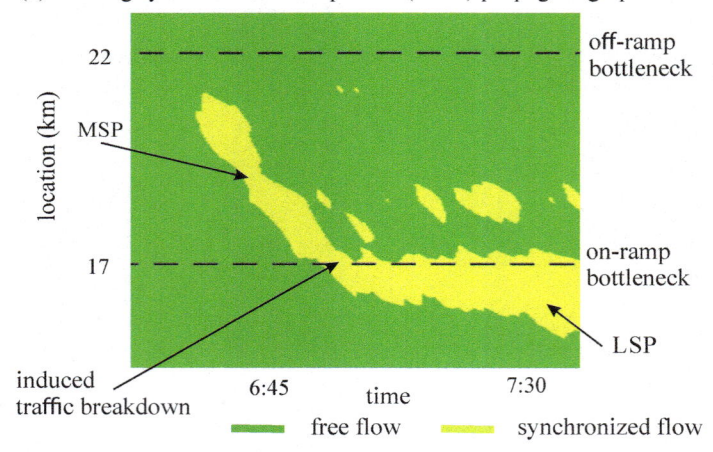

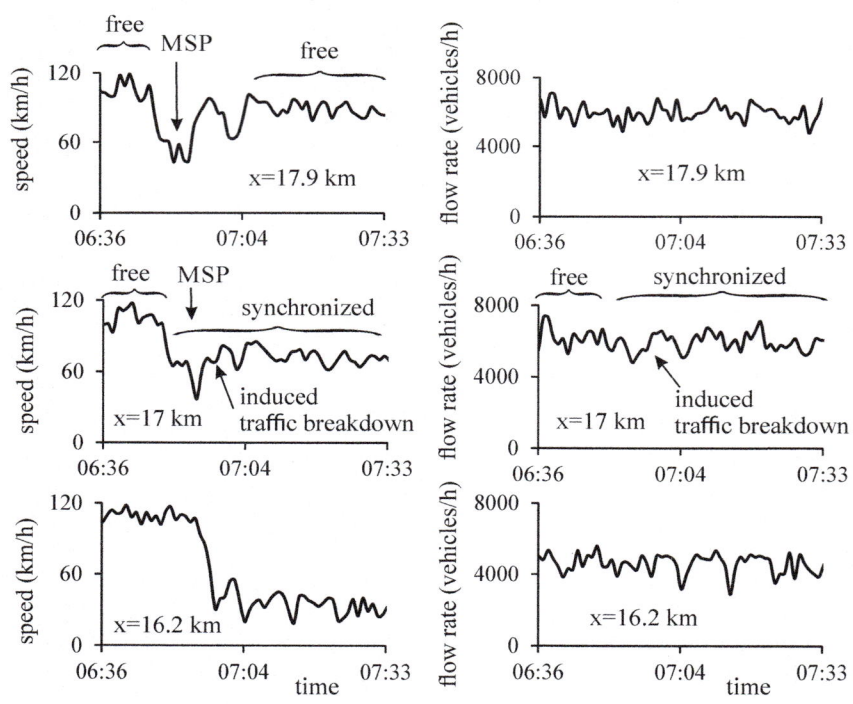

Fig. 11.2 Empirical example of a moving synchronized flow pattern (MSP) occurring at an off-ramp bottleneck: **a** Fragment of Fig. 7.2a related to empirical speed data from April 20, 1998; data are presented in space and time through the use of some averaging method described in Sect. C.2 of [27]. **b** Time dependencies of the average speed (left) and total flow rate (right) measured at different road locations. LSP—a localized synchronized flow pattern. Adapted from [14, 15]

11.1 Synchronized Flow Patterns (SPs)

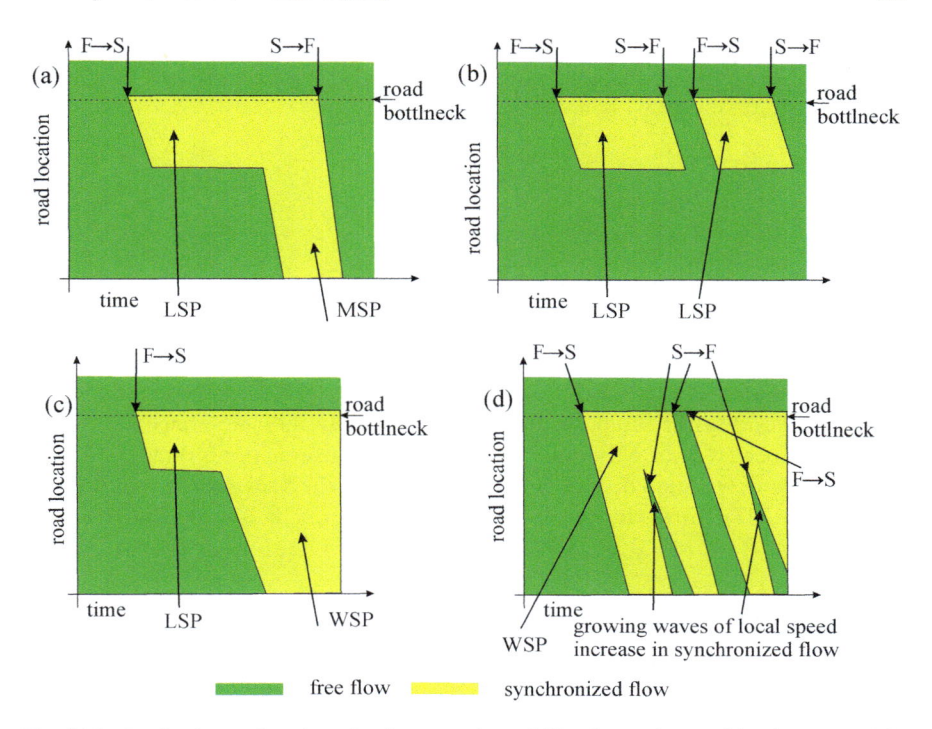

Fig. 11.3 Qualitative explanation of a diverse variety of SPs observed at road bottlenecks. **a** LSP transforms into MSP. **b** Due to S→F transitions at bottleneck, a random sequence of LSPs occurs. **c** LSP transforms into WSP. **d** Complex WSP occurring due to S→F transitions within synchronized flow. Green—free flow, yellow—synchronized flow. Arrows F→S and S→F mark F→S and S→F transitions, respectively

11.1.3 Diverse Variety of SPs at Road Bottlenecks

In empirical traffic data, a diverse variety of SPs is observed. The diverse variety of SPs at a bottleneck can result from both a change in the flow rate at the bottleneck over time and the random occurrence of S→F transitions.

A complex SP can consist of both different alternations of the basic types of the SPs (LSP, MSP, and WSP) with free flow intervals at the bottleneck and transformations of one SP into another SP occurring over time.[3] For example, a complex SP can include a transformation of an LSP into an MSP (Fig. 11.3a) or random alternations of different LSPs occurring due to S→F transitions at the bottleneck (Fig. 11.3b), or else a transformation of an LSP into an WSP (Fig. 11.3c).

---

[3] An empirical example of a complex SP can be seen at the off-ramp bottleneck in Fig. 7.1b.

In some other cases, due to S→F transitions that occur either at the bottleneck or upstream of the bottleneck random alternations between synchronized flow and free flow occur within an initial WSP (Fig. 11.3d).[4]

11.1.4 Boomerang Effect

The three-phase traffic theory predicts that an MSP moving upstream can also emerge spontaneously at a moving bottleneck (MB) (Fig. 11.4a). In this case, the emergence of the MSP can be considered a "boomerang effect".[5]

The boomerang effect is as follows.[6] First, a local speed decrease in free flow at the MB moves downstream at the MB speed. Suddenly, an F→S transition (traffic breakdown) occurs at the MB. Later, a return S→F transition occurs at the MB. The synchronized flow region departs from the location of the MB: An MSP emerges at the MB. The MSP moves further upstream of the MB, i.e., in the opposite direction of the motion of the MB (Fig. 11.4a).

> The initial *downstream* propagation of a local speed decrease in free flow at an MB with the subsequent emergence of an MSP that begins to propagate *upstream* can be considered the boomerang effect in traffic flow.

11.1.5 MSP Propagating in Direction of Traffic Flow

The three-phase traffic theory also predicts that an F→S transition (traffic breakdown) occurring at the MB can result in an SP localized at the MB. In the coordinate system moving at the MB speed, the SP should be considered an LSP. However, the MB moves downstream. Thus, for an observer at a road, the LSP should be classified

---

[4] Growing speed waves of a local speed increase in synchronized flow shown in Fig. 11.3d are caused by an S→F instability discussed in Sect. A.5 of Appendix A.

[5] The boomerang effect was firstly theoretically predicted from a study of a standard traffic flow model [25, 26]. As explained in Sect. C.3, in such models no F→S transition in the metastable free flow can occur. For this reason, in the traffic flow models the boomerang effect is as follows: First, an initial local speed decrease propagates downstream in free flow. Then, within the local speed decrease in free flow a wide moving jam emerges (F→J transition); the jam propagates upstream. Because the spontaneous F→J transition is not observed in real traffic (see explanations in Sect. A.10), we will not consider this theoretical boomerang effect any more.

[6] The boomerang effect has been predicted in numerical simulations with a three-phase traffic flow model (Fig. 2c of [24]).

11.1 Synchronized Flow Patterns (SPs)

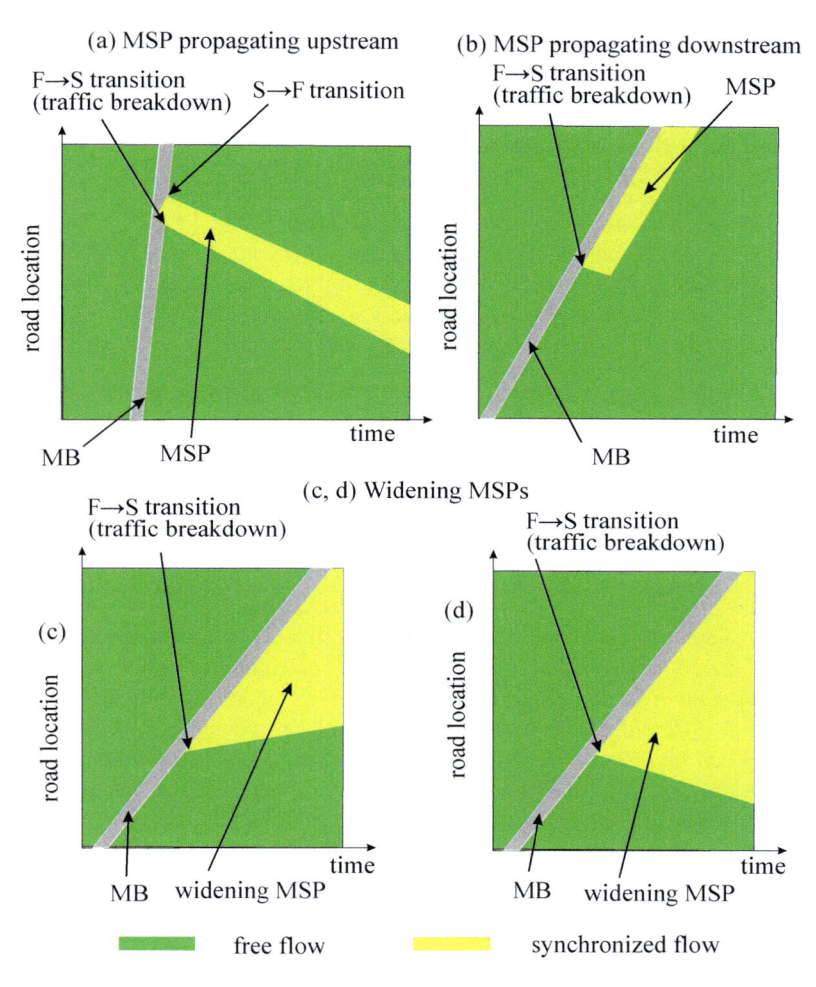

Fig. 11.4 Qualitative explanation of moving synchronized flow patterns (MSPs) occurring at a moving bottleneck (MB): **a** MSP propagating upstream (boomerang effect). **b** MSP propagating downstream. **c, d** Widening MSPs. Green—free flow, yellow—synchronized flow

as an MSP moving downstream. This explains our definition of the SP as the MSP moving in the direction of traffic flow at the MB speed (Fig. 11.4b).[7]

---

[7] An MSP moving downstream has been predicted in numerical simulations with a three-phase traffic flow model (Fig. 2b of [24]).

11.1.6 Basic Types of SPs at Moving Bottleneck

We have defined a moving synchronized flow pattern (MSP) as a localized pattern of synchronized flow that propagates outside road bottlenecks. There can be the following types of MSPs at MBs (Fig. 11.4):

- An MSP that has emerged at an MB moves upstream (Fig. 11.4a).
- An MSP that has emerged at an MB moves downstream (Fig. 11.4b).
- At an MB, an MSP can emerge that downstream front is fixed at the MB, whereas the width of this MSP (in the longitudinal direction) is increasing over time (Fig. 11.4c, d). For this reason, we call such an MSP as a *widening MSP*.[8]

While MSPs moving upstream that emerge at road bottlenecks have been observed in real traffic (Fig. 11.2), MSPs emerging at MBs (Fig. 11.4) should still be considered as theoretical predictions of the three-phase traffic theory.

11.1.7 Diverse Variety of SPs at Moving Bottleneck

The three-phase traffic theory predicts also that similar to a road bottleneck (Fig. 11.3) a diverse variety of SPs can occur at an MB. For example, rather than a single MSP moving downstream (Fig. 11.4b), due to S→F transitions at the MB a random sequence of different MSPs moving downstream can occur (Fig. 11.5a) or S→F transitions can be responsible for alternations between synchronized flow and free flow within a widening MSP (Fig. 11.5b). However, currently there are not enough traffic data for a study of such complex SPs at the MB.

11.2 General Congested Traffic Patterns (GPs)

We define a *general congested traffic pattern (GP)* as a spatiotemporal congested pattern within which congested traffic consists of the synchronized flow and wide moving jam traffic phases.

---

[8] Examples of widening MSPs predicted in the three-phase traffic theory are shown in Fig. 5b, c of [24]. It should be noted that in the system coordinate moving at the MB speed, the widening MSP should be considered an WSP. However, the MB moves downstream. Thus, for an observer at a road, the WSP should be classified as a widening MSP that downstream front moves in the direction of traffic flow at the MB speed. This explains our definition of the MSP as the widening MSP (Figs. 11.4c, d).

11.2 General Congested Traffic Patterns (GPs)

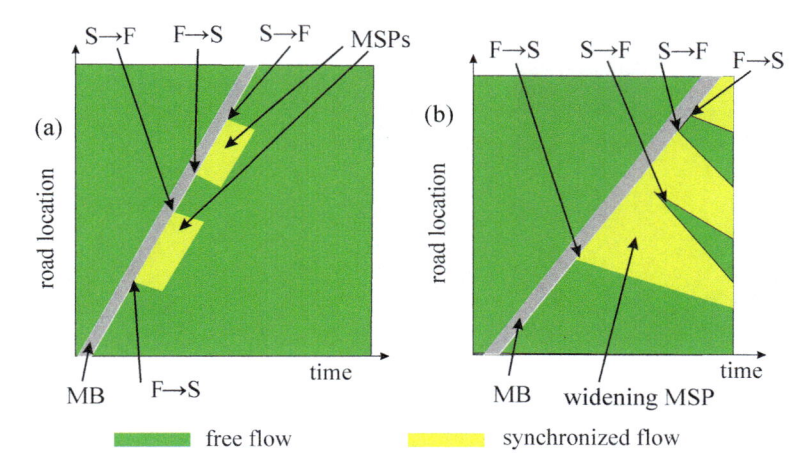

Fig. 11.5 Qualitative explanation of a diverse variety of SPs occurring at an MB: **a** Due to S→F transitions at the MB, a random sequence MSPs propagating downstream occurs. **b** Complex widening MSP occurring due to S→F transitions within synchronized flow. Green—free flow, yellow—synchronized flow. Arrows F→S and S→F mark F→S and S→F transitions, respectively

11.2.1 Basic Types of GPs at Road Bottlenecks

In Fig. 11.6, we present qualitatively the basic types of GPs observed in real traffic:

(i) An GP can emerge that can be considered an MSP within which one (as shown in Fig. 11.6a) or a few growing moving jams emerge. The GP can be called a moving GP (MGP). An empirical example of an MGP will be shown in Fig. A.13 of Appendix A.
(ii) Firstly, an WSP emerges at a road bottleneck. Later, growing moving jams emerge spontaneously in the synchronized flow of the WSP. Consequently, the initial WSP transforms into an GP (Fig. 11.6b). Empirical examples of such GPs are shown in Figs. 5.1 and 10.1.
(iii) Firstly, an LSP emerges at the bottleneck. Later, one or several growing moving jams emerge spontaneously in the synchronized flow of the LSP. Consequently, the initial LSP transforms into an GP (Fig. 11.6c).[9]

11.2.2 Diverse Variety of GPs at Road Bottlenecks

It should be noted that GPs qualitatively shown in Fig. 11.6 are related to a rough simplification of real GPs. In particular, within synchronized flow of an GP, an S→F

---

[9] Empirical examples of such GPs have been shown in the book [15].

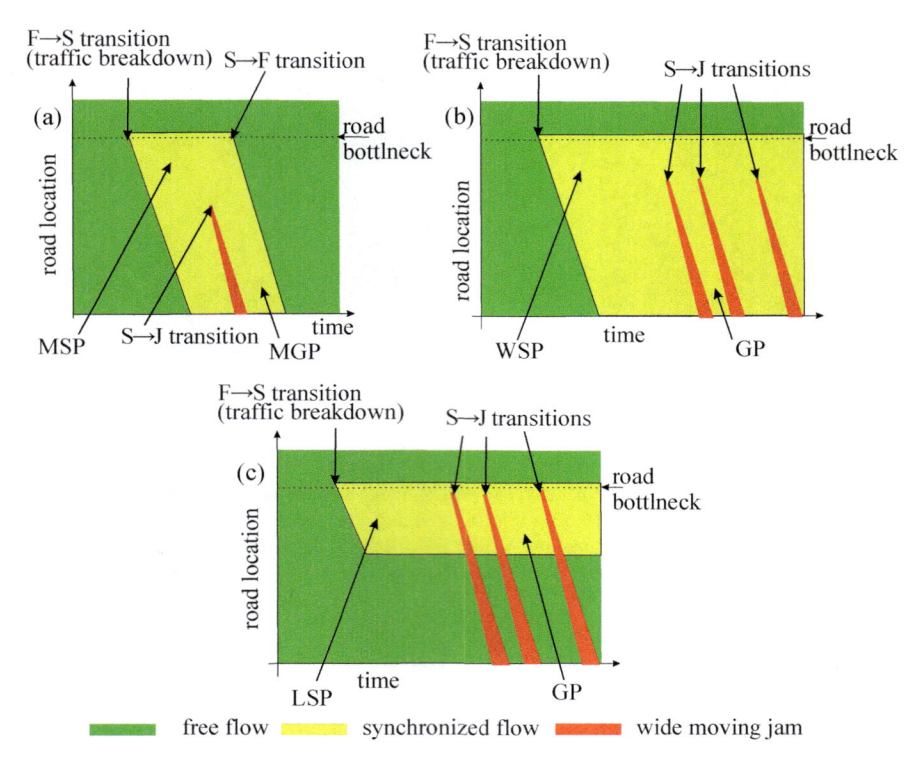

Fig. 11.6 Qualitative illustration of basic types of GPs observed at road bottlenecks in empirical traffic data: **a** Moving GP (MGP) that can be considered an MSP within which a wide moving jam emerges. **b** WSP transforms over time into GP. **c** LSP transforms over time into GP. SP—synchronized flow pattern, MSP—moving SP, MGP—moving GP, WSP—widening SP, LSP—localized SP

transition can occur. The S→F transition results from an S→F instability. The S→F instability is a growing speed wave of a local speed increase in synchronized flow; the subsequent growth of this speed wave leads to the S→F transition.[10]

Moreover, in the three-phase traffic theory it has been found that there is a spatiotemporal competition between S→F and S→J instabilities. Due to the competition between S→F and S→J instabilities there can be two different spatiotemporal traffic phenomena[11]:

- An initial S→J instability can lead to a sequence of S→J→S→F transitions.
- An initial S→F instability can lead to a sequence of S→F→S→J transitions.

---

[10] A more detailed consideration of the S→F instability will be made in Sect. A.5 of Appendix A.

[11] The competition between S→F and S→J instabilities has been disclosed and studied in [20]. A more detailed consideration of the competition between S→F and S→J instabilities will be made in Sect. A.9 of Appendix A.

11.2 General Congested Traffic Patterns (GPs)

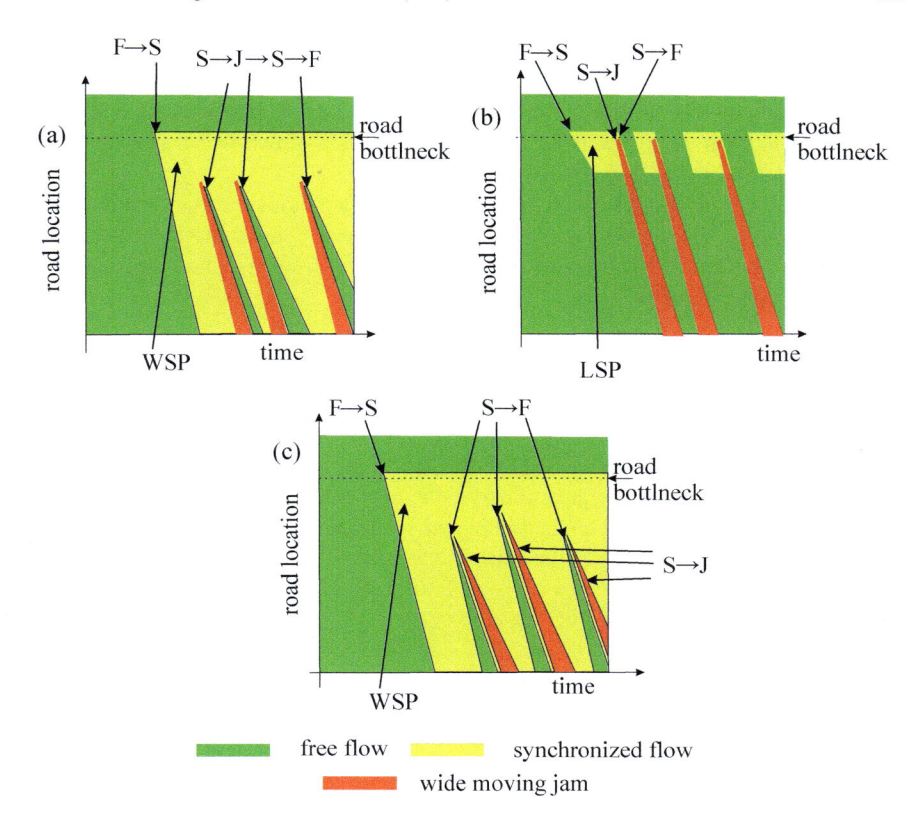

Fig. 11.7 Qualitative illustration of a diverse variety of GPs resulting from S→J→S→F transitions (**a**, **b**) and S→F→S→J transitions (**c**): **a** Downstream of wide moving jams in the GP shown in Fig. 11.6b free flow regions appear. **b** Due to S→F transitions downstream of wide moving jams in the GP shown in Fig. 11.6c, after a wide moving jam emerges within an LSP, free flow returns at the bottleneck. **c** Due to S→J transitions downstream of free flow regions in the WSP shown in Fig. 11.3d, after a region of free flow emerges within an WSP, a wide moving jam emerges downstream of the free flow region. Arrows S→J, F→S, and S→F mark S→J, F→S, and S→F transitions, respectively. LSP—localized SP, WSP—widening SP

The competition between S→F and S→J instabilities results in the occurrence of a diverse variety of GPs, in which the three phases F, S, and J can randomly alternate each other in space and time. Three examples of such complex GPs are qualitatively shown in Fig. 11.7:

- The GP shown in Fig. 11.6b transforms into an GP shown in Fig. 11.7a. This transformation is explained by S→F transitions that are realized downstream of the wide moving jams.
- In Fig. 11.7b, as in Fig. 11.6c, due to an F→S transition an LSP emerges at the bottleneck; then, a wide moving jam due to an S→J transition occurs within the LSP. However, contrary to the GP in Fig. 11.6c, in Fig. 11.7b downstream of the

wide moving jam an S→F transition is realized. As a result of this S→F transition, free flow returns at the bottleneck. Later, a new F→S transition occurs at the bottleneck with the resulting formation of a new LSP, then a new wide moving jam emerges within this new LSP, this jam causes a new S→F transition at the bottleneck, and so on (Fig. 11.7b).

- In Fig. 11.7c, as in Fig. 11.3d, firstly due to S→F transitions the associated regions of free flow emerge within synchronized flow. Then, contrary to the case shown in Fig. 11.3d, due to S→J transitions downstream of the free flow regions wide moving jams emerge.

There can be GPs that can be considered a combination of the GPs shown in Fig. 11.7.

Rather than only one road bottleneck, as shown in Figs. 11.6 and 11.7, there is often a sequence of adjacent road bottlenecks on real highways. In this case, a spatiotemporal structure of an GP can be more complex than the spatiotemporal structure of the GPs shown in Figs. 11.6 and 11.7. Empirical studies of the GPs show that in all cases wide moving jams emerge due to F→S→J transitions, whereas the spatiotemporal structure of GPs is associated with random alternations of the three phases F, S, and J in space and time. Empirical examples of complex spatiotemporal congested patterns at sequences of road bottlenecks are shown in Figs. 2.1, 7.1, and 7.2 as well as in Fig. A.13 of Appendix A.[12]

11.2.3 Basic Types of GPs at Moving Bottlenecks

A diverse variety of GPs has been theoretically predicted also to occur at a moving bottleneck (MB). Some of the basic types of GPs at the MB are shown in Fig. 11.8. It should be noted that in contrast with a road bottleneck that permanently exists on highway, the MB appears often unexpectedly (for example, due to the merging of a slow vehicle from an on-ramp onto the main road). Probably for this reason, there are not enough available empirical traffic data up to now to prove the existence of GPs occurring at the MB (Fig. 11.8).[13]

In Sect. 11.1.4, we have mentioned that the initial *downstream* propagation of a local speed decrease in free flow at an MB with the subsequent emergence of an MSP propagating *upstream* can be considered the boomerang effect in traffic flow. The upstream propagation of each of the GPs shown in Fig. 11.8a–d results also from the initial downstream propagation of a local speed decrease in free flow at an MB.

---

[12] A detailed consideration of empirical GPs and other complex spatiotemporal congested patterns at a sequence of road bottlenecks can be found in [22], Chaps. 2 and 9–14 of the book [15] as well as in [17–19]. In [22], it has also been shown that on multi-lane highways very different SPs and GPs can occur in different highway lanes; however, a discussion of such extremely complex traffic congested patterns is out of scope of this book.

[13] Simulations of GPs at MBs have first been made in [24]. Currently, we know only one empirical example of an GP that has occurred at an MB [12] that is related to the GP shown in Fig. 11.8a.

11.2 General Congested Traffic Patterns (GPs)

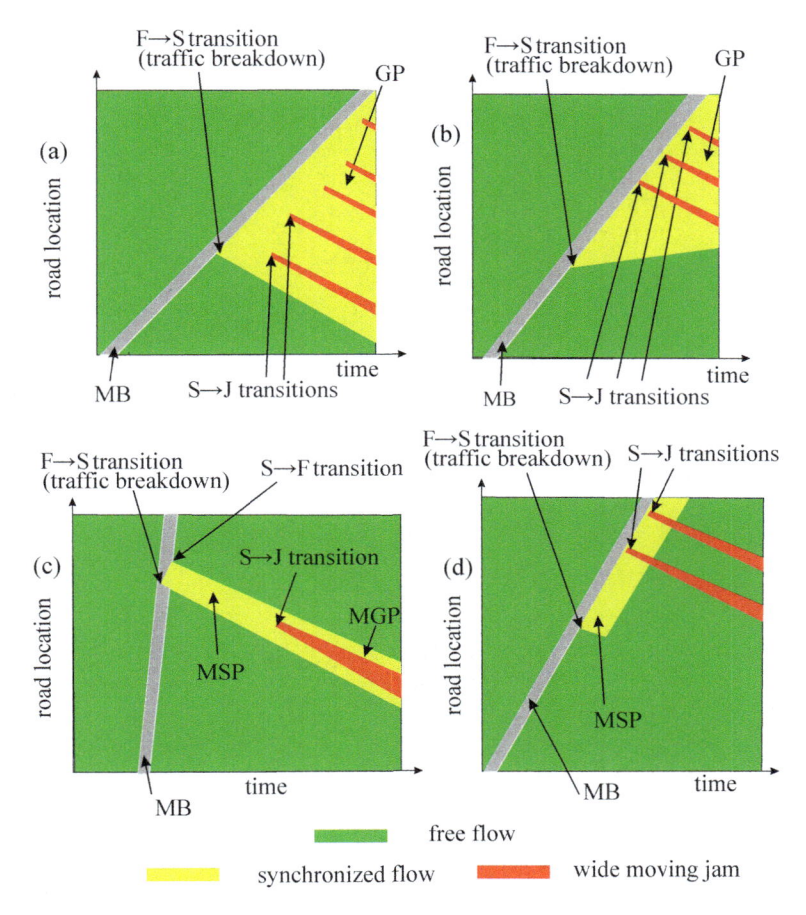

Fig. 11.8 Qualitative illustration of basic types of GPs at a moving bottleneck (MB): **a**, **b** Widening MSPs at MBs transform over time into GPs. **c** MSP at MB transforms over time into MGP. **d** MSP that moves downstream transforms over time into GP

Therefore, each of the GPs shown in Fig. 11.8a–d can be considered the boomerang effect.

11.3 Empirical Microscopic Structure of Wide Moving Jam and Mega-Jam Phenomenon

For a simplification of the above qualitative consideration of the spontaneous emergence of wide moving jams in synchronized flow, we have assumed that the average speed within the wide moving jam is equal to zero. In reality, there can be a complex

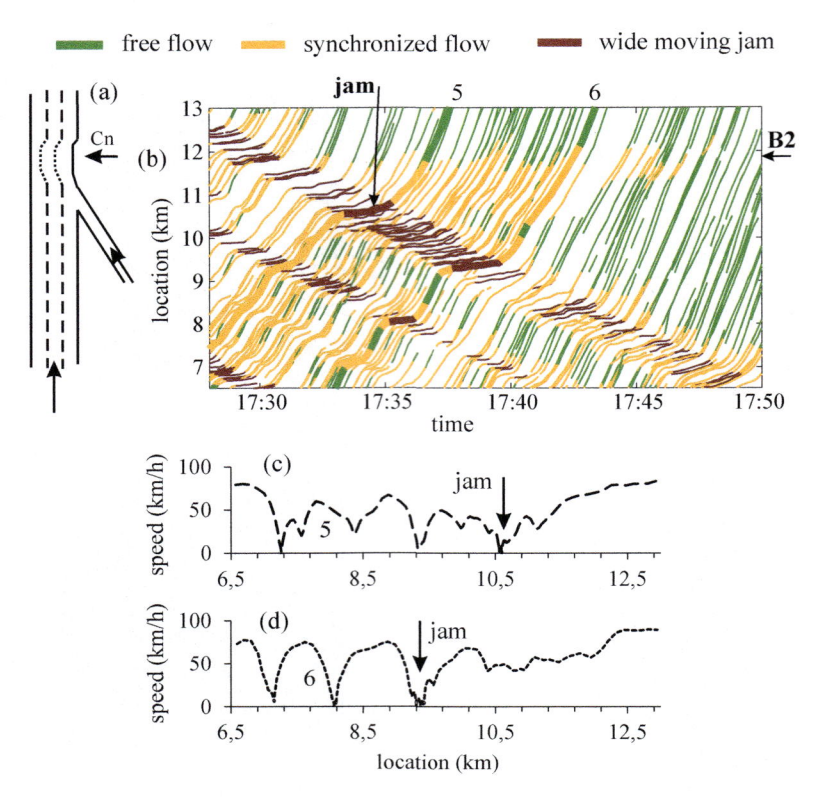

Fig. 11.9 Empirical example of the microscopic structure of a wide moving jam those emergence is shown in Fig. 10.1: **a, b** A part of highway schema **a** taken from Fig. 2.1a and a fragment of vehicle trajectories **b** taken from Fig. 2.2; the wide moving jam studied in Fig. 10.1 is labeled by "jam". **c, d** Road location dependencies of speeds of probe vehicles 5 and 6 those trajectories are marked by bold curves in (**b**). Bottleneck B2 has been explained in caption of Fig. 2.1

spatiotemporal microscopic structure within empirical wide moving jams. Some features of this spatiotemporal microscopic jam structure are discussed in this section.

In accordance with the definition of a wide moving jam (Sect. 2.1.3), the wide moving jam shown in Fig. 11.9 propagates through upstream bottleneck B2 while maintaining the mean velocity of the downstream jam front (labeled by "jam" in Fig. 11.9b). While the width (in the flow direction) of the jam increases over time (trajectories 5 and 6 in Fig. 11.9c, d), a complex spatiotemporal microscopic structure is built within the jam (vehicle trajectory 6 in Fig. 11.10a, b).

To explain the empirical microscopic jam structure, we introduce the term a *flow interruption interval within a wide moving jam*. This term is related to a long enough time interval (longer than about 2–3 s) within which at least a few vehicles are in a standstill (interval labeled by "flow interruption interval" in Fig. 11.10b) or the

11.3 Empirical Microscopic Structure of Wide Moving Jam ... 159

Fig. 11.10 Empirical microscopic structure of wide moving jam: **a, b** Empirical time dependencies of the speed of vehicle 6 that propagates through the wide moving jam labeled by "jam" in Fig. 11.9b. In (**a, b**), the speed of vehicle 6 is presented in two different time scales

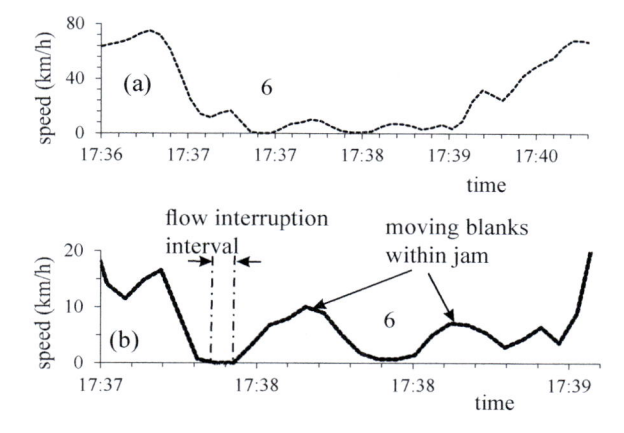

Fig. 11.11 Qualitative explanation of mega-jam phenomenon at heavy road bottleneck: The mega-jam (red region) consists of a random sequence of flow interruption intervals and moving blanks as those within the wide moving jam in Fig. 11.10b. HB—heavy bottleneck

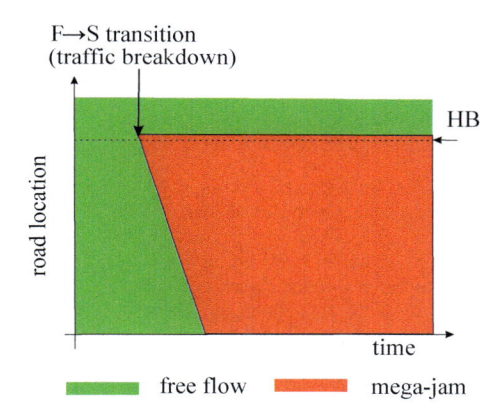

vehicles are moving with a negligible low speed in comparison with the speed in the jam inflow and jam outflow.

Between flow interruption intervals within the wide moving jam, there is the vehicle motion within the wide moving jam that is associated with so called *moving blanks* (labeled by "moving blanks" in Fig. 11.10b). A blank within a wide moving jam is a region with no vehicles. A moving blank within the wide moving jam is the blank, which moves upstream due to the vehicle motion within the jam.[14]

---

[14] This vehicle motion within the jam occurs with long time headway (about 1.5–7 s in empirical data). A *time headway* is a net time gap between two vehicles following each other. When at a time instant the vehicle speed v is larger than zero and the net space distance between two following enough other vehicles (call as *space gap*) is equal to g, then time headway denoted by $\tau^{(net)}$ is equal to $\tau^{(net)} = g/v$. A more detailed consideration of time headway and space gap between vehicles will be done in Appendix A.

> The empirical microscopic structure of a wide moving jam consists of spatiotemporal alternations of flow interruption intervals and moving blanks.

On highways, a *heavy bottleneck* can appear. The heavy bottleneck is a bottleneck that limits the average flow rate within a congested pattern upstream of the bottleneck to such small flow-rate values at which all wide moving jams in congested traffic merge into a mega-wide moving jam (mega-jam for short). The heavy bottleneck can occur through bad weather conditions or accidents or heavy road works.

A mega-jam is a wide moving jam with an extremely large width (in the flow direction) growing often over time (qualitatively a mega-jam is illustrated in Fig. 11.11).[15] As the microscopic structure of a wide moving jam, the empirical microscopic structure of the mega-jam consists of spatiotemporal alternations of flow interruption intervals and moving blanks.

References

1. R.E. Chandler, R. Herman, E.W. Montroll, Oper. Res. **6**, 165–184 (1958)
2. L.C. Edie, Oper. Res. **2**, 107–138 (1954)
3. L.C. Edie, Oper. Res. **9**, 66–77 (1961)
4. L.C. Edie, R.S. Foote, in *Highway Research Board Proceedings*, vol. 37 (HRB, National Research Council, Washington D.C., 1958), pp. 334–344
5. L.C. Edie, R.S. Foote, in *Highway Research Board Proceedings*, vol. 39 (HRB, National Research Council, Washington D.C., 1960), pp. 492–505
6. L.C. Edie, R. Herman, T.N. Lam, Transp. Sci. **14**, 55–76 (1980)
7. D.C. Gazis, *Traffic Theory* (Springer, Berlin, 2002)
8. D.C. Gazis, R. Herman, R.B. Potts, Oper. Res. **7**, 499–505 (1959)
9. D.C. Gazis, R. Herman, R.W. Rothery, Oper. Res. **9**, 545–567 (1961)
10. D.L. Gerlough, M.J. Huber, *Traffic Flow Theory Special Report 165* (Transportation Research Board, Washington D.C., 1975)
11. R. Herman, E.W. Montroll, R.B. Potts, R.W. Rothery, Oper. Res. **7**, 86–106 (1959)
12. W. Huang, Y. Dülgar, H. Rehborn, B.A. Bernhardt, J. Xu, in *Proceedings of the TRB 100th Annual Meeting*, TRB Paper TRBAM-21-01-261 (TRB, Washington D.C., 2021)
13. B.S. Kerner, Phys. Rev. Lett. **81**, 3797–3800 (1998)
14. B.S. Kerner, Phys. Rev. E **65**, 046138 (2002)
15. B.S. Kerner, *The Physics of Traffic* (Springer, Berlin, 2004)
16. B.S. Kerner, J. Phys. A: Math. Theor. **41**, 215101 (2008)
17. B.S. Kerner, *Introduction to Modern Traffic Flow Theory and Control* (Springer, Berlin, 2009)
18. B.S. Kerner, *Breakdown in Traffic Networks* (Springer, Berlin, 2017)
19. B.S. Kerner, *Complex Dynamics of Traffic Management*, Encyclopedia of complexity and systems science series, ed. by B.S. Kerner (Springer, New York, 2019), pp. 387–500
20. B.S. Kerner, Phys. Rev. E **100**, 012303 (2019)
21. B.S. Kerner, S.L. Klenov, Phys. Rev. E **68**, 036130 (2003)
22. B.S. Kerner, S.L. Klenov, Phys. Rev. E **80**, 056101 (2009)
23. B.S. Kerner, S.L. Klenov, Transp. Res. Rec. **2124**, 67–77 (2009)

---

[15] A theory of the mega-jam phenomenon has been developed in [16].

References

24. B.S. Kerner, S.L. Klenov, J. Phys. A: Math. Theor. **43**, 425101 (2010)
25. B.S. Kerner, P. Konhäuser, Phys. Rev. E **50**, 54–83 (1994)
26. B.S. Kerner, P. Konhäuser, M. Schilke, Phys. Rev. E **51**, 6243–6246 (1995)
27. B.S. Kerner, H. Rehborn, R.-P. Schäfer, S.L. Klenov, J. Palmer, S. Lorkowski, N. Witte, Phys. A **392**, 221–251 (2013)
28. E. Kometani, T. Sasaki, J. Oper. Res. Soc. Jpn. **2**, 11–26 (1958)
29. E. Kometani, T. Sasaki, Oper. Res. **7**, 704–720 (1959)
30. E. Kometani, T. Sasaki, Oper. Res. Soc. Jpn. **3**, 176–190 (1961)
31. E. Kometani, T. Sasaki, in *Theory of Traffic Flow*. ed. by R. Herman (Elsevier, Amsterdam, 1961), pp. 105–119
32. M. Koshi, M. Iwasaki, I. Ohkura, in *Proceedings of the 8th International Symposium on Transportation and Traffic Theory*, ed. by V.F. Hurdle (University of Toronto Press, Toronto, 1983), pp. 403
33. G.F. Newell, Oper. Res. **9**, 209–229 (1961)
34. G.F. Newell, in *Proceedings of the Second International Symposium on Traffic Road Traffic Flow* (OECD, London, 1963), pp. 73–83
35. G.F. Newell, Transp. Res. B **36**, 195–205 (2002)
36. H. Rehborn, M. Koller, S. Kaufmann, *Data-Driven Traffic Engineering: Understanding of Traffic and Applications Based on Three-Phase Traffic Theory* (Elsevier, Amsterdam, 2021)
37. J. Treiterer, *Investigation of Traffic Dynamics by Aerial Photogrammetry Techniques*, Ohio State University Technical Report PB 246 094 (Columbus, Ohio, 1975)
38. J. Treiterer, J.A. Myers, in *Proceedings of the 6th International Symposium on Transportation and Traffic Theory*, ed. by D.J. Buckley (A.H. & AW Reed, London, 1974), pp. 13–38
39. J. Treiterer, J.I. Taylor, Highw. Res. Rec. **142**, 1–12 (1966)

Chapter 12
Discussion and Outlook

12.1 Kuhn's Structure of Scientific Revolutions in Application to Transportation Science

Since 1962 when Kuhn's theory[1] about the structure of scientific revolutions was published, there had been many critical discussions about the question whether Kuhn's theory exhibits a generic character. Many of philosophers of science nowadays believe that Kuhn's theory is incorrect at least for generic cases. It seems that between these philosophers of science the consensus appears to be that Kuhn's view on the science development is an artifact of posterior investigations and it does not reflect the process of science in the generic case properly. In Kuhn's view, a paradigm shift is a change of the perception rather than a cumulative process in the scientific field. However, many philosophers of science believe that view of science as a cumulative process matches twentieth-century science very well.

In contrast to the above-mentioned criticism of many philosophers of science on Kuhn's view of the science development, the author's experience with the development of traffic and transportation science made since 1992 and up to now leads to conclusion that Kuhn's theory predicted almost exactly the events that happened in the traffic and transportation research community as the response to the three-phase traffic theory and the author's criticism on standard traffic theories and models. For this reason, a discussion of the application of Kuhn's structure of scientific revolutions to traffic and transportation science could be interesting to understand why the empirical fundamental of transportation science (empirical nucleation nature of traffic breakdown (F$\rightarrow$S transition)) appears to be ignoring in the traffic and transportation research community up to now and what development of traffic and transportation science we might expect in the future.

In Kuhn's theory about a scientific revolution in a scientific field, there are the following features:

---

[1] We refer here to the 4th edition of the famous Kuhn's book [28] firstly published in 1962.

- *Normal science* (Sect. 12.1.1).
- *Crisis* in the field (Sect. 12.1.2).
- *Empirical anomaly* in the field that is the origin of the crisis (Sect. 12.1.3).
- *Response to crisis*, i.e., the emergence of new theory (Sect. 12.1.4).
- *Incommensurability* of the old (standard) theory and the new theory (Sect. 12.1.5).
- *Paradigm shift* in the field (Sect. 12.1.6).
- *Response to the new theory* in the scientific community working in the field (Sect. 12.1.7).

12.1.1 Normal Science: Cumulative Process in Standard Traffic and Transportation Science

Since 1920s–1930s and up to now there is a cumulative process of empirical and theoretical studies of empirical traffic breakdown and resulting traffic congestion made in standard traffic and transportation science. We should note that the term *standard science* used in this book is related to the term *normal science* in Kuhn's terminology.

We should emphasize that standard traffic flow theories and models have had a great impact on the understanding of many empirical traffic phenomena (Sect. 5.5). Moreover, many ideas of the standard traffic flow theories are very important and they have been used in the three-phase traffic theory (Sect. 1.1).

- Without the cumulative process of empirical and theoretical studies of empirical traffic breakdown and resulting traffic congestion made in standard traffic and transportation science, the three-phase traffic theory could not been introduced and developed.

The cumulative process that accordingly to Kuhn can also be called "puzzle solving" in standard traffic and transportation science is related to the perception of vehicular traffic that is as follows (Sect. 5.1). At any given time instant, there is a particular value of stochastic highway capacity at a highway bottleneck that determines the occurrence of traffic breakdown: When at a time instant the flow rate exceeds the capacity value at this time instant, traffic breakdown must occur at the bottleneck.

As we have shown in the book, this standard perception of vehicular traffic contradicts basically to the empirical nucleation nature of traffic breakdown (F→S transition) at the bottleneck. We have also found that understanding real traffic is *not* possible, if the empirical nucleation nature of traffic breakdown (F→S transition) at the bottleneck is ignored, as made in standard traffic and transportation science.

12.1 Kuhn's Structure of Scientific Revolutions in Application ...

> The cumulative process in standard traffic and transportation science, in which the empirical nucleation nature of traffic breakdown (F→S transition) at the bottleneck is ignored, cannot lead to understanding real traffic. This critical conclusion about standard traffic and transportation science is independent of the duration of this cumulative process.

12.1.2 Crisis: Failure of Engineering Applications of Standard Traffic Theories

In 1992, as the author began to work in traffic and transportation science at the Daimler Company (Stuttgart, Germany), one of the main questions that we should answer was as follows: Are there traffic models that can be used for effective traffic control, management, and the evaluation of applications of intelligent transportation systems (ITS)? The basic importance for the author to answer this question was the encounter with the following empirical result obtained by traffic engineers working at the Heusch/Boesefeldt GmbH (Aachen, Germany):

- All numerous attempts to apply standard traffic models for traffic control and management of real traffic failed.

> At least at the beginning of 1990s, practical traffic engineers knew that standard traffic models for traffic control and management developed in the traffic and transportation research community failed by their applications in the real world. This is the *crisis in standard traffic and transportation science*.

12.1.3 Anomaly: Empirical Induced Traffic Breakdown at Bottleneck

The question raised after the author began to study empirical traffic data has been as follows:

- Is there an empirical traffic phenomenon that can be considered the fundamental empirical basis of transportation science?

In this book, we have shown that this empirical traffic phenomenon does exist: The phenomenon is the empirical nucleation nature of traffic breakdown (F→S transition) at a highway bottleneck (Chap. 6). The empirical nucleation nature of traffic breakdown (F→S transition) is proved by the the evidence of the empirical induced traffic breakdown at the bottleneck. However, *none* of standard traffic theories and models that are the theoretical basis for traffic control and management as well as for the evaluation of ITS applications can explain the empirical nucleation nature of empirical traffic breakdown (F→S transition) at the bottleneck.

> The empirical induced traffic breakdown (F→S transition) at a highway bottleneck can be considered an empirical anomaly for standard traffic and transportation science. This empirical anomaly can explain the crisis in traffic and transportation science.

The basic reason for the criticism of the standard traffic theories is the ignoring of the evidence of the empirical traffic phenomenon "empirical induced traffic breakdown" in standard traffic and transportation science. The evidence of the empirical traffic phenomenon "empirical induced traffic breakdown" at a highway bottleneck leads to the basic conclusion that real traffic breakdown (F→S transition) at the bottleneck exhibits the empirical nucleation nature. In its turn, the empirical nucleation nature of traffic breakdown (F→S transition) at the bottleneck changes fundamentally the perception of vehicular traffic. The consequence of the change in the perception of vehicular traffic is understanding real traffic with the use of the three-phase traffic theory.

12.1.4 Response to Crisis: Emergence of Three-Phase Traffic Theory

The response to the crisis in traffic science is the emergence of the three-phase traffic theory (Chaps. 7 and 8). The main reason for this theory is the explanation of the empirical nucleation nature of traffic breakdown (F→S transition) at a highway bottleneck. Based on studies of empirical traffic data, it has been found that in congested traffic there are two traffic phases: (i) synchronized flow and (ii) wide moving jam. The speed in synchronized flow is lower than the speed in free flow. Basic features of the synchronized flow traffic phase (Chaps. 8 and 9) are responsible for the empirical nucleation nature of traffic breakdown (F→S transition) at the bottleneck.

> The three-phase traffic theory has been introduced to explain the empirical nucleation nature of traffic breakdown (F→S transition) at a highway bottleneck.

12.1.5 Incommensurability of Standard Traffic Theories with Three-Phase Traffic Theory

It turned out that the three-phase traffic theory is incommensurable with any standard traffic flow theory. The term *incommensurability* has been introduced by Kuhn to explain a paradigm shift in a field of science. The nucleation nature of traffic breakdown (F→S transition) at a bottleneck in a traffic network is the basic feature of the three-phase traffic theory. *None* of the standard traffic and transportation theories can explain the empirical nucleation nature of traffic breakdown (F→S transition) at the bottleneck.

> The three-phase traffic theory is incommensurable with standard traffic and transportation theories because the empirical nucleation nature of the F→S transition at the bottleneck has *no sense* for the standard traffic and transportation theories.

12.1.6 Paradigm Shift in Traffic and Transportation Science

> The empirical nucleation nature of traffic breakdown (F→S transition) at a highway bottleneck explained in the three-phase traffic theory changes fundamentally the meaning of stochastic highway capacity.

The paradigm shift in traffic and transportation science is the fundamental change in the meaning of stochastic highway capacity. This is because the meaning of stochastic highway capacity is the basis for the development of any method for traffic control, management, and organization of a traffic network as well as for many other ITS applications. The crucial difference between the understanding of standard stochastic highway capacity made in the standard traffic and transportation theories and the understanding of stochastic highway capacity made in the three-phase traffic theory is as follows:

- The old paradigm in traffic and transportation science can been formulated as follows: *At any time instant*, there is a *particular value* of stochastic highway capacity. When at a time instant the flow rate at the bottleneck exceeds the capacity at this time instant, then traffic breakdown must occur at the bottleneck.
- The new paradigm in traffic and transportation science formulated in the three-phase traffic theory is as follows: *At any time instant*, there is a *range of stochastic highway capacities*. When at a time instant the flow rate at the bottleneck is within the range of stochastic highway capacities at this time instant, then traffic break-

down at the bottleneck can occur with some probability but traffic breakdown does not necessarily occur.
- This basic change in the meaning of stochastic highway capacity results from the empirical nucleation nature of traffic breakdown at a bottleneck.

We have already mentioned that in Kuhn's view, a paradigm shift is a change of the perception rather than a cumulative process in a scientific field. The experience of the author made during last 25 years of research in traffic and transportation science confirms Kuhn's view.

- The ignoring of the empirical nucleation nature of traffic breakdown (F→S transition) in standard traffic and transportation science is related to the perception of vehicular traffic, in which at any given time instant there should be a particular value of stochastic highway capacity that determines the occurrence of traffic breakdown.
- In contrast with this perception of vehicular traffic made in standard traffic and transportation science, as we have shown in this book, the three-phase traffic theory predicts that only through the acceptance of the empirical nucleation nature of traffic breakdown (F→S transition), which leads to a range of stochastic highway capacities existing at any given time instant, understanding real traffic is possible.

> In accordance with Kuhn's view on a paradigm shift in a scientific field, the paradigm shift in traffic and transportation science is a change of the perception of stochastic highway capacity rather than a cumulative process of subsequent studies of vehicular traffic in which the empirical nucleation nature of traffic breakdown (F→S transition) is ignored.

12.1.7 Response of Traffic and Transportation Research Community

In a discussion of the question about a conversion to the new paradigm, Kuhn wrote "How, then, are scientists brought to make this transposition (to the new paradigm)? Part of the answer is that they are very often not".[2] To the same question, Max Planck wrote "A new scientific truth does not triumph by convincing its opponents and making them see the light, but rather because its opponents eventually die, and a new generation grows up that is familiar with it".[3] The author's experience with the response of the traffic and transportation research community on the emergence

---

[2] See pages 149 and 150 of [28].

[3] See [37].

12.1 Kuhn's Structure of Scientific Revolutions in Application ...

of the three-phase traffic theory confirms general conclusions about the conversion to the new paradigm made in Kuhn's book.

> Most of the scientists working in standard traffic and transportation science cannot accept the three-phase traffic theory because the scientists see that a good share of the traffic papers they have published do not contribute much of anything to understanding real traffic or how to reduce congestion as it has been explained in the three-phase traffic theory.

As mentioned in Chap. 1, the standard traffic and transportation theories that failed when applied in the real world are currently the methodologies of teaching programs in most universities and the subject of publications in most transportation research journals and scientific conferences. This state of the traffic and transportation research community persists: Although the empirical proof of the empirical nucleation nature of traffic breakdown (F→S transition) at a bottleneck is not denied, the empirical proof appears to be simply ignoring in the traffic and transportation research community. Such a behavior of scientists is confirmed by the historical analysis made by Kuhn. Kuhn wrote "the transfer of allegiance from paradigm to paradigm is a conversion experience that cannot be forced", "paradigm change cannot be justified by proof", and "the source of resistance (of research community) is the assurance that the older paradigm will ultimately solve all its problems that nature can be shoved into the box of the paradigm provides".[4] Thus, it could be understandable why it is very difficult for most traffic and transportation researchers to realize there is an empirical traffic phenomenon that call into question basically all fundamentals of the standard traffic flow theories and models as well as to accept the three-phase traffic theory solving the problem. In accordance with Kuhn's analysis, the empirical proof of the three-phase traffic theory has no significance for the acceptance of this theory by the traffic and transportation research community.

It could be assumed that the failure of standard methodologies of traffic and transportation science for reliable traffic management can be explained by a very long time interval between the development of the standard traffic theories made in the 1950s–1960s and the understanding of the empirical nucleation nature of traffic breakdown at road bottlenecks made at the end of the 1990s. During this long time interval, several generations of traffic researches developed a huge number of standard traffic flow models.

What could be the future development of traffic science expected from Kuhn's historical analysis? The three-phase traffic theory works in the real world. In contrast, applications of standard traffic theories failed in the real world. Due to a particular importance of the development of reliable traffic management for the economy of the world, the author could expect that the acceptance of the three-phase traffic theory

---

[4] See pages 150 and 151 of [28].

will come firstly from *practical* traffic engineers: For the practical traffic engineers, it is mostly important that a traffic theory works in real traffic applications.[5]

12.2 Can Autonomous Driving Improve Traffic?

12.2.1 Mixed Traffic Flow

The term *mixed traffic flow* means vehicular traffic consisting of random distributed human driving and autonomous (automated) driving vehicles. It is commonly assumed that future vehicular traffic is mixed traffic flow.[6] An autonomous driving vehicle is a self-driving vehicle that can move without a driver. Autonomous driving is realized through the use of an automated system in a vehicle: The automated system has control over the vehicle in traffic flow. For this reason, autonomous driving vehicle is also called automated driving (or automatic driving) vehicle.

It should be noted that in the engineering science the terms *autonomous driving* and *automated driving* are not synonymous. There are two reasons for this. While an autonomous driving vehicle should be able to move without a driver in the vehicle, there are several different levels of automation associated with automated driving. The levels include, for example, a level of "conditional automation" in which the driver must be present to provide any corrections when needed and a level of "full automation" in which the automated vehicle system is in complete control of the vehicle and human presence is no longer needed. Additionally, in contrast with autonomous driving vehicle that moves fully autonomous from other vehicles, it is often assumed that automated driving can be supported by so-called cooperative driving that can be realized through a diverse variety of cooperative automated systems like vehicle-to-vehicle (V2V) communication (ad hoc vehicle networks) and vehicle-to-infrastructure (V2X) communication.

It is generally assumed that autonomous and cooperative driving vehicles can improve mixed vehicular traffic, while enhancing traffic safety and comfort as well

---

[5] This expectation is confirmed by a recent book by Rehborn et al. [40] written by practical traffic engineers working in industry. Additionally, we should also mention that accordingly Max Planck's conclusion that "... a new generation grows up that is familiar with" the three-phase traffic theory, already now there are many research papers made in the framework of the three-phase traffic theory (see, e.g., [6, 12–17, 27, 29, 38, 41, 46, 50–52, 54–56]).

[6] There is a huge number of publications devoted to mixed traffic flow (see, e.g., [1–5, 7, 8, 11, 18–21, 30–36, 39, 43–45, 47, 48]). In particular, there exist a large series of papers by the well-known and massive "Automated Highway System" project involving the US government and a large number of transportation researchers [1, 2], EU projects [11], and projects made in Germany [3]. A consortium of researches all over the world performed extensive and pioneering research into autonomous and automated driving vehicle systems (see references to these extensive research, for example, in reviews and books by Ioannou [18], Ioannou and Sun [19], Ioannou and Kosmatopoulos [20], Shladover [43], Rajamani [39], Meyer and Beiker [36], Bengler et al. [4], and Van Brummelen et al. [47]).

12.2 Can Autonomous Driving Improve Traffic?

as increase highway capacity. Here a question arises: How can this assumption be proven? It must be emphasized that to understand real traffic consisting only of human driving vehicles, there is a sufficient amount of empirical spatiotemporal traffic data measured in real field traffic. *Understanding real traffic* presented in this book is based on results of the analysis of these empirical spatiotemporal traffic data.

In contrast to real traffic consisting only of human driving vehicles, currently there are *no* empirical datasets for mixed traffic flow consisting of autonomous driving and human driving vehicles that can be used for understanding real mixed traffic. Therefore, traffic researchers have to use simulation models of mixed traffic flow for the evaluation of features of mixed traffic. On the one hand, this is the reason why in this book we have not discussed in detail whether autonomous driving can improve traffic: The book is entirely devoted to understanding real traffic based on measurements of real field spatiotemporal traffic data. On the other hand, the most simulation results made in the traffic and transportation research community for the evaluation of mixed traffic are based on the use of standard models for traffic flow of human driving vehicles. As emphasized many times in this book, standard models for traffic flow of human driving vehicles widely used in the traffic and transportation research community are invalid for the evaluation of most ITS applications.[7] In particular, these standard models are invalid for the evaluation of the effect of autonomous driving and cooperative driving on highway capacity of mixed traffic flow. We have proven that for a valid evaluation of the effect of autonomous driving and cooperative driving on highway capacity in mixed traffic flow a traffic flow model in the framework of the three-phase traffic theory that can show the nucleation nature of traffic breakdown (F→S transition) at a highway bottleneck should be used.[8]

As mentioned above, it is generally assumed that autonomous driving vehicles should considerably enhance highway capacity. To discuss this point, firstly recall that highway capacity is limited by traffic breakdown at a bottleneck. In the near future, we could expect mixed traffic flow in which the share of autonomous driving vehicles is very small. In the case of a very small share of the autonomous vehicles, almost any autonomous vehicle is surrounded by long sequences of human driving vehicles. In other words, the autonomous vehicle can be considered a "single" autonomous vehicle moving between human driving vehicles. Therefore, the question arises whether and how a "single" autonomous vehicle moving between human driving vehicles can effect on conditions of traffic breakdown.

Conditions of traffic breakdown depend on the amplitude of a local speed decrease occurring in free flow at a bottleneck (Chap. 8): At the same other conditions, the larger the local speed decrease in free flow at the bottleneck caused by a vehicle moving through the bottleneck, the larger the probability that the vehicle causes traffic breakdown at the bottleneck. In other words, to enhance highway capacity,

---

[7] This criticism can be found in Chaps. 1 and 5 as well as in Appendix C. In more detail, the criticism of the evaluation of mixed traffic flow with the use of standard models for traffic flow of human driving vehicles has been considered in Refs. [23, 24, 26].

[8] See papers [23–26].

autonomous vehicles should lead to a smaller local speed decrease at the bottleneck occurring when the autonomous vehicles travel in free flow through the bottleneck.

Unfortunately, for the classical (standard) idea of the dynamic behavior of autonomous vehicles dominated in the literature[9], the opposite case is realized: as we have proven,[10] the standard dynamics of autonomous vehicles does lead to a larger local speed decrease at the bottleneck than that caused by human driving vehicles.

- Rather than enhanced highway capacity, the standard autonomous driving vehicles can deteriorate mixed traffic, while provoking traffic breakdown at the bottleneck.

To avoid this deterioration of traffic system, the author introduced a strategy of autonomous driving in the framework of the three-phase theory called TPACC[11]:

(i) In contrast with the standard autonomous driving vehicle, the TPACC vehicle can lead to a smaller local speed decrease in free flow at a bottleneck.
(ii) In mixed traffic flow with TPACC vehicles, the probability of traffic breakdown at the bottleneck can be reduced.

This discussion about the effect of autonomous vehicles on traffic breakdown in mixed traffic allows us to assume that future systems for autonomous driving should be developed whose rules are consistent with those of human driving vehicles. Otherwise, we could expect that autonomous driving vehicles can be considered as "obstacles" for drivers.

12.2.2 Can Vehicular Traffic Consisting of 100% Autonomous Vehicles Be Real Option in Near Future?

In Sect. 12.2.1, we have focused on mixed traffic flow with a small share of autonomous driving vehicles. It should be emphasized that dynamic rules of autonomous vehicles can be developed that are totally different from the dynamic behavior of human driving vehicles. In other words, for traffic flow consisting of 100% of autonomous driving vehicles the dynamic rules of autonomous driving vehicle should not necessarily be consistent with the dynamic behavior of human driving vehicles. One of the consequences is that in traffic flow consisting of 100% of autonomous driving vehicles highway capacity can be considerably larger in comparison with highway capacity of mixed traffic flow. Therefore, a question can arise: why does the effect of an autonomous driving vehicle on traffic breakdown and

---

[9] Standard models of autonomous vehicles can be found in [1, 2, 4, 5, 7, 8, 10, 18–21, 30–36, 39, 43–45, 47, 48].

[10] See papers [23, 24, 26].

[11] TPACC–Three-traffic-Phase Adaptive Cruise Control. One of the most important features of TPACC is the existence of the indifferent zone for car-following of the three-phase traffic theory. The indifferent zone for car-following of the three-phase traffic theory will be defined and considered in Sect. A.1.1 of Appendix A. A theory of the effect of TPACC vehicles on traffic breakdown at a bottleneck can be found in [23, 24, 26].

12.2 Can Autonomous Driving Improve Traffic?

capacity of mixed traffic flow discussed in Sect. 12.2.1 is important for future vehicular traffic? To answer this question, we should discuss whether and when traffic flow consisting of 100% of autonomous driving vehicles in a traffic network without human driving vehicles could be possible to expect. First, we assume that in the future all vehicles are autonomous driving vehicles only. We could expect that there is at least one reason that might prevent the realization of this case:

- *Service costs for autonomous vehicles* could be many times larger than those for conventional vehicles driven by humans. Indeed, the check of systems for autonomous driving should be made more frequently than it is needed for the case for a manual driving vehicle. This is because a sudden failure of a system for autonomous driving in a vehicle can lead to an accident with catastrophic consequences for both passengers of the autonomous driving vehicle and passengers of several other following vehicles in traffic flow. Such a frequent check of systems for autonomous driving could lead to enormous service costs that could be paid by only a (small) part of vehicle owners.[12]

If in the future some of the vehicles moving in traffic networks are manual driving ones, then we can assume that two independent of each other traffic networks might be developed: (i) One network is dedicated to autonomous driving vehicles only. (ii) Another network is dedicated to human driving vehicles only. We could expect that there is at least one reason that might prevent the realization of this case:

- *Extremely high costs* for the development of two independent of each other traffic networks.

To explain this statement, we should recall that traffic congestion might be prevented, if either highways with much enough lanes were build or many enough parallel highways for some travel routes in a network were build. However, due to the extremely high road-repairing costs this well-known idea could not be realized. Indeed, after 20–30 years almost each highway should be repaired. The result is the extremely high road-repairing costs and/or very many road works that act as highway bottlenecks for traffic breakdown.

It is clear that the organization of high-speed highway lanes, or some separated roads, or else *a part* of a traffic network dedicated to connected autonomous vehicles is possible.[13] In this case, in another network part in which human driving vehicles can move, mixed traffic flow is realized. Therefore, it could be expected that mixed traffic flow will remain the reality also for future vehicular traffic. In this case, as mentioned in Sect. 12.2.1, already a single autonomous driving vehicle whose dynamics is based on the standard approach can provoke traffic breakdown at the bottleneck. To avoid such a deterioration of traffic system, autonomous driving vehicles should learn some

---

[12] To explain this statement, we should note that the mean time headway (see the definition of the term *time headway* in Appendix A) in vehicular traffic can reach values 1–2 s or less. Even for an autonomous driving vehicle in which there is a possibility for driver control of the vehicle, none of the passengers is able to take the vehicle under control during the short time interval 1–2 s.

[13] The organization of high-speed highway lanes dedicated to connected autonomous vehicles has been studied, for example, in [9, 42, 49, 53].

important features of the driver behaviors in the car-following, for example, as it has been done in TPACC vehicles.

Understanding real traffic presented in this book based on a study of real field traffic data with the use of the three-phase traffic theory will be the basis for the understanding of future mixed traffic flow consisting of random distributed human driving and automated driving vehicles.

References

1. Automated Highway Systems, http://www.seminarsonly.com/Civil_Engineering/automated-highway-systems.php
2. Automated Highway Systems, https://seminarprojects.blogspot.de/2012/01/detailed-report-on-automated-highway.html
3. Automatisches Fahren, http://www.tuvpt.de/index.php?id=foerderung000
4. K. Bengler, K. Dietmayer, B. Farber, M. Maurer, Ch. Stiller, H. Winner, IEEE Intell. Transp. Syst. Mag. **6**, 6–22 (2014)
5. L.C. Davis, Phys. Rev. E **69**, 066110 (2004)
6. L.C. Davis, Phys. A **387**, 6395–6410 (2008)
7. L.C. Davis, Phys. A **405**, 128–139 (2014)
8. L.C. Davis, Phys. A **451**, 320–332 (2016)
9. L.C. Davis, Phys. A **555**, 124743 (2020)
10. L.C. Davis, Phys. A **562**, 125402 (2021)
11. European Roadmap Smart Systems for Automated Driving (2015), https://www.smart-systems-integration.org/public/documents/
12. D.-J. Fu, Q.-L. Li, R. Jiang, B.-H. Wang, Phys. A **559**, 125075 (2020)
13. K. Gao, R. Jiang, S.-X. Hu, B.-H. Wang, Q.-S. Wu, Phys. Rev. E **76**, 026105 (2007)
14. S. He, W. Guan, L. Song, Phys. A **389**, 825–836 (2010)
15. X.-J. Hu, X.-T. Hao, H. Wang, Z. Su, F. Zhang, Phys. A **545**, 123725 (2020)
16. X.-J. Hu, F. Zhang, J. Lu, M.-Y. Liu, Y.-F. Ma, Q. Wan, Phys. A **527**, 121176 (2019)
17. X.-j. Hu, H. Liu, X. Hao, Z. Su Z. Yang, Phys. A **563**, 125495 (2021)
18. P.A. Ioannou (ed.), *Automated Highway Systems* (Plenum Press, New York, 1997)
19. P.A. Ioannou, J. Sun, *Robust Adaptive Control* (Prentice Hall Inc, Upper Saddle River, 1996)
20. P.A. Ioannou, E.B. Kosmatopoulos, in *Wiley Encyclopedia of Electrical and Electronics Engineering*, ed. by J.G. Webster (Wiley, New York, 2000), https://doi.org/10.1002/047134608X.W1002
21. P.A. Ioannou, C.C. Chien, IEEE Trans. Veh. Technol. **42**, 657–672 (1993)
22. B.S. Kerner, Procedia Comput. Sci. **130**, 785–790 (2018)
23. B.S. Kerner, Phys. Rev. E **97**, 042303 (2018)
24. B.S. Kerner, in *Complex Dynamics of Traffic Management*, ed. by B.S. Kerner. Encyclopedia of Complexity and Systems Science Series (Springer, New York, 2019), pp. 343–385
25. B.S. Kerner, in *Complex Dynamics of Traffic Management*, ed. by B.S. Kerner. Encyclopedia of Complexity and Systems Science Series (Springer, New York, 2019), pp. 195–283
26. B.S. Kerner, Phys. A **562**, 125315 (2021)
27. S. Kokubo, J. Tanimoto, A. Hagishima, Phys. A **390**, 561–568 (2011)
28. T.S. Kuhn, *The Structure of Scientific Revolutions*, 4th edn. (The University of Chicago Press, Chicago, 2012)

References

29. H.K. Lee, R. Barlović, M. Schreckenberg, D. Kim, Phys. Rev. Lett. **92**, 238702 (2004)
30. W. Levine, M. Athans, IEEE Trans. Autom. Control **11**, 355–361 (1966)
31. C.-Y. Liang, H. Peng, Veh. Syst. Dyn. **32**, 313–330 (1999)
32. C.-Y. Liang, H. Peng, J.S.M.E. Jnt, J. Ser. C **43**, 671–677 (2000)
33. T.-W. Lin, S.-L. Hwang, P. Green, Saf. Sci. **47**, 620–625 (2009)
34. J.-J. Martinez, C. Canudas-do-Wit, IEEE Trans. Control Syst. Technol. **15**, 246–258 (2007)
35. M. Maurer, J.Ch. Gerdes, B. Lenz, H. Winner (eds.), *Autonomes Fahren* (Springer, Berlin, 2015)
36. G. Meyer, S. Beiker, *Road Vehicle Automation* (Springer, Berlin, 2014)
37. M. Planck, *Scientific Autobiography and Other Papers*, trans. by F. Gaynor (Williams & Norgate LTD, London, 1950), pp. 33–34
38. Y.-S. Qian, X. Feng, J.-W. Zeng, Phys. A **479**, 509–526 (2017)
39. R. Rajamani, in *Vehicle Dynamics and Control*. Mechanical Engineering Series (Springer, Boston, 2012)
40. H. Rehborn, M. Koller, S. Kaufmann, *Data-Driven Traffic Engineering: Understanding of Traffic and Applications Based on Three-Phase Traffic Theory* (Elsevier, Amsterdam, 2021)
41. F. Rempe, P. Franeck, U. Fastenrath, K. Bogenberger, Transp. Res. C **85**, 644–663 (2017)
42. X. Shan, P. Hao, K. Boriboonsomsin, G. Wu, M. Barth, X. Chen, Transp. Res. A **118**, 25–37 (2018)
43. S.E. Shladover, Veh. Syst. Dyn. **24**, 551–595 (1995)
44. D. Swaroop, J.K. Hedrick, IEEE Trans. Autom. Control **41**, 349–357 (1996)
45. D. Swaroop, J.K. Hedrick, S.B. Choi, IEEE Trans. Veh. Technol. **50**, 150–161 (2001)
46. J.-F. Tian, C.-Q. Zhu, R. Jiang, in *Complex Dynamics of Traffic Management*, ed. by B.S. Kerner. Encyclopedia of Complexity and Systems Science Series (Springer, New York, 2019), pp. 313–342
47. J. Van Brummelen, M. O'Brien, D. Gruyer, H. Najjaran, Transp. Res. C **89**, 384–406 (2018)
48. P. Varaiya, IEEE Trans. Autom. Control **38**, 195–207 (1993)
49. J.-P. Wang, H.-J. Huang, X. Ban, Phys. A **524**, 354–361 (2019)
50. J.J. Wu, H.J. Sun, Z.Y. Gao, Phys. Rev. E **78**, 036103 (2008)
51. H. Yang, J. Lu, X.-J. Hu, J. Jiang, Phys. A **392**, 4009–4018 (2013)
52. H. Yang, X. Zhai, C. Zheng, Phys. A **509**, 567–577 (2018)
53. L. Ye, T. Yamamoto, Phys. A **512**, 588–597 (2018)
54. J.-W. Zeng, Y.-S. Qian, F. Lv, F. Yin, L. Zhu, Y. Zhang, D. Xu, Phys. A **574**, 125918 (2021)
55. J.-W. Zeng, Y.-S. Qian, S.-B. Yu, X.-T. Wei, Phys. A **530**, 121567 (2019)
56. H.-T. Zhao, L. Lin, C.-P. Xu, Z.-X. Li, X. Zhao, Phys. A **553**, 124213 (2020)

Appendix A
Characteristics of Synchronized Flow in Three-Phase Traffic Theory

The objective of Appendix A is a more detailed consideration of features of synchronized flow observed in empirical data and explained in the three-phase traffic theory.[1]

A.1 Two-Dimensional Region of Steady States of Synchronized Flow

In the three-phase traffic theory, based on empirical data measured in synchronized flow[2] it has been assumed that steady states of synchronized flow cover a two-dimensional (2D) region in the flow–density plane (dashed region in Fig. A.1). To explain this hypothesis, we should firstly define the term *steady state* of synchronized flow.

We consider a vehicle following the preceding vehicle that moves at a time-independent speed v_ℓ in synchronized flow. We assume that the vehicle speed v is equal to v_ℓ (Fig. A.2a):

$$v = v_\ell. \qquad (A.1)$$

The net distance between these two vehicles following each other is called a space gap denoted by g. We denote the vehicle length by d. For a simplification of explanations, we assume that all vehicles exhibit the same parameters, in particular, the same vehicle length. Furthermore, we assume that each of the vehicles in synchronized flow moves at the same space gap g and the same speed satisfying (A.1). Such a

---

[1] In this appendix, results discussed in Sects. A.1–A.4 are based on Refs. [1–15], results of Sects. A.5–A.7 are based on Refs. [3–40], results presented in Sect. A.8 are based on Refs. [15, 17, 22, 29, 30], results presented in Sect. A.9 are based on Refs. [26, 27], and results presented in Sect. A.10 are based on Ref. [15].

[2] Readers can find examples of the traffic data in Sect. 10.7 of the book [15].

178   Appendix A: Characteristics of Synchronized Flow in Three-Phase Traffic Theory

Fig. A.1 Qualitative illustration of the hypothesis of three-phase traffic theory about 2D region of steady states of synchronized flow (dashed region) in the flow–density plane; curve F for free flow is taken from Fig. 6.3b. F—free flow; S—steady states of synchronized flow

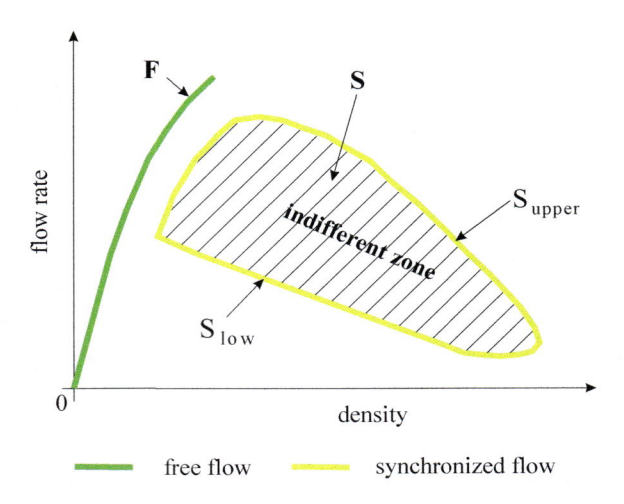

hypothetical state of synchronized flow can be called as a steady state of synchronized flow.

To understand the sense and features of the 2D region of steady states of synchronized flow (Fig. A.1), we should consider the correspondence between the space gap g and the vehicle density ρ. We assume that in a steady state of synchronized flow on a road section of length M (km) there are exactly N vehicles ($N \gg 1$) (Fig. A.3). From the definition of the vehicle density (Sect. 1.1), we get

$$\rho = N/M \text{ (vehicles/km)}. \tag{A.2}$$

For the steady state of synchronized flow (Fig. A.3), we get

$$N(d + g) = 1000M, \tag{A.3}$$

where we assume that the space gap g and the vehicle length d are measured in the units of meter. From (A.2) and (A.3), we get[3]

$$\rho = \frac{1000}{d + g} \text{ (vehicles/km)}. \tag{A.4}$$

Clearly, rather than steady states of synchronized flow, states of synchronized flow are dynamic ones in real traffic. However, to understand real synchronized flow, firstly hypothetical steady states of synchronized flow should be discussed.

There are the low boundary S_{low} and the upper boundary S_{upper} of the 2D states of synchronized flow (Fig. A.1): Below the boundary S_{low} and above the boundary

---

[3] At $N \gg 1$ in (A.2), formula (A.4) is also valid for traffic flow with different values of the space gap between vehicles that have different lengths, if we replace in (A.3) the constant values g and d by their mean values, respectively.

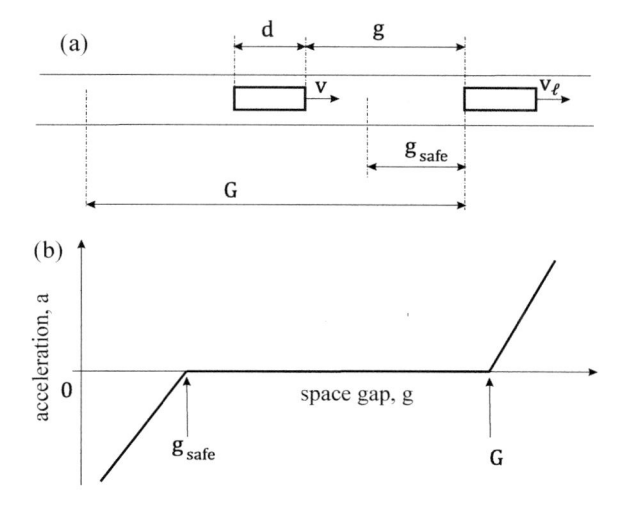

Appendix A: Characteristics of Synchronized Flow in Three-Phase Traffic Theory

indifferent zone in car-following:

$a = 0$ at $g_{safe} \leq g \leq G$

asymmetric deceleration-acceleration driver behavior:

$$\begin{cases} a < 0 \text{ at } g < g_{safe} \\ a > 0 \text{ at } g > G \end{cases}$$

Fig. A.2 Qualitative explanation of driver behavior in car-following in the three-phase traffic theory [15]: **a** A schema of car-following within indifferent zone (A.6). **b** Indifferent zone for car-following and asymmetric deceleration–acceleration driver behavior. g is the space gap between two following each other vehicles, d is the vehicle length, g_{safe} is the safe space gap, G is the synchronization space gap, a is the vehicle acceleration (deceleration)

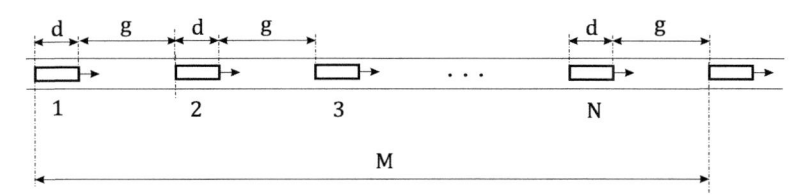

Fig. A.3 Qualitative explanation of relation (A.4) between the space gap g between two following each other vehicles and the vehicle density ρ. d is the vehicle length

$S_{\rm upper}$ no steady states of synchronized flow can be realized, i.e., only dynamic states of synchronized flow can occur. The hypothesis about the 2D region of the steady states of synchronized flow introduced in the three-phase traffic theory[4] is equivalent to the following features of car-following[5]:

- There is an *indifferent zone for car-following*: The vehicle motion is indifferent to the space gap to the preceding vehicle within the 2D region in Fig. A.1 (Sect. A.1.1).
- In its turn, the existence of the indifferent zone for car-following results in a deceleration–acceleration driver behavior introduced in the three-phase traffic theory that can be called as *asymmetric* deceleration–acceleration driver behavior: In dynamic states above the boundary $S_{\rm upper}$ in Fig. A.1 the vehicle decelerates, whereas in dynamic states below the boundary $S_{\rm low}$ in Fig. A.1 the vehicle accelerates (Sect. A.1.2).

A.1.1 Indifferent Zone for Car-Following

The upper boundary $S_{\rm upper}$ in Fig. A.1 is related to safety driving conditions. At the boundary $S_{\rm upper}$, the space gap g to the preceding vehicle is equal to some *safe space gap* (safe gap for short) denoted by $g_{\rm safe}$ (Fig. A.2a). In the states above the boundary $S_{\rm upper}$ in Fig. A.1, the space gap g to the preceding vehicle is smaller than the safe space gap $g_{\rm safe}$. We denote the vehicle acceleration (deceleration) by a. The definition of the safe space gap $g_{\rm safe}$ is as follows: At $g < g_{\rm safe}$, a driver decelerates to avoid the collision with the preceding vehicle ($a < 0$ at $g < g_{\rm safe}$ in Fig. A.2b).

At the low boundary $S_{\rm low}$ in Fig. A.1, the space gap g to the preceding vehicle is equal to some *synchronization space gap* (synchronization gap for short) denoted by G (Fig. A.2a). The definition of the synchronization space gap G in synchronized flow is as follows: In the states below the boundary $S_{\rm low}$ in Fig. A.1, within which the space gap g to the preceding vehicle is larger than the synchronization space gap G, the driver accelerates while following the preceding vehicle ($a > 0$ at $g > G$ in Fig. A.2b). From the indifferent zone for car-following introduced in the three-phase traffic theory, we get

$$G > g_{\rm safe}. \tag{A.5}$$

The indifferent zone for car-following is defined as follows. There is a range of space gaps between the synchronized space gap G and the safe space gap $g_{\rm safe}$

$$g_{\rm safe} \leq g \leq G. \tag{A.6}$$

The space-gap range (A.6) defines the indifferent zone for car-following introduced in the three-phase traffic theory. Within the space-gap range (A.6), condition (A.1)

---

[4] See papers [2–14] and the book [15].
[5] Recall that the term *car-following* means the dynamic behavior of a vehicle that follows the preceding vehicle.

Appendix A: Characteristics of Synchronized Flow in Three-Phase Traffic Theory 181

is satisfied. This means that in a steady state of synchronized flow under conditions (A.6) the vehicle acceleration or deceleration is equal to zero ($a = 0$ in Fig. A.2b): In the steady state of synchronized flow, the vehicle moves at the time-independent speed v (A.1) independent of the space gap to the preceding vehicle as long as conditions (A.6) are satisfied.

A.1.2 Asymmetric Deceleration–Acceleration Driver Behavior

In the three-phase traffic theory, the asymmetric deceleration–acceleration driver behavior is defined as follows: In dynamic states of synchronized flow that are above the boundary S_{upper} in Fig. A.1, the vehicle decelerates, whereas in dynamic states below the boundary S_{low} in Fig. A.1 the vehicle accelerates. This definition means that the following conditions should be satisfied in car-following:

$$a < 0 \text{ at } g < g_{\mathrm{safe}}, \tag{A.7}$$

$$a > 0 \quad \text{at } g > G. \tag{A.8}$$

Condition (A.7) means that the vehicle decelerates when its space gap g to the preceding vehicle becomes smaller than the safe gap g_{safe}. Condition (A.8) means that the vehicle accelerates when its space gap g to the preceding vehicle becomes larger than the synchronization gap G.

In addition to the space gap g, traffic researchers use very often time headway (net time gap) to the preceding vehicle denoted by $\tau^{(\mathrm{net})}$:

$$\tau^{(\mathrm{net})} = g/v, \tag{A.9}$$

where it is assumed that the vehicle speed $v > 0$. Time headway (A.9) determines the time it takes for the vehicle to move at the speed v the distance that is equal to the space gap g to the preceding vehicle. In accordance with (A.9), the dependence of vehicle acceleration (deceleration) a on time headway $\tau^{(\mathrm{net})}$ is qualitatively the same as that shown in Fig. A.2b for the gap dependence of the vehicle acceleration $a(g)$. Thus, the asymmetric deceleration–acceleration driver behavior (Fig. A.2b) can also be formulated as follows[6]:

- Drivers accept on average shorter time headway by deceleration than time headway accepting by acceleration.

Therefore, the asymmetric deceleration–acceleration driver behavior (A.7), (A.8) can be also rewritten through the following equivalent formulas:

---

[6] See Ref. [38] in which this definition for the asymmetric deceleration–acceleration driver behavior has been used for empirical and theoretical studies of the driver behavior in synchronized flow and wide moving jams (see also Sect. B.3 for more details).

$$a < 0 \text{ at } \tau^{(\text{net})} < \tau_{\text{safe}}, \tag{A.10}$$

$$a > 0 \text{ at } \tau^{(\text{net})} > \tau_{\text{G}}, \tag{A.11}$$

where τ_{safe} is a safe time headway that is equal to $\tau_{\text{safe}} = g_{\text{safe}}/v$, τ_{G} is the synchronization time headway $\tau_{\text{G}} = G/v$,

$$\tau_{\text{G}} > \tau_{\text{safe}}, \tag{A.12}$$

and we assume that the vehicle speed $v > 0$.

A.2 Origin of Indifferent Zone for Car-Following and Asymmetric Deceleration–Acceleration Driver Behavior

> The 2D region of steady states of synchronized flow (Fig. A.1) is the origin of the indifferent zone for car-following and the asymmetric deceleration–acceleration driver behavior occurring at the boundaries of the indifferent zone.

To explain this statement, firstly, we consider a correspondence between a given vehicle speed v and a point (ρ, q) in the flow–density plane (colored black circle in Fig. A.4). In accordance with formula (5.1), we get $v = q/\rho$. Therefore, any point on a solid line in Fig. A.4 that intersects the beginning of the coordinate system in the flow–density plane ($\rho = 0, q = 0$) and the given point (ρ, q) is related to the same average speed $v = q/\rho$. Tangent of the angle α between the solid line in Fig. A.4 and the density axis is equal to $\tan \alpha = q/\rho$, and therefore we get $v = \tan \alpha = q/\rho$. Thus, the slope of any line $\tan \alpha$ intersecting the beginning of the coordinate system in the flow–density plane is equal to the speed v for any point (ρ, q) that lies at this line in the flow–density plane.

In accordance with Fig. A.4, in Fig. A.5a that is taken from Fig. A.1, we have drawn two lines related to two different constant vehicle speeds $v = v_1$ and $v = v_2$, where $v_2 > v_1$. We denote in Fig. A.5a the vehicle densities at intersection points 1 and 3 with boundary S_{low} by ρ_1 and ρ_3, respectively. Correspondingly, we denote the vehicle densities at intersection points 2 and 4 with boundary S_{upper} by ρ_2 and ρ_4, respectively. Then, Fig. A.5b results from Fig. A.5a as follows.

We draw two vertical lines $v = v_1$ and $v = v_2$ in the space gap–speed plane (Fig. A.5b). As follows from Fig. A.4 and formula (A.4), we get the following results: (i) the density ρ_1 in Fig. A.5a determines the synchronization gap $G = G_1$ in the intersection point 1 of the line $v = v_1$ with a speed dependence of the synchronization gap $G(v)$ in Fig. A.5b; (ii) the density ρ_2 in Fig. A.5a determines the safe gap $g_{\text{safe}} = g_{\text{safe 1}}$ in the intersection point 2 of the line $v = v_1$ with a speed dependence

Fig. A.4 Qualitative explanation of the correspondence between a given speed v (solid line) and a point (ρ, q) (colored black circle) in the flow–density plane

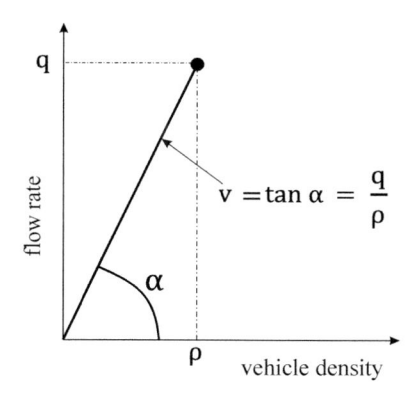

$$v = \tan \alpha = \frac{q}{\rho}$$

of the safe gap $g_{\text{safe}}(v)$ in Fig. A.5b; (iii) the density ρ_3 in Fig. A.5a determines the synchronization gap $G = G_2$ in the intersection point 3 of the line $v = v_2$ with the dependence $G(v)$ in Fig. A.5b; (iv) the density ρ_4 in Fig. A.5a determines the safe gap $g_{\text{safe}} = g_{\text{safe 2}}$ in the intersection point 4 of the line $v = v_2$ with the dependence $g_{\text{safe}}(v)$ in Fig. A.5b. We see that $\rho_1 > \rho_3$ (Fig. A.5a) and, therefore, accordingly to formula (A.4) $G_2 > G_1$ (Fig. A.5b). Correspondingly, $\rho_2 > \rho_4$ (Fig. A.5a) and, therefore, according to formula (A.4) $g_{\text{safe 2}} > g_{\text{safe 1}}$ (Fig. A.5b). From this qualitative analysis of the hypothesis about 2D steady states of synchronized flow, it follows that both $G(v)$ and $g_{\text{safe}}(v)$ are increasing speed functions (Fig. A.5b). Moreover, because $\rho_2 > \rho_1$ and $\rho_4 > \rho_3$ (Fig. A.5a), according to formula (A.4), we get $G_1 > g_{\text{safe 1}}$ and $G_2 > g_{\text{safe 2}}$ (Fig. A.5a). Thus, the hypothesis about 2D steady states of synchronized flow proves that conditions (A.5) and (A.12) are valid for any given speed v in steady states of synchronized flow, when $v > 0$.

In its turn, points 1–4 in the space gap–speed plane (Fig. A.5b) determine, respectively, the associated points 1–4 related to the space gap dependencies of the vehicle acceleration (Fig. A.5c, d): Points 1 and 2 are related to $v = v_1$ as shown in Fig. A.5c whereas points 3 and 4 are related to $v = v_2$ as shown in Fig. A.5d. In particular, in Fig. A.5b–d within the space-gap range for the indifferent zone for car-following (A.6), at $v = v_\ell$ (A.1) the vehicle acceleration is equal to zero: $a = 0$.

Outside the indifferent zone for car-following, as explained above, formulas (A.7) and (A.8) define the asymmetric deceleration–acceleration behavior of human driving vehicles as shown in Fig. A.5c, d. In Fig. A.5b, this asymmetric deceleration–acceleration behavior of human driving vehicles is symbolically shown by the label "acceleration" with down arrows at $g > G$ (A.7) as well as by the label "deceleration" with up arrows at $g < g_{\text{safe}}$ (A.8).[7]

---

[7] Features of the indifferent zone for car-following and asymmetric deceleration–acceleration driver behavior of the three-phase traffic theory [2–15], in particular, formulas (A.5)–(A.12) have been firstly incorporated in the mathematical microscopic stochastic [29] and deterministic traffic flow models [32] of Kerner and Klenov (see simplified explanations of the mathematical incorporation of the indifferent zone for car-following and asymmetric deceleration–acceleration driver behavior in the Kerner–Klenov models in Chap. 11 of the book [17]).

184   Appendix A: Characteristics of Synchronized Flow in Three-Phase Traffic Theory

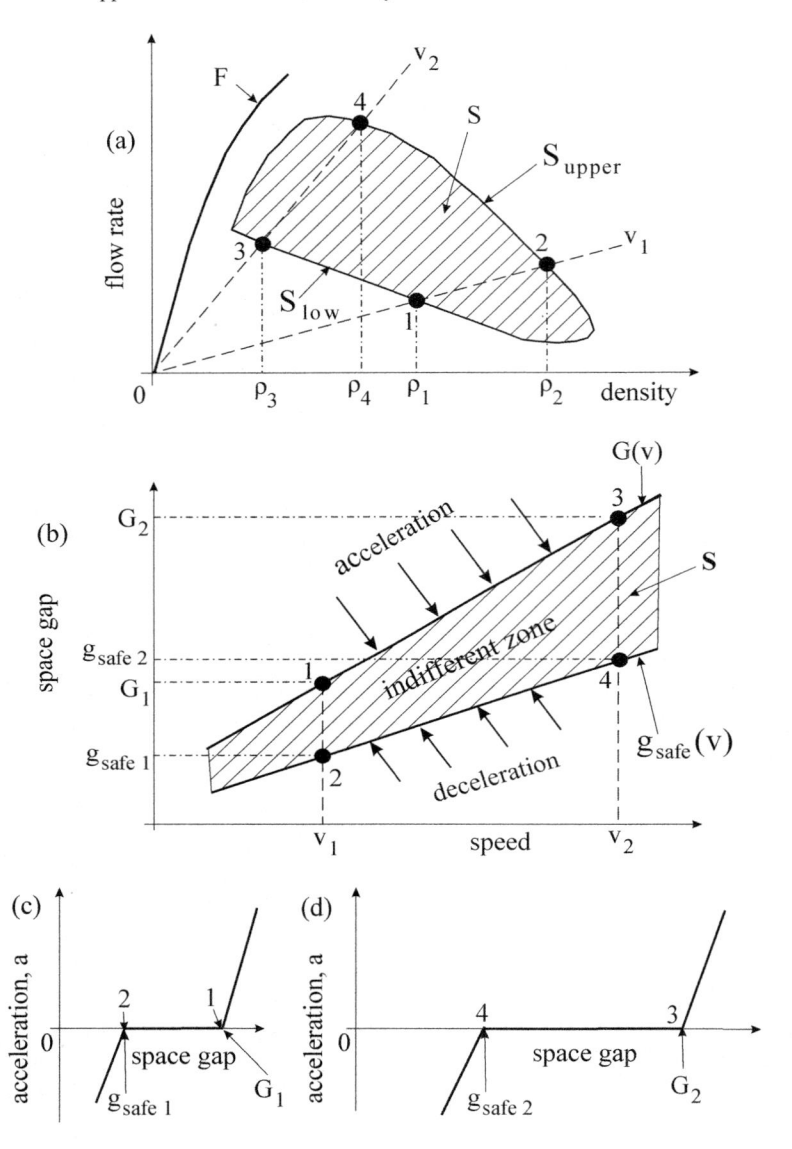

Fig. A.5 Qualitative explanations of the origin of the indifferent zone for car-following and asymmetric deceleration–acceleration driver behavior: **a** 2D region of steady states of synchronized flow (dashed 2D region) in the flow–density plane taken from Fig. A.1. **b** A part of the steady states of synchronized flow (dashed 2D region) in (**a**) that are shown in the space gap–speed plane; $g_{\mathrm{safe}}(v)$ is the speed dependence of the safe space gap g_{safe}, $G(v)$ is the speed dependence of the synchronization space gap G. **c, d** Vehicle acceleration (deceleration) as space-gap functions that illustrate indifferent zones for car-following related to space-gap range (A.6) for $v = v_1$ (**c**) and for $v = v_2$ (**d**) as well as the associated asymmetric deceleration–acceleration driver behavior outside the indifferent zones; constant speeds v_1 and v_2 are marked in (**a, b**) by associated dashed lines; points 1–4 are the same as those in (**a, b**)

Fig. A.6 Qualitative explanation of driver speed adaptation in the speed–space gap plane (dashed region for synchronized flow is taken from Fig. A.5b) in the three-phase traffic theory. S—synchronized flow. g_{safe} is the safe space gap, G is the synchronization space gap

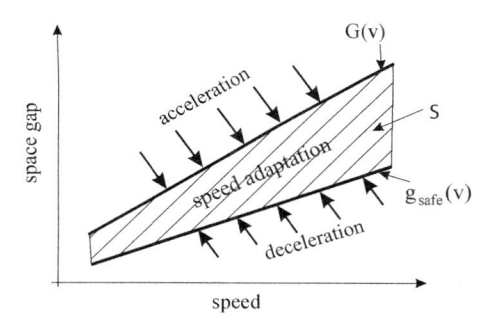

> The 2D region of steady states of synchronized flow (Fig. A.1) of the three-phase traffic theory is equivalent to the indifferent zone for car-following and the asymmetric deceleration–acceleration driver behavior occurring at the boundaries of the indifferent zone (Fig. A.5b–d).

A.3 Driver Speed Adaptation within Indifferent Zone for Car-Following

Rather than steady states of synchronized flow, dynamic states of synchronized flow are realized in real traffic. However, as we will see below, the consideration of the 2D steady states of synchronized flow made in Sects. A.1 and A.2 is very important for the understanding of many features of dynamic states of synchronized flow.

One of the features of the dynamic states of synchronized flow is the driver speed adaptation effect. In Sect. 8.1.1, we have already briefly considered driver speed adaptation that occurs when a driver approaches a slower moving preceding vehicle moving at a speed v_ℓ and the driver cannot pass this slow vehicle (Fig. 8.1a). In the three-phase traffic theory, it has been assumed that the driver speed adaptation occurs within the indifferent zone of car-following (A.6) (Fig. A.6)[8]: The driver begins to

---

[8] The indifferent zone in car-following and asymmetric deceleration–acceleration driver behavior resulting from the hypothesis of the three-phase traffic theory about the 2D region of synchronized flow states (Fig. A.5a–d) are also the theoretical basic for autonomous driving in the framework of the three-phase traffic theory called TPACC that was introduced by the author in 2004 (see references to TPACC inventions in [28]). TPACC has already been mentioned in Sect. 12.2. In particular, the speed adaptation of TPACC occurs within the indifferent zone of TPACC given by formula (A.6). Speed adaptation effect of TPACC means that if the TPACC vehicle cannot still pass the slow-moving preceding vehicle, then under conditions (A.6) the TPACC vehicle acceleration (deceleration) $a^{(TPACC)}$ changes over time in accordance with formula

$$a^{(TPACC)} = K_{\Delta v}(v_\ell - v) \text{ at } g_{safe} \leq g \leq G, \tag{A.13}$$

decelerate adapting the speed v to the speed v_ℓ of the preceding vehicle within the synchronization space gap G to the preceding vehicle, i.e., when the space gap to the preceding vehicle satisfies condition $g \leq G$. This speed synchronization between vehicles explains the meaning of the term *synchronization space gap*.

The driver speed adaptation effect means that if the vehicle cannot pass the slow-moving preceding vehicle, then under conditions (A.6) the vehicle acceleration (deceleration) a changes over time in accordance with formulas:

$$\begin{aligned} a &> 0 \text{ if } v < v_\ell, \\ a &= 0 \text{ if } v = v_\ell, \\ a &< 0 \text{ if } v > v_\ell. \end{aligned} \qquad (A.15)$$

Conditions (A.6) and (A.15) mean that when the space gap $g \leq G$, the vehicle accelerates, if it is slower than the preceding vehicle and the vehicle decelerates if it is faster than the preceding vehicle. This speed adaptation (A.15) does not depend on the space gap g to the preceding vehicle as long as the space gap satisfies conditions for the indifferent zone (A.6).

A.4 Driver Over-Acceleration within Indifferent Zone for Car-Following

To escape from the car-following of the slow-moving preceding vehicle, the driver searches for the opportunity to accelerate. In Sect. 8.1.1, we have already called vehicle acceleration from the car-following of the slow-moving preceding vehicle as over-acceleration. Driver over-acceleration within the indifferent zone (A.6) satisfies condition

$$a > 0 \quad \text{when } g_{\text{safe}} \leq g \leq G, \text{ even if } v \geq v_\ell. \qquad (A.16)$$

Condition for driver over-acceleration (A.16) contradicts basically conditions (A.15) for speed adaptation: Through over-acceleration, a driver can also accelerate when the vehicle speed v is equal to or it even higher than the speed of the preceding vehicle v_ℓ.

It must be noted that in real traffic flow the speed adaptation and over-acceleration effects appear usually in their dynamic competition. For this reason, a separate consideration of the over-acceleration effect made here is a rough simplification of the reality. As we will show in Sect. A.5, the competition of the speed adaptation and

---

where $K_{\Delta v}$ is a dynamic coefficient ($K_{\Delta v} > 0$). The asymmetric deceleration–acceleration behavior of TPACC means that

$$a^{(\text{TPACC})} < 0 \text{ at } g < g_{\text{safe}}, \quad a^{(\text{TPACC})} > 0 \text{ at } g > G, \qquad (A.14)$$

where $G > g_{\text{safe}}$.

over-acceleration exhibits the nucleation character that is the origin of the nucleation nature of traffic breakdown at a bottleneck.

A.5 Growing Wave of Local Speed Increase in Synchronized Flow (S→F Instability)

We have already mentioned that in real traffic there are no steady states of synchronized flow: All states of synchronized flow are dynamic states. Therefore, a question arise: why have the steady states of synchronized flow been considered in detail in Sects. A.1 and A.2? The answer on this question is as follows: The existence of the indifferent zone for car-following associated with the hypothesis about the 2D region of steady states of synchronized flow (dashed 2D region in Fig. A.1) changes qualitatively the understanding of driver behavior. We will show in this section that within the indifferent zone (A.6) a competition between driver speed adaptation (Sect. A.3) and over-acceleration (Sect. A.4) is realized that explains in more detail the empirical nucleation nature of traffic breakdown at a bottleneck (Sect. 8.2).

A.5.1 Decay of Local Increase in Speed in Initially Homogeneous Synchronized Flow

We assume that there is an initially homogeneous state of synchronized flow on a single-lane road, in which all vehicles move at the same time-independent synchronized flow speed (vehicle 1 in Fig. A.7). Furthermore, we assume that at a time instant through over-acceleration a vehicle (vehicle 2 in Fig. A.7) accelerates. The speed of the preceding vehicle that is equal to the speed in the initial synchronized flow is less than the speed of vehicle 2. For this reason, vehicle 2 decelerates later due to speed adaptation to the lower speed of the preceding vehicle within the indifferent zone (A.6): The initial increase in the speed of vehicle 2 caused by over-acceleration is short (short time increase in the speed of vehicle 2 in Fig. A.7b).

This short increase in the speed of vehicle 2 can lead to the occurrence of a speed wave of a local speed increase in synchronized flow: Due to the speed increase of vehicle 2, vehicles that follow vehicle 2 can also accelerate. However, due to speed adaptation, the speed increase caused by the acceleration of the vehicles following vehicle 2 is also short (short time increase in the speed of vehicles 3 and 4 in Fig. A.7b): All following vehicles must reduce the speed to the lower speed in the initial synchronized flow. This effect of the acceleration of the vehicles following vehicle 2 with the subsequent deceleration of these vehicles causes the emergence of a speed wave of a local speed increase in synchronized flow that is labeled by dashed-dotted curves in Fig. A.7a (vehicles 3 and 4 in Fig. A.7 illustrate this behavior leading to the emergence of the speed wave).

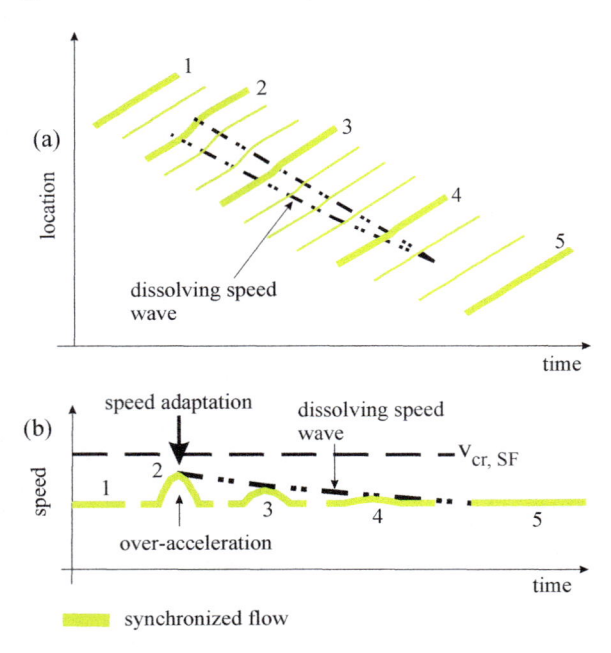

Fig. A.7 Qualitative illustration of dissolving speed wave of a local speed increase in synchronized flow: **a** Parts of trajectories of some of the vehicles (five of the vehicles are marked by numbers 1–5) in the vicinity of the local speed increase that boundaries are marked by dashed-dotted curves (vehicles 1 and 5 move in homogeneous synchronized flow; vehicles 2–4 move through the local speed increase). **b** Time dependencies of speeds of vehicles 1–5 in (**a**) during decay of the local speed increase; dashed-dotted curve marks the decay of the local speed increase over time; thick down arrow "speed adaption" and thin up arrow "over-acceleration" show symbolically a spatiotemporal competition between driver speed adaptation and over-acceleration within the local speed increase in synchronized flow; dashed horizontal line is qualitatively related to critical speed $v_{\text{cr, SF}}$ for the S→F instability (see Sect. A.5.2). Yellow curves—synchronized flow

To understand the result of the competition between over-acceleration and speed adaptation shown in Fig. A.7, we recall that the mean time delay in over-acceleration in synchronized flow is long enough (Fig. 8.2). Therefore, for a small enough local speed increase in synchronized flow (vehicle 2 in Fig. A.7), the mean time delay in over-acceleration within the local speed increase is approximately equal to that in synchronized flow. Due to the long enough mean time delay in over-acceleration, speed adaptation overcomes on average over-acceleration. This causes the decay of the local speed increase resulting in a dissolving speed wave of the local speed increase in synchronized flow (Fig. A.7).

A.5.2 Critical Speed for S→F Instability

Contrary to synchronized flow, in free flow the mean time delay in over-acceleration is short (Fig. 8.2). Therefore, when a local speed increase in synchronized flow is large enough (vehicle 2 in Fig. A.8), the mean time delay in over-acceleration within the local speed increase in synchronized flow can be as short as in free flow. Due to a short mean time delay in over-acceleration, over-acceleration overcomes on average speed adaptation within the local speed increase: The spatiotemporal competition between driver speed adaptation (A.15) and over-acceleration (A.16) leads to the speed growth within the local speed increase in synchronized flow (Fig. A.8b). As a result of this competition, a growing speed wave of the local speed increase in synchronized flow is realized. This growing wave has been called an S→F instability. A continuous development of the S→F instability leads to a transition from synchronized flow to free flow (S→F transition) (Fig. A.8).

There should be a critical speed for the S→F instability: The critical speed denoted by $v_{cr,\ SF}$ separates cases of the decay and the growth of a local speed increase in synchronized flow (dashed horizontal lines $v_{cr,\ SF}$ in Figs. A.7b and A.8b):

- When the speed within the local speed increase in synchronized flow is less than the critical speed $v_{cr,\ SF}$ (this is related to a small speed increase for vehicle 2 in Fig. A.7), speed adaptation (thick down arrow "speed adaption" in Fig. A.7b) overcomes on average over-acceleration (thin up arrow "over-acceleration" in Fig. A.7b). As a result, the local speed increase decays over time resulting in a dissolving speed wave of the local speed increase in synchronized flow (Fig. A.7).
- Contrarily, when the speed within the local speed increase in synchronized flow is equal to or larger than the critical speed $v_{cr,\ SF}$ (this is related to a large speed increase for vehicle 2 in Fig. A.8), over-acceleration (thick up arrow "over-acceleration" in Fig. A.8b) overcomes on average speed adaptation (thin down arrow "speed adaption" in Fig. A.8b). As a result, the local speed increase in synchronized flow grows over time resulting in a growing speed wave of the local speed increase in synchronized flow, i.e., an S→F instability is realized (Fig. A.8).

A.5.3 Nucleation Nature of S→F Instability

The existence of the critical speed $v_{cr,\ SF}$ for the occurrence of the S→F instability means that the S→F instability exhibits the nucleation character: A local speed increase in synchronized flow initiates the S→F instability only if the speed within the local speed increase is equal to or larger than the critical speed $v_{cr,\ SF}$. Such a local speed increase in synchronized flow can be considered a *nucleus* for the S→F instability (a large local speed increase in synchronized flow related to vehicle trajectory 2 in Fig. A.8 is an example of a nucleus for the S→F instability). Contrarily, when a local speed increase in synchronized flow occurs, within which the speed is

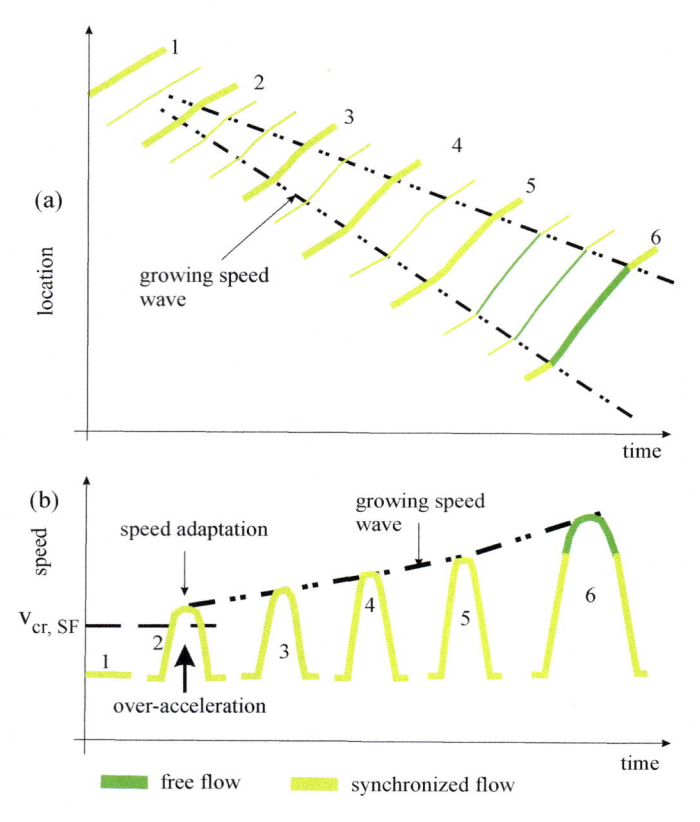

190   Appendix A: Characteristics of Synchronized Flow in Three-Phase Traffic Theory

Fig. A.8 Qualitative explanation of S→F instability: **a** Parts of vehicle trajectories (six of the trajectories are marked by numbers 1–6) in the vicinity of a local speed increase in synchronized flow that boundaries are marked by dashed-dotted curves (vehicle 1 moves in homogeneous synchronized flow; vehicles 2–6 move through the local speed increase). **b** Time dependencies of speeds of vehicles 1–6 in (**a**); dashed-dotted curve marks the growth of the local speed increase over time; thin down arrow "speed adaption" and thick up arrow "over-acceleration" show symbolically the spatiotemporal competition between speed adaptation and over-acceleration occurring within the local speed increase; dashed horizontal line is qualitatively related to critical speed $v_{\text{cr, SF}}$ for the S→F instability. Green curves—free flow, yellow curves—synchronized flow

less than the critical speed $v_{\text{cr, SF}}$, then the dissolving wave of the local speed increase is realized in synchronized flow (Fig. A.7).

> The cause of the nucleation nature of the S→F instability is the discontinuous character of over-acceleration together with the spatiotemporal competition between over-acceleration and speed adaptation occurring within the local speed increase in synchronized flow.

A.6 Why Does Nucleation Nature of S→F Instability Govern Nucleation Nature of Traffic Breakdown?

The competition between speed adaptation and over-acceleration is responsible for the nucleation nature of the S→F instability (Sect. A.5). The same competition is also responsible for the nucleation nature of traffic breakdown (F→S transition) at a bottleneck (Sect. 8.2.1): when the speed within a local speed decrease in free flow at the bottleneck is larger than the critical speed $v_{cr,\ FS}^{(B)}$ for the F→S transition, over-acceleration overcomes on average speed adaptation. As a result, no traffic breakdown occurs at the bottleneck. This effect of over-acceleration within the local speed decrease in free flow that prevents traffic breakdown can also be considered an S→F instability occurring within the local speed decrease in free flow at the bottleneck. Thus, the nucleation nature of the S→F instability associated with the discontinuous character of over-acceleration (Fig. 8.2) governs the nucleation nature of traffic breakdown at the bottleneck.

> The nucleation nature of an S→F instability at a highway bottleneck governs the nucleation nature of an F→S transition at the bottleneck. In other words, nucleation nature of the S→F instability leads to the metastability of free flow with respect to an F→S transition (traffic breakdown) at the bottleneck.

A.7 Main Prediction of Three-Phase Traffic Theory

The main prediction of the three-phase traffic theory is as follows: The cause of the empirical nucleation nature of traffic breakdown (F→S transition) at a highway bottleneck is the discontinuous character of over-acceleration together with the spatiotemporal competition between over-acceleration and speed adaptation (Chap. 8).

The following traffic phenomena result from the main prediction of the three-phase traffic theory:

- There is an S→F instability in synchronized flow. The S→F instability is a growing speed wave of a local speed increase in synchronized flow. The growth of this speed wave leads to an S→F transition. The S→F instability exhibits the nucleation nature: only a large enough local speed increase in synchronized flow can lead to the S→F instability, whereas a small enough local speed increase in synchronized flow decays.
- The nucleation nature of the S→F instability governs the metastability of free flow with respect to the F→S transition at a bottleneck. In its turn, the metastability of free flow with respect to the F→S transition at the bottleneck explains the empirical nucleation nature of traffic breakdown.

> The cause of both the nucleation nature of traffic breakdown (F→S transition) at the bottleneck and the nucleation nature of the S→F instability is the discontinuous character of over-acceleration together with the spatiotemporal competition between over-acceleration and speed adaptation.

A.8 Characteristics of Traffic Breakdown and Wide Moving Jam Emergence

In this section, we consider the critical speed $v^{(B)}_{\text{cr, FS}}$ for traffic breakdown (F→S transition) at a bottleneck discussed in Sect. 8.2 as well as the critical speed $v_{\text{cr, SJ}}$ for the S→J instability in synchronized flow discussed in Sect. 10.2.2 of the main text in more detail. This consideration allows us to find some important characteristics of (i) traffic breakdown in free flow (Sect. A.8.1), (ii) moving jam emergence in synchronized flow (Sect. A.8.2), and (iii) a sequence of F→S→J transitions (Sect. A.8.3).

A.8.1 Z-Characteristic for Traffic Breakdown (F→S Transition) at Bottleneck

The critical speed $v^{(B)}_{\text{cr, FS}}$ for traffic breakdown (F→S transition) at a bottleneck should be an increasing function of the flow rate q in free flow at the bottleneck within the flow-rate range $C_{\min} \leq q < C_{\max}$ (7.6) of the metastability of free flow with respect to an F→S transition (Fig. A.9). To show this, we introduce a flow-rate function

$$\Delta v_{\text{FS}}(q) = v^{(B)}_{\text{free}}(q) - v^{(B)}_{\text{cr, FS}}(q), \qquad (\text{A.17})$$

where Δv_{FS} is the speed difference between the minimum speed $v^{(B)}_{\text{free}}$ within the average local speed decrease in free flow at the bottleneck and the critical speed $v^{(B)}_{\text{cr, FS}}$ for the F→S transition (Fig. A.9a).

The larger the flow rate q in free flow at the bottleneck, the larger the density within the average local speed decrease in free flow at the bottleneck. Obviously, the larger the vehicle density in free flow, the larger the critical speed $v^{(B)}_{\text{cr, FS}}$ for the F→S transition. Therefore, the larger the flow rate q, the smaller the speed difference $\Delta v_{\text{FS}} = v^{(B)}_{\text{free}} - v^{(B)}_{\text{cr, FS}}$ (A.17) (Fig. A.9b). We can see that when the flow rate q is close to the maximum highway capacity $C_{\max}$, the speed difference Δv_{FS} (A.17) is close to zero. Respectively, the larger the difference $C_{\max} - q$ is, the larger should be the speed difference Δv_{FS} (A.17).

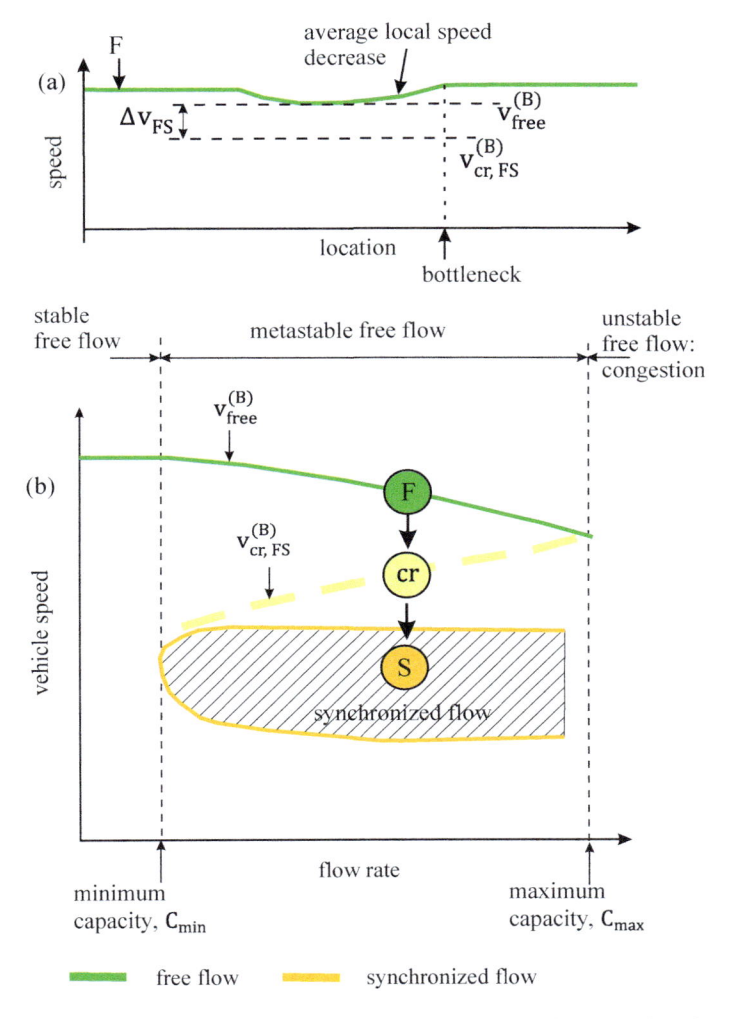

Appendix A: Characteristics of Synchronized Flow in Three-Phase Traffic Theory 193

Fig. A.9 Qualitative explanation of Z-characteristic for F→S transition at bottleneck: **a** Average local speed decrease at bottleneck (solid curve) taken from Fig. 8.5a showing the speed difference Δv_{FS} (A.17) for some given flow rate q within the flow rate range $C_{\min} \leq q < C_{\max}$ (7.6); $v_{\mathrm{free}}^{(B)}$ is the minimum speed within the average local speed decrease in free flow at bottleneck; $v_{\mathrm{cr, FS}}^{(B)}$ is the critical speed for F→S transition. **b** Qualitative Z-characteristic for traffic breakdown (F→S transition). In (**b**), the flow-rate dependence of the speed $v_{\mathrm{free}}^{(B)}$ is taken from Fig. 8.3b; curve labeled by $v_{\mathrm{cr, FS}}^{(B)}$ is the flow-rate dependence of the critical speed $v_{\mathrm{cr, FS}}^{(B)}$ for F→S transition; dashed two-dimensional (2D) region "synchronized flow" taken from Fig. 8.3b is related to synchronized flow at the bottleneck occurring due to traffic breakdown; down arrow from a state of free flow "F" to a state "cr" on curve $v_{\mathrm{cr, FS}}^{(B)}$ shows symbolically the emergence of a nucleus for an F→S transition, whereas down arrow from the state "cr" to a state "S" of synchronized flow shows symbolically the development of the F→S transition (traffic breakdown) at the bottleneck

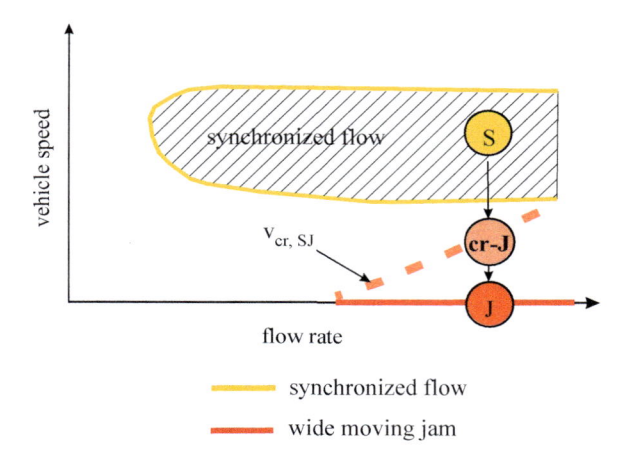

[FIGURE: Qualitative Z-characteristic for S→J transition. Axes: vehicle speed (vertical) vs flow rate (horizontal). Shows 2D region of synchronized flow (hatched), state S with down arrow to cr-J on dashed curve $v_{\mathrm{cr,\,SJ}}$, then down arrow to state J on wide moving jam line.]

Fig. A.10 Qualitative Z-characteristic for S→J transition. 2D region of synchronized flow (labeled by "synchronized flow") is taken from Fig. A.9b. Down arrow from a state of synchronized flow "S" to a state "cr-J" on curve of the flow-rate dependence of the critical speed $v_{\mathrm{cr,\,SJ}}$ for an S→J instability shows symbolically the emergence of a nucleus for the S→J instability, whereas down arrow from the state "cr-J" to a state "J" related to a wide moving jam shows symbolically the wide moving jam emergence through the S→J instability. For simplicity, we neglect a possible spatiotemporal jam structure (Sect. 11.3) assuming that the speed within the jam is equal to zero

The flow-rate function of the minimum speed $v_{\mathrm{free}}^{(B)}$ within the average local speed decrease in free flow at the bottleneck together with the flow-rate dependence of the critical speed $v_{\mathrm{cr,\,FS}}^{(B)}$ for an F→S transition and the 2D region of synchronized flow states (labeled by "synchronized flow") are built a so-called Z-characteristic for traffic breakdown (F→S transition) at the bottleneck (Fig. A.9b).

A.8.2 Z-Characteristic for S→J Transition

The critical speed $v_{\mathrm{cr,\,SJ}}$ within a local speed decrease in synchronized flow required for the S→J instability discussed in Sect. 10.2.2 of the main text should be an increasing function of the flow rate (dashed curve for the flow-rate dependence of the critical speed $v_{\mathrm{cr,\,SJ}}$ in Fig. A.10). Indeed, the density within synchronized flow increases on average, when the flow rate increases. The larger the density in synchronized flow, the larger should be the critical speed for the S→J instability: If the density of synchronized flow is large enough, then already a small enough local speed decrease in synchronized flow can be a nucleus for the S→J instability.

The 2D region of synchronized flow states (labeled by "synchronized flow" in Fig. A.10) that is the same as that shown in Fig. A.9b, the critical speed for the S→J instability as the flow-rate function (dashed curve labeled by $v_{\mathrm{cr,\,SJ}}$) together with states within a wide moving jam (wide moving jam states labeled by "J") are built a so-called Z-characteristic for the S→J transition (Fig. A.10).

A.8.3 Double Z-Characteristic for Phase Transitions

Now we draw the Z-characteristic for F→S transition (traffic breakdown) at a highway bottleneck (Fig. A.9b) *together* with the Z-characteristic for S→J transition in synchronized flow resulted from traffic breakdown at the bottleneck (Fig. A.10). Then, we come to a so-called double Z-characteristics (2Z-characteristic) for phase transitions in traffic flow (Fig. A.11). The 2Z-characteristic for phase transitions (Fig. A.11) consists of the following traffic states:

(i) states for free flow $v_{\text{free}}^{(B)}$,
(ii) states for the critical speed $v_{\text{cr, FS}}^{(B)}$ for the F→S transition,
(iii) a 2D region of synchronized flow states,
(iv) states for the critical speed $v_{\text{cr, SJ}}$ for the S→J instability,
(v) states within wide moving jams.

Down arrows in the 2Z-characteristic for phase transitions (Fig. A.11) present symbolically the sequence of the F→S→J transitions discussed in Sect. 10.1 as follows: Firstly, a nucleus for an F→S transition occurs in a free flow state "F" at a bottleneck (down arrow from "F" to "cr"). Then, the nucleus development leads to the F→S transition at the bottleneck. In its turn, the F→S transition results in the propagation of synchronized flow upstream of the bottleneck (down arrow from a state "cr" on dashed curve for the flow-rate dependence of the critical speed $v_{\text{cr, FS}}^{(B)}$ to a state "S" of synchronized flow). Later, a nucleus for an S→J instability occurs in synchronized flow (down arrow from the state "S" to a state "cr-J" on dashed curve of the flow-rate dependence of the critical speed $v_{\text{cr, SJ}}$). Finally, the development of the S→J instability causes an S→J transition (down arrow from the state "cr-J" to a state "J" related to the wide moving jam phase).

A.9 Competition of S→F and S→J Instabilities

In synchronized flow, there can randomly occur either a local speed increase or a local speed decrease. In Sect. A.5, we have already explained that if a large enough local speed *increase* occurs in synchronized flow, that is, a nucleus for an S→F instability, then the S→F instability is realized leading to the emergence of free flow (S→F transition). In Sects. 10.2 and A.8.2, we have considered the opposite case, when a large enough local speed *decrease* occurs in synchronized flow, that is, a nucleus for an S→J instability. In this case, the S→J instability is realized leading to the emergence of a wide moving jam in the synchronized flow (S→J transition).

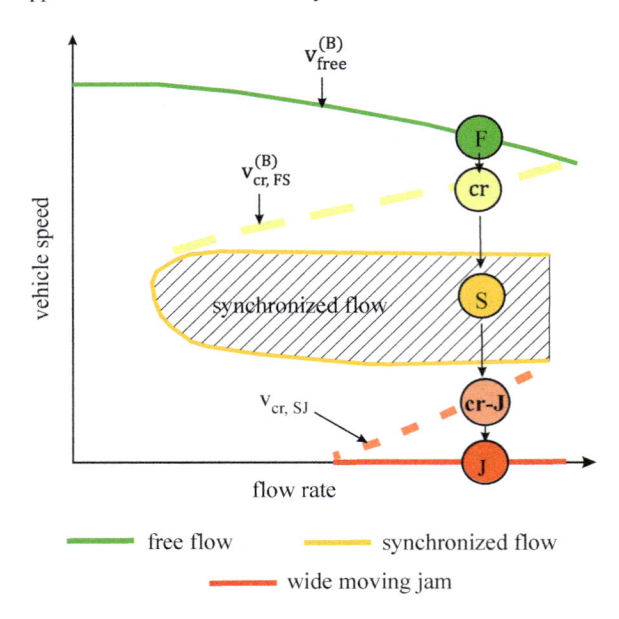

Fig. A.11 Qualitative double Z-characteristic (2Z-characteristic) for phase transitions in traffic flow shown in the speed–flow plane. Curves of the flow-rate dependencies of the speed $v_{\text{free}}^{(B)}$ within the average local speed decrease in free flow at bottleneck and of the critical speed $v_{\text{cr, FS}}^{(B)}$ for F→S transition, 2D states of synchronized flow as well as the related states "F", "cr" and "S" are taken from Fig. A.9b, whereas curve of the flow-rate dependence of the critical speed $v_{\text{cr, SJ}}$ for S→J transition, states related to wide moving jams as well as the related states "S", "cr-J", and "J" are taken from Fig. A.10

A.9.1 Qualitative Explanation of Competition of S→F and S→J Instabilities

Thus, we could expect that in the *same state of synchronized flow* one of two qualitatively different nucleation effects can randomly occur[9]:

1. If a nucleus for an S→F instability appears randomly in the synchronized flow, then the S→F instability is realized (Fig. A.12a).
2. If, in contrast, a nucleus for an S→J instability appears randomly in the synchronized flow, then the S→J instability (Fig. A.12b) is realized.

Qualitatively the spatiotemporal competition of S→F and S→J instabilities is shown in Fig. A.12. In a state of synchronized flow (vehicle trajectory 1 in Fig. A.12a), a local speed increase can randomly occur (trajectory 2 in Fig. A.12a). We assume that the speed within the local speed increase is larger than the critical speed $v_{\text{cr, SF}}$ for the S→F instability (Fig. A.12a). This means that this local speed increase is a

---

[9] This assumption is indeed confirmed in simulations with a microscopic stochastic three-phase traffic flow model [27].

Appendix A: Characteristics of Synchronized Flow in Three-Phase Traffic Theory 197

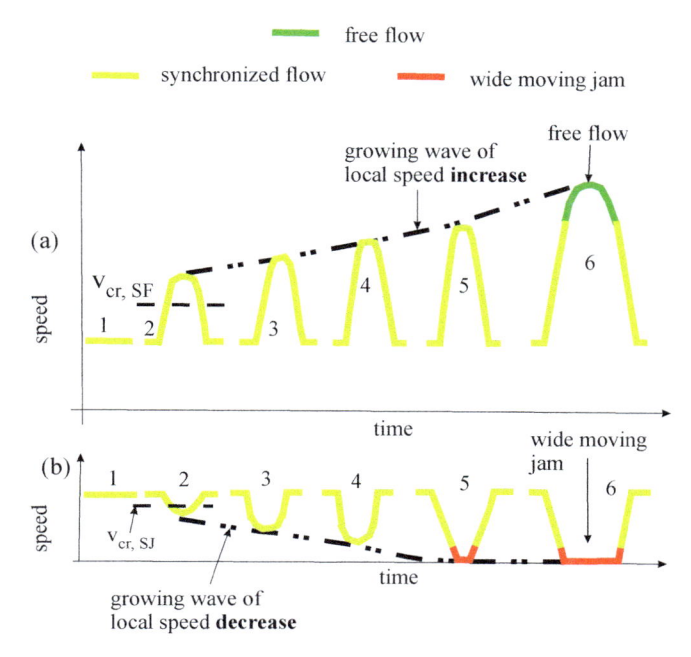

Fig. A.12 Qualitative explanation of competition of S→F and S→J instabilities: **a** Fragments of time functions of speed of some vehicles 1–6 that illustrate the development of S→F instability taken from Fig. A.8b. **b** Fragments of time functions of speed of some vehicles 1–6 that illustrate the development of S→J instability taken from Fig. 10.2b. Dashed horizontal line in (**a**) is qualitatively related to the critical speed $v_{\text{cr, SF}}$ for the S→F instability. Dashed horizontal line in (**b**) is qualitatively related to the critical speed $v_{\text{cr, SJ}}$ for the S→J instability, where $v_{\text{cr, SJ}} < v_{\text{cr, SF}}$ (A.18)

nucleus for the S→F instability. Due to the development of the S→F instability, the S→F transition occurs (trajectory 6 in Fig. A.12a).

Rather than a local speed *increase*, in the same state of synchronized flow (vehicle 1 in Fig. A.12a and vehicle 1 in Fig. A.12b is the same vehicle), a local speed *decrease* can randomly occur within which the speed is less than the critical speed $v_{\text{cr, SJ}}$ for the S→J instability (trajectory 2 in Fig. A.12b). This means that this local speed decrease is a nucleus for the S→J instability. Due to the development of the S→J instability, the S→J transition occurs (trajectory 6 in Fig. A.12b).

In the three-phase traffic theory, it is assumed that the following condition

$$v_{\text{cr, SJ}} < v_{\text{cr, SF}} \quad (A.18)$$

is satisfied (Fig. A.12). A qualitative presentation of the competition of the S→F and S→J instabilities (Fig. A.12) emphasizes a *non-predictive behavior* of synchronized flow: Both the nucleus occurrence for the S→F instability (Fig. A.12a) and the nucleus occurrence for the S→J instability (Fig. A.12b) are random events that can be realized in the same state of synchronized flow.

> The three-phase traffic theory predicts that there are states of synchronized flow within which in dependence on whether a large enough local speed increase or a large enough local speed decrease occurs randomly, respectively, either an S→F instability or an S→J instability is realized.

A.9.2 Empirical Alternations of Regions of Free Flow, Synchronized Flow, and Wide Moving Jams

In Sect. 11.2.2, we have already mentioned that spatiotemporal competition of S→F and S→J instabilities (Fig. A.12) can explain empirical alternations of regions of free flow, synchronized flow, and wide moving jams[10] observed in real traffic. Here we discuss in more detail some of empirical traffic phenomena that can be explained through the competition of the S→F and S→J instabilities (Fig. A.13):

- *MSP formation with subsequent MSP transformation into an MGP*: Due to the S→F instability at off-ramp bottleneck B1, a moving synchronized flow pattern (MSP) has occurred (labeled by "MSP" in Fig. A.13b). During the MSP upstream propagation, S→J instabilities occur within the MSP leading to the formation of a moving general congested traffic pattern (MGP) (labeled by "MGP" in Fig. A.13b) (compare with a qualitative explanation of an MGP in Sect. 11.2). Within the MGP, two wide moving jams emerge (these jams are symbolically shown by down arrows "J" on the location dependence of the speed of vehicle 1 in Fig. A.13c).
- *Alternations of synchronized flow and free flow at off-ramp bottleneck B1*: Moving through the bottleneck location, vehicle 2 is within synchronized flow (Fig. A.13b, d). Later, vehicle 3 moves in free flow due to the empirical S→F transition at the bottleneck (Fig. A.13b, e). Due to the subsequent F→S transition at the bottleneck, vehicle 4 moves again in synchronized flow (Fig. A.13b, f).
- *Alternations of free flow, synchronized flow, and wide moving jams*: Such complex alternations are observed by a driver of vehicle 5 moving through the whole congested pattern (Fig. A.13b, g).[11]

---

[10] Theoretical predictions of the complexity of spatiotemporal alternations of regions of free flow, synchronized flow, and wide moving jams that can occur in traffic due to the competition of S→F and S→J instabilities can be found in [26, 27].

[11] It should be emphasized that for a detailed empirical analysis of theoretical microscopic features of S→F and S→J instabilities as well as their competition, rather than only trajectories of some probe vehicles (Fig. A.13b), microscopic vehicle speeds as time-space functions for all vehicles are needed. Such empirical microscopic data in which traffic breakdown (F→S transition) has been measured at highway bottlenecks are not currently available. We hope that such microscopic

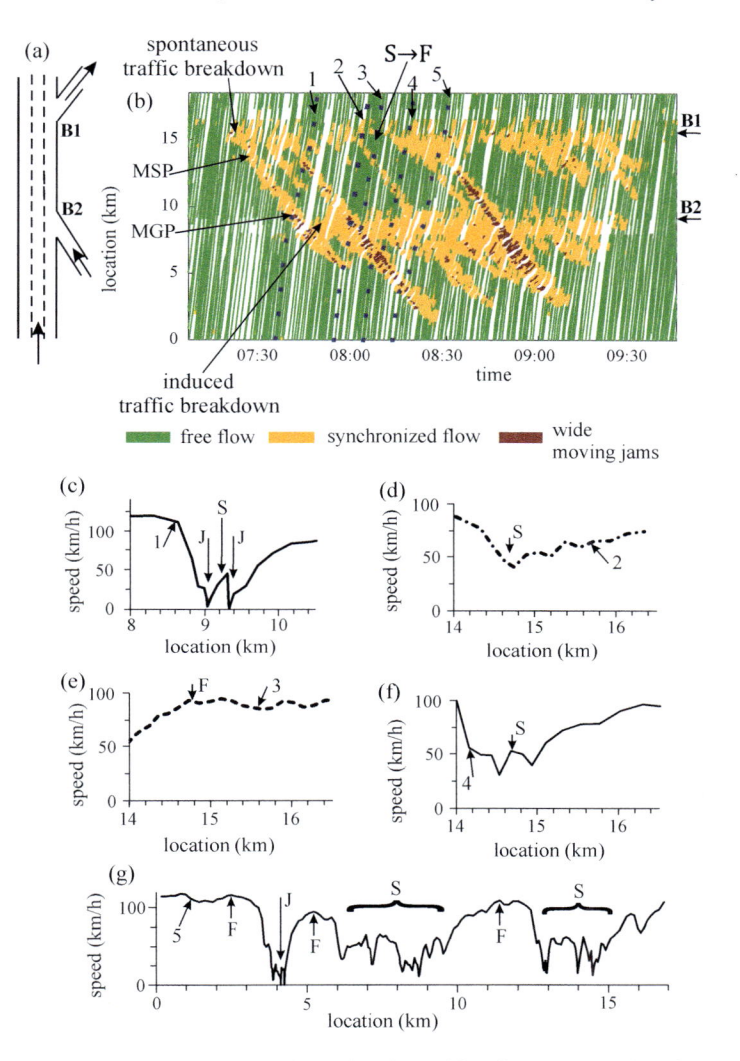

Appendix A: Characteristics of Synchronized Flow in Three-Phase Traffic Theory

Fig. A.13 Empirical example of alternations of regions of free flow, synchronized flow, and wide moving jams (probe vehicle data were measured on three-lane highway A5-South in Germany on May 08, 2017). **a** Schema of a highway section with off-ramp bottleneck B1 within intersection "Nordwestkreuz Frankfurt" (at about 15 km) and on-ramp bottleneck B2 within at intersection "Bad Homburger Kreuz" (at about 9 km). **b** Trajectories of probe vehicles. **c–g** Road-location functions of the speed of probe vehicles 1–5. In **c–g**, probe vehicles 1–5 are, respectively, the same vehicles 1–5 marked by dotted curves in (**b**); F—free flow; S—synchronized flow; J—wide moving jam. In (**b**), arrow S→F shows the time instant of one of the S→F transitions at bottleneck B1. MSP—moving synchronized flow pattern, MGP—moving general congested pattern. In (**b**), there is empirical induced traffic breakdown at bottleneck B2 labeled by "induced traffic breakdown"; the empirical induced traffic breakdown exhibits qualitatively the same features as that shown in Fig. 6.1b and discussed in Sect. 6.1. Adapted from [27]

> A spatiotemporal competition of S→F and S→J instabilities can explain empirical alternations of regions of free flow, synchronized flow, and wide moving jams observed in real traffic.

A.10 Why Is Spontaneous Emergence of Moving Jams Not Observed in Real Free Flow?

We have already mentioned (Sect. 2.4) that no spontaneous emergence of moving jams is observed in free flow. However, the classical traffic flow instability discussed in Sect. 10.2 can *theoretically* occur in free flow (the classical traffic flow in free flow is called an F→J instability). The continuous development of the F→J instability leads to the spontaneous emergence of a wide moving jam in free flow (F→J transition). Moreover, as traffic breakdown (F→S transition), the F→J transition should exhibit the nucleation nature.[12] Therefore, the question arises:

- Although the spontaneous emergence of the wide moving jam in free flow (F→J transition) could be theoretically possible, why is the F→J transition not observed in real free flow?

To answer this question, we should compare a critical nucleus required for traffic breakdown (F→S transition) with a critical nucleus required for the spontaneous emergence of a wide moving jam in free flow (F→J transition). As explained in Sect. 8.2.2, the speed within the critical nucleus required for traffic breakdown is equal to the critical speed $v_{\mathrm{cr,\,FS}}^{(B)}$ for the F→S transition (Fig. A.14a, b). Respectively, the speed within the critical nucleus required for the spontaneous emergence of the wide moving jam in free flow is equal to the critical speed for the F→J instability denoted by $v_{\mathrm{cr,\,FJ}}$ (Fig. A.14c, d).

It must be emphasized that a theoretical analysis of the occurrence of the classical traffic flow instability in free flow and synchronized flow shows that the critical speed $v_{\mathrm{cr,\,FJ}}$ for an F→J instability is approximately the same as the critical speed $v_{\mathrm{cr,\,SJ}}$ for S→J instability (Fig. A.14c, d). As can be seen from Fig. A.11, for any flow rate q the critical speed $v_{\mathrm{cr,\,SJ}}$ for the S→J instability is considerably lower than the critical speed $v_{\mathrm{cr,\,FS}}^{(B)}$ for the F→S transition: $v_{\mathrm{cr,\,SJ}}(q) \ll v_{\mathrm{cr,\,FS}}^{(B)}(q)$. As mentioned above, $v_{\mathrm{cr,\,FJ}}(q) \approx v_{\mathrm{cr,\,SJ}}(q)$; therefore, we get (Fig. A.14d)

$$v_{\mathrm{cr,\,FJ}}(q) \ll v_{\mathrm{cr,\,FS}}^{(B)}(q). \tag{A.19}$$

---

empirical data will be available in the near future. We believe that a study of empirical microscopic features of F→S, S→F, and S→J transitions will be one of the important tasks for future traffic research.

[12] The analysis of this section is based on Refs. [1, 2, 4, 6, 15].

Appendix A: Characteristics of Synchronized Flow in Three-Phase Traffic Theory 201

(a, b) Nucleation of traffic breakdown (F→S transition)

(c, d) Nucleation of wide moving jam in free flow (F→J transition)

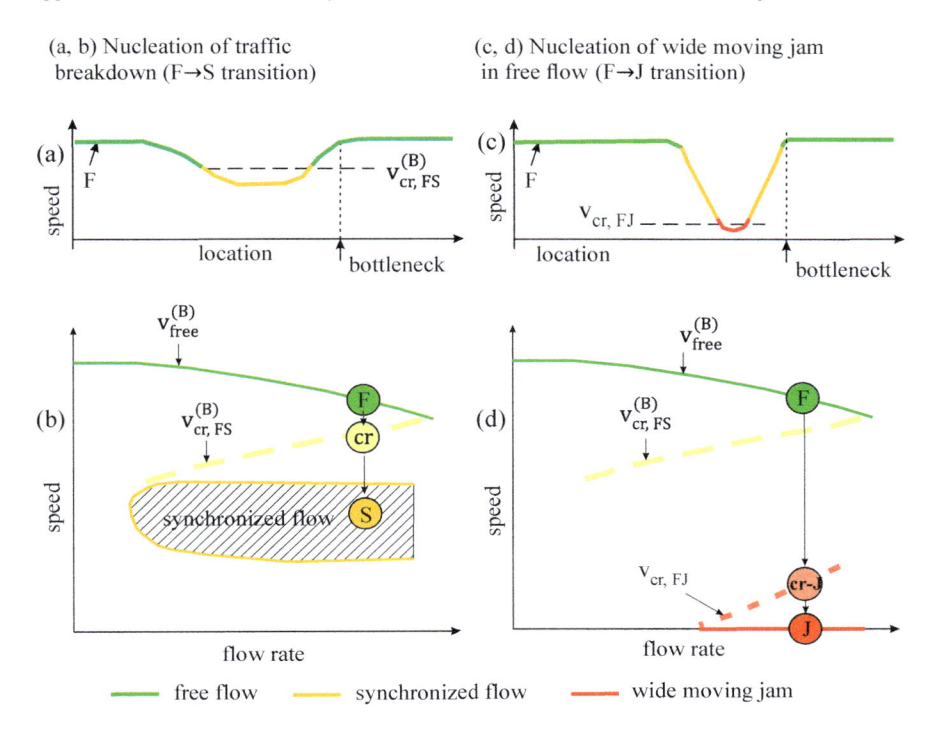

Fig. A.14 Qualitative comparison of condition for F→S transition (**a, b**) with condition for F→J transition (**c, d**): **a** A local speed decrease in free flow at a bottleneck (taken from Fig. 8.7c) that is a nucleus for F→S transition because the speed within the local speed decrease is less than the critical speed $v_{cr,\,FS}^{(B)}$. **b** Z-characteristics taken from Fig. A.9b. **c** A local speed decrease in free flow that is a nucleus for F→J instability because the speed within the local speed decrease is less than the critical speed $v_{cr,\,FJ}$. **d** A part of 2Z-characteristics taken from Fig. A.11 in which 2D states of synchronized flow are omitted and the flow-rate dependence of the critical speed for S→J instability $v_{cr,\,SJ}$ is replaced by the flow-rate dependence of the critical speed for F→J instability $v_{cr,\,FJ}$. In (**b, d**), curves labeled by $v_{free}^{(B)}$ are related to the flow-rate dependence of the minimum speed $v_{free}^{(B)}$ within the average local speed decrease in free flow at the bottleneck (Sect. 5.2)

A nucleus for traffic breakdown (F→S transition) is a local speed decrease in free flow at a bottleneck, when the speed within this local speed decrease is equal to or less than the critical speed $v_{cr,\,FS}^{(B)}$ (Fig. A.14a). The emergence of the nucleus for the F→S transition is symbolically shown by down arrow from a free flow state "F" to a state "cr" on the flow-rate dependence of the critical speed $v_{cr,\,FS}^{(B)}$ (Fig. A.14b). Contrarily, a nucleus for the F→J instability is a local speed decrease in free flow, when the speed within this local speed decrease is equal to or less than the critical speed $v_{cr,\,FJ}$ (Fig. A.14c). The emergence of the nucleus for the F→J instability is symbolically shown by down arrow from the free flow state "F" to a state "cr-J" on the flow-rate dependence of the critical speed $v_{cr,\,FJ}$ (Fig. A.14d). Condition (A.19) explains why a spontaneous F→J transition (symbolically shown in Fig. A.14d by down arrow from the state "cr-J" to a state "J" related to a wide moving jam) is not observed in real traffic as follows.

> In the same state of free flow, the probability of a random occurrence of a very large local speed decrease in free flow needed for the spontaneous F→J transition is *negligible* in comparison with the probability of a random occurrence of a considerably smaller local speed decrease in free flow needed for the spontaneous F→S transition (traffic breakdown).

We can make the following conclusions:

- For any given flow rate in free flow at the bottleneck, the critical speed $v_{\text{cr, FJ}}$ for the F→J instability is considerably lower than the critical speed $v_{\text{cr, FS}}^{(B)}$ for the F→S transition (A.19) (Fig. A.14).
- Therefore, a considerably larger local speed decrease in free flow is needed for the F→J instability (Fig. A.14c) in comparison with a smaller local speed decrease that is needed for the spontaneous F→S transition (traffic breakdown) (Fig. A.14a).
- Because the probability of the occurrence of a nucleus for the F→J instability is negligible in comparison with the probability of the occurrence of a nucleus for the spontaneous F→S transition, the classical traffic flow instability (F→J instability) and the resulting F→J transition are *not* observed in real free flow.
- Rather than moving jams emerge spontaneously in free flow, the moving jams result from a sequence of the F→S→J transitions, as explained in Chap. 10.

> Traffic breakdown in real free flow at a bottleneck is the F→S transition, *not* the F→J transition.

References

1. B.S. Kerner, Phys. Rev. Lett. **81**, 3797–3800 (1998)
2. B.S. Kerner, in *Proceedings of the 3rd Symposium on Highway Capacity and Level of Service*, ed. by R. Rysgaard (Road Directorate, Copenhagen, Ministry of Transport – Denmark 1998), pp. 621–642
3. B.S. Kerner, Transp. Res. Rec. **1678**, 160–167 (1999)
4. B.S. Kerner, in *Transportation and Traffic Theory*, ed. by A. Ceder (Elsevier Science, Amsterdam, 1999), pp. 147–171
5. B.S. Kerner, Phys. World **12**, 25–30 (August, 1999)
6. B.S. Kerner, J. Phys. A: Math. Gen. **33**, L221–L228 (2000)
7. B.S. Kerner, in *Traffic and Granular Flow '99: Social, Traffic and Granular Dynamics*, ed. by D. Helbing, H.J. Herrmann, M. Schreckenberg, D.E. Wolf (Springer, Heidelberg, 2000), pp. 253–284

8. B.S. Kerner, Transp. Res. Rec. **1710**, 136–144 (2000)
9. B.S. Kerner, Netw. Spat. Econ. **1**, 35–76 (2001)
10. B.S. Kerner, Transp. Res. Rec. **1802**, 145–154 (2002)
11. B.S. Kerner, Math. Comput. Model. **35**, 481–508 (2002)
12. B.S. Kerner, in *Traffic and Transportation Theory in the 21st Century*, ed. by M.A.P. Taylor (Elsevier Science, Amsterdam, 2002), pp. 417–439
13. B.S. Kerner, Phys. Rev. E **65**, 046138 (2002)
14. B.S. Kerner, Phys. A **333**, 379–440 (2004)
15. B.S. Kerner, *The Physics of Traffic*, (Springer, Berlin, 2004)
16. B.S. Kerner, J. Phys. A: Math. Theor. **41**, 215101 (2008)
17. B.S. Kerner, *Introduction to Modern Traffic Flow Theory and Control*, (Springer, Heidelberg, 2009)
18. B.S. Kerner, Phys. Rev. E **85**, 036110 (2012)
19. B.S. Kerner, Phys. A **392**, 5261–5282 (2013)
20. B.S. Kerner, Elektrotech. Inf. **132**, 417–433 (2015)
21. B.S. Kerner, Phys. A **450**, 700–747 (2016)
22. B.S. Kerner, *Breakdown in traffic networks*, (Springer, Berlin, 2017)
23. B.S. Kerner, Phys. Rev. E **97**, 042303 (2018)
24. B.S. Kerner, in *Complex Dynamics of Traffic Management*, ed. by B.S. Kerner, Encyclopedia of Complexity and Systems Science Series (Springer, New York, 2019), pp. 21–77
25. B.S. Kerner, in *Complex Dynamics of Traffic Management*, ed. by B.S. Kerner, Encyclopedia of Complexity and Systems Science Series (Springer, New York, 2019), pp. 195–283
26. B.S. Kerner, in *Complex Dynamics of Traffic Management*, ed. by B.S. Kerner, Encyclopedia of Complexity and Systems Science Series (Springer, New York, 2019), pp. 387–500
27. B.S. Kerner, Phys. Rev. E **100**, 012303 (2019)
28. B.S. Kerner, Phys. A **562**, 125315 (2021)
29. B.S. Kerner, S.L. Klenov, J. Phys. A: Math. Gen. **35**, L31–L43 (2002)
30. B.S. Kerner, S.L. Klenov, Phys. Rev. E **68**, 036130 (2003)
31. B.S. Kerner, S.L. Klenov, J. Phys. A: Math. Gen. **37**, 8753–8788 (2004)
32. B.S. Kerner, S.L. Klenov, J. Phys. A: Math. Gen. **39**, 1775–1809 (2006)
33. B.S. Kerner, S.L. Klenov, Phys. Rev. E **80**, 056101 (2009)
34. B.S. Kerner, S.L. Klenov, Transp. Res. Rec. **2124**, 67–77 (2009)
35. B.S. Kerner, S.L. Klenov, G. Hermanns, M. Schreckenberg, Phys. A **392**, 4083–4105 (2013)
36. B.S. Kerner, S.L. Klenov, A. Hiller, J. Phys. A: Math. Gen. **39**, 2001–2020 (2006)
37. B.S. Kerner, S.L. Klenov, A. Hiller, Nonlinear Dyn. **49**, 525–553 (2007)
38. B.S. Kerner, S.L. Klenov, A. Hiller, H. Rehborn, Phys. Rev. E **73**, 046107 (2006)
39. B.S. Kerner, S.L. Klenov, M. Schreckenberg, Phys. Rev. E **84**, 046110 (2011)
40. B.S. Kerner, S.L. Klenov, M. Schreckenberg, Phys. Rev. E **89**, 052807 (2014)

Appendix B
Empirical Features of Wide Moving Jams

In Chap. 10, we have explained that empirical wide moving jams emerge in real traffic through a sequence of the F→S→J transitions. The objective of Appendix B is a discussion of some empirical features of wide moving jams that are important for a deeper understanding real congested traffic.

B.1 Empirical Characteristic Parameters of Jam Propagation: Line *J*

Results of empirical studies show that after a wide moving jam has emerged, the jam propagates upstream through any states of free flow and synchronized flow as well as through any bottleneck while maintaining the mean velocity of the jam downstream front denoted by v_g. At the downstream jam front, vehicles accelerate while escaping from the jam to traffic flow (either free flow or synchronized flow) downstream. The empirical feature of the jam to maintain the mean velocity of the downstream jam front v_g can be considered a *characteristic parameter of the propagation of the wide moving jam* (Fig. B.1). As already mentioned in Sect. 2.1.3, this characteristic feature of the wide moving jam is called the jam feature [J]. Another characteristic parameter of the wide moving jam propagation is the vehicle density within the jam denoted by $\rho_{\max}$. The definition of empirical characteristic parameters of the wide moving jam propagation is as follows[13]:

- Empirical characteristic parameters of the wide moving jam propagation are the parameters that at given traffic parameters (weather, share of trucks, etc.) are the same for different wide moving jams; the characteristic parameters of a wide

---

[13] Characteristic parameters of the wide moving jam propagation on highways have theoretically been predicted in [17–19] and studied in [5, 16, 20]. The characteristic parameters of the wide moving jam propagation have been found in empirical traffic data in Refs. [21, 22].

Fig. B.1 Qualitative explanation of empirical characteristic parameters of wide moving jam propagation. Schematic representation of a wide moving jam at a fixed time instant. Spatial distribution of the speed v, flow rate q, and the density ρ in the wide moving jam, which propagates through a homogeneous state of free flow. v_g is the mean velocity of the downstream jam front. Empirical examples of the characteristic parameters of wide moving jam propagation can be found in [21, 22] as well as in the books [8, 27]

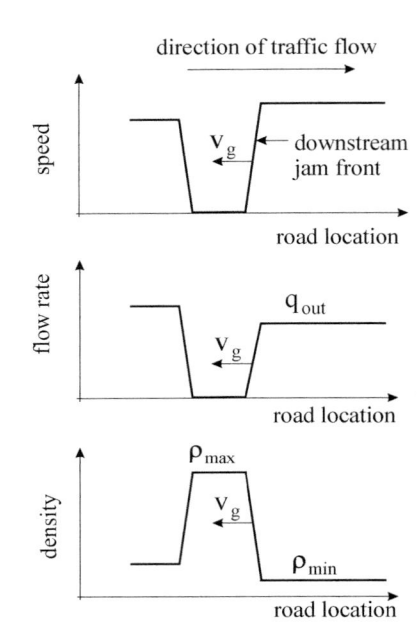

moving jam do not depend on initial traffic variables upstream of the jam and they remain during the wide moving jam propagation.

Empirical studies of the wide moving jam propagation show also that when free flow is formed downstream of the jam, the flow rate in the jam outflow denoted by q_{out} and the vehicle density in this free flow denoted by ρ_{min} are also the characteristic parameter of the wide moving jam propagation (Figs. B.1 and B.2).

The empirical characteristic parameters of the wide moving jam propagation shown in Fig. B.1 can be presented in the flow–density plane by a line J (Fig. B.2).

- Line J is defined through its left (ρ_{min}, q_{out}) and right coordinates (ρ_{max}, 0) that are given by the characteristic parameters of the wide moving jam propagation (Fig. B.1) shown in the flow–density plane (Fig. B.2).

B.2 Jam Absorption Effect

The important empirical feature of the line J is as follows. The line J divides the free flow states and 2D states of synchronized flow into two different classes of traffic flow states (Fig. B.2)[14]:

---

[14] This feature of the line J was predicted in the three-phase traffic theory in 1998 [7] (for a review see the books [8, 9, 11]).

Appendix B: Empirical Features of Wide Moving Jams

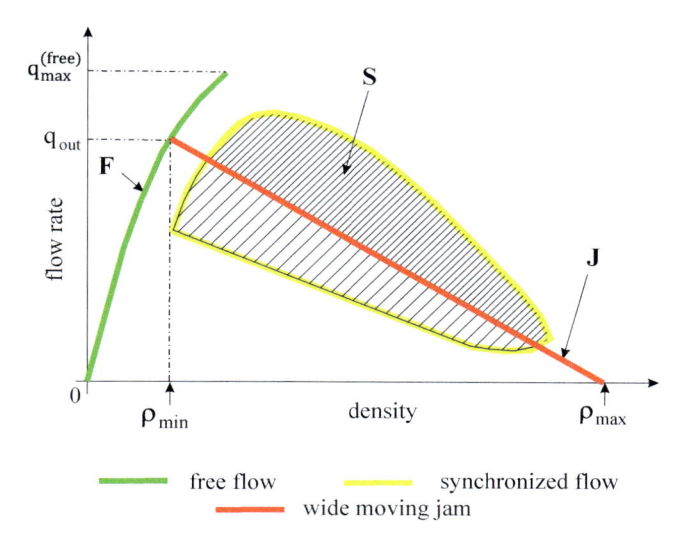

Fig. B.2 Qualitative presentation of the three traffic phases in the flow–density plane. Line J (red line) presents the characteristic parameters of the wide moving jam propagation. Curve F for free flow and 2D region for steady states of synchronized flow (dashed region S) are taken from Fig. A.1. $q_{\max}^{\rm (free)}$ is the maximum flow rate that is possible in free flow. Adapted from [8]

1. States on and above the line J. These states of free flow and synchronized flow are metastable with respect to wide moving jam emergence.
2. States below the line J. These states of free flow and synchronized flow are stable with respect to wide moving jam emergence: no wide moving jams can persist in these states.

Thus, in synchronized flow states below the line J no wide moving jams can persist over time. Therefore, if drivers change their behavior in accordance with features of these synchronized flow states, all moving jams dissolve and only synchronized flow remains. This jam dissolution can be called *jam absorption effect* (Fig. B.3).[15]

To understand features of the synchronized flow states below the line J (Fig. B.2), in which all moving jams dissolve, we consider a characteristic space gap between two following each other vehicles denoted by g_J (Fig. B.3) that is related to a synchronized flow state on the line J in the flow–density plane (Fig. B.2). To see that the characteristic space gap g_J is an increasing function $g_J(v)$ of the speed v in synchronized flow (Fig. B.3a), we use explanations of the reconstruction of Fig. A.5b from Fig. A.5a of Sect. A.2: If, as made in Fig. A.5a, a line of a constant speed v is

---

[15] The jam absorption effect has theoretically been studied in detail on highways in [10]. We should have mentioned that jam absorption effect can also occur in city traffic [14] in which this effect has indeed been observed in empirical probe vehicle data [13]. A detailed empirical study of the jam absorption effect that has confirmed theoretical results of Refs. [10, 14] has been made with the use of empirical vehicle trajectories measured by aerial traffic observations at traffic signal in city traffic by Kaufmann et al. [6]. A consideration of real city traffic is out of scope of this book (for more details, see [27]).

208　　　　　　　　　　　　　Appendix B: Empirical Features of Wide Moving Jams

(a)

— synchronized flow　　— wide moving jam

jam absorption effect:
no moving jams can persist

$G(v)$

$g_J(v)$

$g_{safe}(v)$

metastable synchronized flow:
moving jams can emerge

space gap → speed

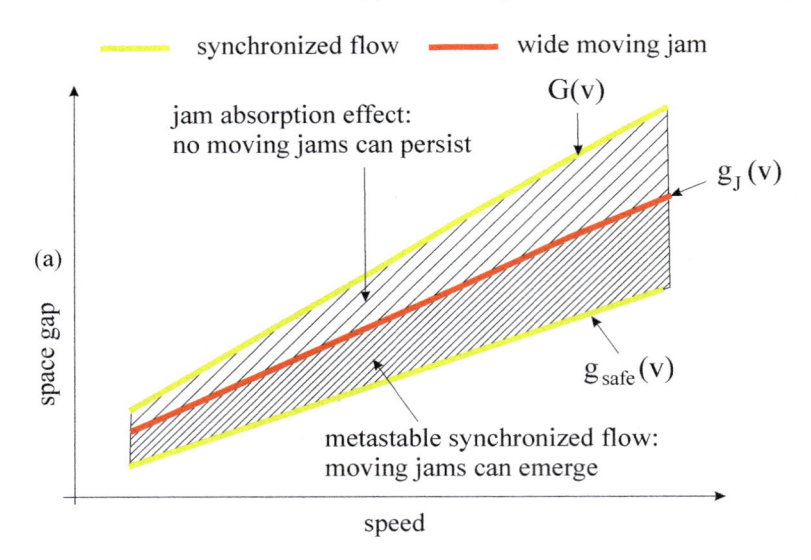

(b) moving jams can emerge in synchronized flow

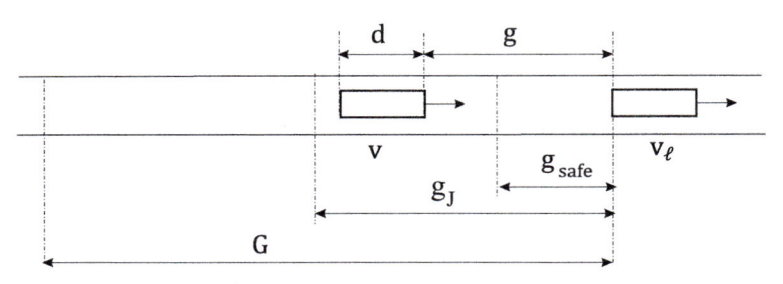

(c) no moving jams can persist in synchronized flow:
jam absorption effect

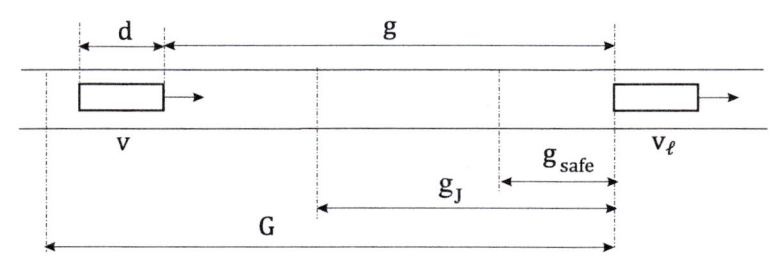

Fig. B.3 Qualitative explanation of jam absorption effect. **a** Two classes of steady states of synchronized flow (dashed 2D regions) in the space gap–speed plane; red curve $g_J(v)$ is related to the line J in Fig. B.2, yellow curves $G(v)$ and $g_{safe}(v)$ are taken from Fig. A.5b. **b, c** Schemes of car-following within indifferent zone (A.6) under condition $g < g_J$ (**b**) and under condition $g > g_J$ (**c**) (B.1) for steady states of synchronized flow satisfying condition (A.1). g is the space gap between two vehicles following each other; G and g_{safe} are, respectively, the synchronization space gap and safe space gap; d is the vehicle length (see Sect. A.1)

drawn in the flow–density plane (not shown in Fig. B.2), then in accordance with (A.4), the vehicle density at the intersection point between the line J and the line of the constant speed v in Fig. B.2 corresponds to the space gap g_J at the intersection point between a vertical line of the same constant speed v (not shown in Fig. B.3a) and the speed dependence $g_J(v)$ in Fig. B.3a. The speed dependence $g_J(v)$ in the speed–space gap plane (Fig. B.3a) is equivalent to the line J for synchronized flow states in the flow–density plane (Fig. B.2).

States below the line J in the flow–density plane (Fig. B.2), in which no moving jams can persist, are related to the space gaps within indifferent zone (A.6) satisfying condition

$$g(v) > g_J(v). \tag{B.1}$$

In accordance with the above-mentioned feature of line J (Fig. B.2), the speed dependence $g_J(v)$ separates synchronized flow states into two different classes:

(i) In states in the speed–space gap plane that are above the speed dependence $g_J(v)$ (B.1), no moving jams can persist.
(ii) In states in the speed–space gap plane that are on and below the speed dependence $g_J(v)$, moving jams can emerge and can persist.

If vehicles maintain on average a large enough space gap[16] that satisfies conditions (A.6), (B.1) then all moving jams in synchronized flow dissolve.[17] Through this jam absorption effect, GPs discussed in Sect. 11.2 can transform into one of the SPs discussed in Sect. 11.1.

To understand the physics of the jam absorption effect, we consider condition $g(v) < g_J(v)$ that is the opposite one to (B.1). We recall that the S→J instability leading to the formation of a wide moving jam (S→J transition) occurs when within a local speed decrease in synchronized flow the over-deceleration effect overcomes on average speed adaptation occurring under conditions (A.6). The over-deceleration occurs, when the space gap becomes smaller than the safe space gap g_{safe}. When $g(v) < g_J(v)$, the driver reaches quickly the safe space gap g_{safe}. This is because the smaller the space gap, the less the effect of speed adaptation within a local speed decrease in synchronized flow. Thus, at $g(v) < g_J(v)$ the over-deceleration can overcome on average the speed adaptation within a large enough local speed decrease in synchronized flow. This results in the S→J instability.

---

[16] The fact that the classical traffic flow instability occurs only when the average space gap in traffic flow is small enough (the vehicle density is large enough) was known already from the classical papers of Herman, Gazis, Rothery, Montroll, Chandler, and Potts [1–4] as well as Kometani and Sasaki [23–26]. In other words, it was clear that no traffic flow instability in traffic flow, in which the average space gap between vehicles is large enough, can occur. Therefore, the consideration of the jam absorption effect made in Sect. B.2 can be considered the application of this well-known feature of the classical traffic flow instability [1–4, 23–26] for 2D region of synchronized flow states of the three-phase traffic theory [10].

[17] In real synchronized flow, the space gaps between vehicles can be very different. In this case, the development of the S→J instability can be interrupted, when already one or a few drivers have large enough space gaps to the associated preceding vehicles [12].

Contrarily, under condition $g(v) > g_J(v)$ (B.1), a driver has enough time for speed adaptation to the speed of the preceding vehicle before the space gap reaches the safe space gap g_{safe}. Due to the speed adaptation, the driver can adapt its speed to the speed of the preceding vehicle without the necessity of over-deceleration to a lower speed than the speed of the preceding vehicle. In this case, the speed adaptation overcomes on average the over-deceleration: no S→J instability can occur and no moving jam can persist in such synchronized flow.

B.3 Jam Loops: Empirical Proof of Asymmetric Deceleration–Acceleration Driver Behavior

We explain[18] that empirical studies of traffic dynamics (Fig. B.4) in which vehicles traveled through a wide moving jam[19] can confirm the hypothesis of the three-phase traffic theory about the asymmetric deceleration–acceleration driver behavior (Sect. A.1.2).

Within a vehicle platoon[20] propagating through the wide moving jam shown in Fig. B.4a,[21] firstly, drivers moved through phase transition points[22] between free flow and synchronized flow upstream of the wide moving jam. Then, the drivers moved through phase transition points between synchronized flow and the wide moving jam.[23] Based on Treiterer's empirical results (Fig. B.4a), it can be found that for a

---

[18] Results of Sect. B.3 are based on Sect. IV.B "Jam propagation loops: Nature of Treiterer's hysteresis phenomenon" of Ref. [15].

[19] First empirical studies of vehicles traveling through a wide moving jam have been made by Treiterer et al. [28–30]. Treiterer et al. observed the wide moving jam (Fig. B.4a) based on aerial photography.

[20] A qualitative explanation of the term *vehicle platoon* has been made in Fig. B.4b.

[21] Based on a study of the empirical propagation of vehicle platoons through the wide moving jam (Fig. B.4a), Treiterer [28] found two hysteresis loops in the flow–density and speed–density planes called by Treiterer as "loop A" and "loop B". Treiterer [28] called this empirical phenomenon as *hysteresis phenomenon* [28]. Treiterer's hysteresis phenomenon was theoretically explained in 2006 [15] by the propagation of vehicle platoons through the three traffic phases F, S, and J as well as through the asymmetric deceleration–acceleration driver behavior (Sect. A.1.2 of Appendix A). Empirical proof for the asymmetric deceleration–acceleration driver behavior of the three-phase traffic theory considered in Sect. B.3 is based on results of [15]. Some years later empirical results of Treiterer [28] and their theoretical explanations made in [15] were rediscovered in many studies of vehicle platoon propagation through empirical moving jams.

[22] The definition of the term *phase transition point* has been made in Sect. 3.1.2.

[23] As explained in [15], in which Treiterer's hysteresis phenomenon has been studied in detail, later the drivers moved through phase transition points between the wide moving jam and synchronized flow. Finally, the drivers moved through phase transition points between synchronized flow and free flow downstream of the wide moving jam. Such a vehicle platoon propagation through the three traffic phases F, S, and J explains Treiterer's hysteresis phenomenon. However, for empirical proof of the asymmetric deceleration–acceleration driver behavior (Sect. A.1 of Appendix A) that is the objective of Sect. B.3, it is sufficient to limit a discussion of vehicle platoon propagation through the wide moving jam phase (J) only. This limitation is related to a fragment of jam loops denoted by

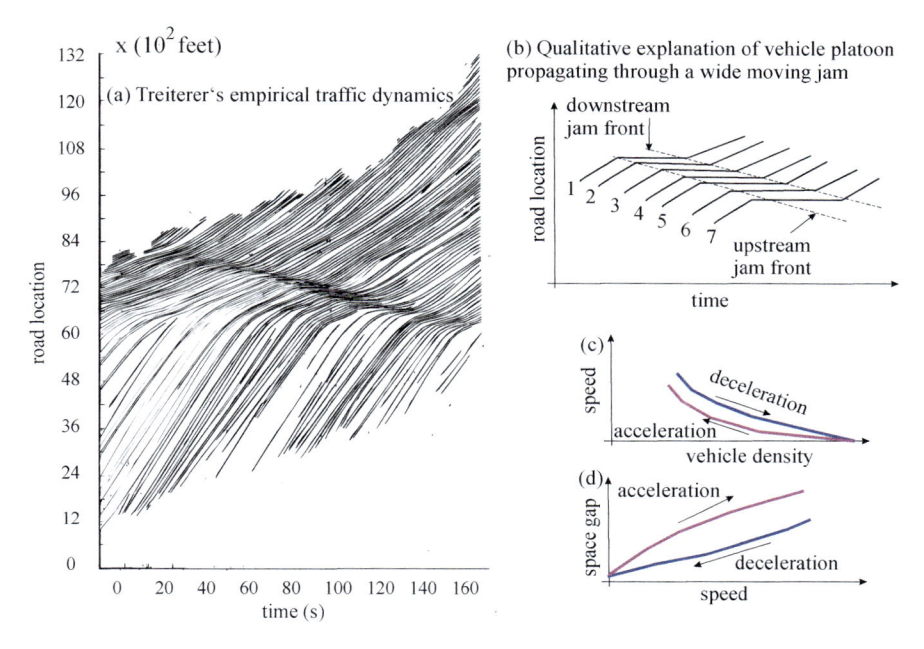

Appendix B: Empirical Features of Wide Moving Jams

Fig. B.4 Empirical proof for asymmetric deceleration–acceleration driver behavior [15]: **a** Treiterer's empirical traffic dynamics derived from aerial photography; adapted from Treiterer [28]. **b** Explanation of the term *vehicle platoon* through a qualitative illustration of a platoon of 7 vehicles (labeled by 1–7) propagating through a wide moving jam. **c, d** Qualitative explanations of empirical and theoretical results of the vehicle platoon analysis in the speed–density (**c**) and space gap–speed planes (**d**). In **c, d**, colored blue curves—vehicles decelerate at the upstream jam front, colored purple curves—vehicles accelerate at the downstream jam front; figures (**c, d**) are qualitative presentations of results shown in Figs. 13 and 14 of [15]

given speed the vehicle density in the vehicle platoon when drivers decelerate at the upstream front of the jam is larger than the density is when the vehicles accelerate at the jam downstream front (Fig. B.4c). In accordance with Fig. A.3 and Formulas (A.4), (A.9), the larger the density in the vehicle platoon, the shorter both the mean space gap g and the mean time headway within the vehicle platoon. Therefore, Fig. B.4c is equivalent to Fig. B.4d in which, instead of the vehicle density in the platoon, the average space gap in the platoon as the speed function is shown. Other empirical studies of moving jams confirm these conclusions.[24]

Treiterer as "loop A" [28]. Explanations of the nature of both "loop A" and "loop B" of Treiterer's empirical data [28] can be found in Sect. IV.B of Ref. [15]. It must be noted that as explained in Sect. 3.1.2, a phase transition point is *not* the phase transition that is responsible for the emergence of a new traffic phase. For this reason, Treiterer's hysteresis phenomenon has a totally different meaning in comparison with traffic hysteresis observed at a highway bottleneck (Sect. 5.3). See also a discussion of interpretations of traffic hysteresis in Sect. C.4 of Appendix C.

[24] See formula (3) and related empirical results of Sect. II.D of [15].

Empirical results confirm the asymmetric deceleration–acceleration driver behavior resulting from the hypothesis about the 2D states of synchronized flow of the three-phase traffic theory: Vehicles decelerating at the upstream jam front accept shorter time headway (smaller space gaps to the preceding vehicle) than time headway (space gaps) of vehicles accelerating at the downstream jam front.

References

1. R.E. Chandler, R. Herman, E.W. Montroll, Oper. Res. **6**, 165–184 (1958)
2. D.C. Gazis, R. Herman, R.B. Potts, Oper. Res. **7**, 499–505 (1959)
3. D.C. Gazis, R. Herman, R.W. Rothery, Oper. Res. **9**, 545–567 (1961)
4. R. Herman, E.W. Montroll, R.B. Potts, R.W. Rothery, Oper. Res. **7**, 86–106 (1959)
5. M. Herrmann, B.S. Kerner, Phys. A **255**, 163–188 (1998)
6. S. Kaufmann, B.S. Kerner, H. Rehborn, M. Koller, S.L. Klenov, Transp. Res. C **86**, 393–406 (2018)
7. B.S. Kerner, Phys. Rev. Lett. **81**, 3797–3800 (1998)
8. B.S. Kerner, *The Physics of Traffic* (Springer, Berlin, 2004)
9. B.S. Kerner, *Introduction to Modern Traffic Flow Theory and Control* (Springer, Heidelberg, 2009)
10. B.S. Kerner, Phys. Rev. E **85**, 036110 (2012)
11. B.S. Kerner, *Breakdown in Traffic Networks* (Springer, Berlin, 2017)
12. B.S. Kerner, in *Complex Dynamics of Traffic Management*, ed. by B.S. Kerner. Encyclopedia of Complexity and Systems Science Series (Springer, New York, 2019), pp. 387–500
13. B.S. Kerner, P. Hemmerle, M. Koller, G. Hermanns, S.L. Klenov, H. Rehborn, M. Schreckenberg, Phys. Rev. E **90**, 032810 (2014)
14. B.S. Kerner, S.L. Klenov, G. Hermanns, P. Hemmerle, H. Rehborn, M. Schreckenberg, Phys. Rev. E **88**, 054801 (2013)
15. B.S. Kerner, S.L. Klenov, A. Hiller, H. Rehborn, Phys. Rev. E **73**, 046107 (2006)
16. B.S. Kerner, S.L. Klenov, P. Konhäuser, Phys. Rev. E. **56**, 4200–4216 (1997)
17. B.S. Kerner, P. Konhäuser, Phys. Rev. E **48**, 2335–2338 (1993)
18. B.S. Kerner, P. Konhäuser, Phys. Rev. E **50**, 54–83 (1994)
19. B.S. Kerner, P. Konhäuser, M. Schilke, Phys. Rev. E **51**, 6243–6246 (1995)
20. B.S. Kerner, P. Konhäuser, M. Schilke, Phys. Lett. A **215**, 45–56 (1996)
21. B.S. Kerner, H. Rehborn, Phys. Rev. E **53**, R1297–R1300 (1996)
22. B.S. Kerner, H. Rehborn, Phys. Rev. E **53**, R4275–R4278 (1996)
23. E. Kometani, T. Sasaki, J. Oper. Res. Soc. Jpn. **2**, 11–26 (1958)
24. E. Kometani, T. Sasaki, Oper. Res. **7**, 704–720 (1959)
25. E. Kometani, T. Sasaki, Oper. Res. Soc. Jpn. **3**, 176–190 (1961)
26. E. Kometani, T. Sasaki, in *Theory of Traffic Flow*, ed. by R. Herman (Elsevier, Amsterdam, 1961), pp. 105–119

27. H. Rehborn, M. Koller, S. Kaufmann, *Data-Driven Traffic Engineering: Understanding of Traffic and Applications based on Three-Phase Traffic Theory* (Elsevier, Amsterdam, 2021)
28. J. Treiterer, *Investigation of Traffic Dynamics by Aerial Photogrammetry Techniques*, Ohio State University Technical Report PB 246 094 (Columbus, Ohio, 1975)
29. J. Treiterer, J.A. Myers, in *Proceedings of the 6th International Symposium on Transportation and Traffic Theory*, ed. by D.J. Buckley (A.H. & AW Reed, London, 1974), pp. 13–38
30. J. Treiterer, J.I. Taylor, Highw. Res. Rec. **142**, 1–12 (1966)

Appendix C
Empirical Induced Traffic Breakdown—Failure of Standard Traffic Theories

In Chap. 5, we have emphasized that standard traffic methodologies and theories can explain many empirical features of vehicular traffic. On the one hand, this might be responsible for the wide acceptance of the standard traffic and transportation theories in the traffic and transportation research community. This might also be the reason for the expectation that the subsequent "puzzle solving" and associated cumulative process in standard traffic and transportation science might solve main problems on the long road to understanding real traffic.

On the other hand, we have also emphasized in Chap. 5 that the explanation of many empirical features of vehicular traffic through the standard traffic theories is the fundamental problem for both understanding real traffic and for many traffic applications in intelligent transportation systems (ITS): The standard traffic and transportation theories failed by their applications in the real world. This is because in standard traffic and transportation science the evidence of empirical induced traffic breakdown at a highway bottleneck is ignored. For this reason, the term *traffic breakdown* used in standard traffic and transportation science and the term *spontaneous traffic breakdown* used in the three-phase traffic theory can be considered synonymous. As we have shown in Chaps. 6 and 7, the evidence of the empirical induced traffic breakdown means that real traffic breakdown (F→S transition) at the bottleneck exhibits the empirical nucleation nature.

> The fundamental requirement for traffic flow theories and models that claim to explain real traffic is as follows: The theories and models should explain the empirical nucleation nature of traffic breakdown (F→S transition) at a highway bottleneck.

The empirical nucleation nature of real traffic breakdown (F→S transition) at the bottleneck has *no* sense for the standard traffic and transportation theories and models. This explains the failure of the standard traffic and transportation theories and models in the real world. It turns out that the evidence of the empirical induced traffic breakdown at the bottleneck changes basically the perception of vehicular traffic as a spatiotemporal dynamic process. This change in the perception of vehicular traffic has resulted in the emergence of the three-phase traffic theory (Chap. 8).

The objective of Appendix C is to explain why the empirical nucleation nature of real traffic breakdown (F→S transition) at the bottleneck has no sense for the standard traffic methodologies and theories.[25]

C.1 Standard Understanding of Stochastic Highway Capacity

Since 1920s–1930s, it was generally accepted in standard traffic and transportation science that highway capacity denoted by C in Fig. C.1 is equal to the maximum flow rate in free flow at which free flow still persists at a bottleneck.[26] This means that if the flow rate in free flow at the bottleneck is equal to or less than highway capacity, i.e.,

$$q \leq C, \qquad (C.1)$$

no traffic breakdown can occur at the bottleneck. Thus, under condition (C.1) free flow is stable.[27] Contrarily, when the flow rate in free flow at the bottleneck exceeds highway capacity, i.e.,

$$q > C, \qquad (C.2)$$

traffic breakdown must occur (arrow in Fig. C.1). The understanding of highway capacity related to Formulas (C.1) and (C.2) can be considered a standard understanding of highway capacity.

In empirical observations, it was found that highway capacity exhibits a probabilistic behavior, in particular, at the same flow rate in free flow at a bottleneck traffic breakdown can occur but it should not necessarily occur.[28] The empirical stochastic highway capacity denoted by $C(t)$ has been explained through the standard understanding of highway capacity as follows[29]:

---

[25] In more detail, the criticism of standard traffic methodologies, theories, and models can be found in review [68].

[26] Although in Sect. 5.1 of the main text we have briefly considered the standard definition of stochastic highway capacity that is generally accepted in standard traffic and transportation science, due to a great importance of this subject, we explain here this capacity definition in more detail.

[27] Here we assume that initially there is no traffic congestion at the bottleneck and there are no other neighborhood highway bottlenecks that can affect on the free flow existence at the bottleneck under consideration. In other words, the bottleneck under consideration can be considered as an isolated bottleneck on a highway section.

[28] The empirical probabilistic character of highway capacity was firstly observed by Elefteriadou et al. [36] in 1995.

[29] The standard understanding of highway capacity is the theoretical basis of "Highway Capacity Manual" [63, 64] used as a textbook for most students in transportation science (see also papers [5–

Fig. C.1 Qualitative explanation of standard understanding of highway capacity. Curve for free flow is the same as that in Fig. 7.4. Arrow shows symbolically a transition from free flow to congested traffic in the flow–density plane. C—standard highway capacity

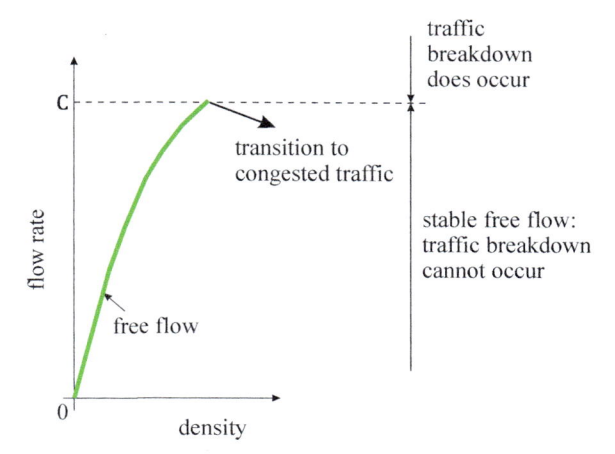

Fig. C.2 Qualitative explanation of standard understanding of stochastic highway capacity: Hypothetical fragment of a qualitative time dependence of standard stochastic highway capacity $C(t)$ at a bottleneck

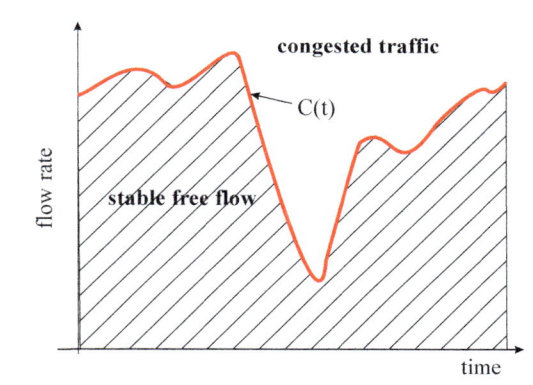

- At *any given time instant*, there is a particular value of highway capacity of free flow at a bottleneck; the value of highway capacity is a stochastic time function.

This definition of stochastic highway capacity is qualitatively illustrated with Fig. C.2. A curve $C(t)$ shows qualitatively a fragment of a stochastic time dependence of highway capacity.

If at a time instant the flow rate $q(t)$ in free flow is equal to or less than highway capacity at this time instant, i.e.,

$$q(t) \leq C(t), \tag{C.3}$$

no traffic breakdown can occur at the bottleneck. This means that all flow rates below the curve $C(t)$ are related to stable free flow with respect to traffic breakdown (dashed

---

10, 15–25, 33, 35, 36, 47, 50–60, 92, 94, 105–109] as well as reviews and books [1, 11–14, 27, 29, 31, 34, 37–40, 44, 48, 49, 61, 65, 66, 75, 84, 87, 91, 93, 100, 103, 104, 110, 112, 114, 116–118, 123, 124, 126]).

region labeled by "stable free flow" in Fig. C.2). Contrarily, if at another time instant the flow rate $q(t)$ in free flow is larger than highway capacity at this time instant, i.e.,

$$q(t) > C(t), \qquad (C.4)$$

traffic breakdown must occur at the bottleneck. This means that at any flow rate at the bottleneck that is above the curve $C(t)$ traffic breakdown must occur (labeled by "congested traffic" in Fig. C.2).

As we have already explained in Sects. 6.4 and 7.3, the empirical proof for the distinct decision between different theories of traffic breakdown at a highway bottleneck is *the empirical induced traffic breakdown* at a highway bottleneck:

1. The empirical induced traffic breakdown at the bottleneck found in real field traffic data proves the existence of a range of stochastic highway capacities and the metastability of free flow with respect to traffic breakdown (F→S transition) at the bottleneck within this capacity range (Chap. 7).
2. The empirical induced traffic breakdown at the bottleneck has *no sense* for the standard understanding of stochastic highway capacity. This proves that the standard understanding of stochastic highway capacity is invalid for real vehicular traffic.

In the standard understanding of stochastic highway capacity is assumed that at any given time instant there is a *particular value* of stochastic highway capacity $C(t)$. In contrast, in the three-phase traffic theory it is assumed that at any given time instant there is a *range* of stochastic highway capacities. A comparison of Fig. C.2 for the standard understanding of stochastic highway capacity with Fig. 7.6 for a range of stochastic highway capacities of the three-phase traffic theory explains the fundamental difference between these two definitions of stochastic highway capacity.

In Fig. 7.6, there is a range of the flow rates at the bottleneck between the minimum highway capacity and the maximum highway capacity (7.9). Within this range of the flow rates, free flow is in a metastable state with respect to traffic breakdown (F→S transition) at the bottleneck. This means that when the flow rate at a given time instant is within the range of stochastic highway capacities (7.9), traffic breakdown can occur but it should not necessarily occur at this time instant. In contrast with the capacity range (Fig. 7.6) of the three-phase traffic theory, for the standard understanding of stochastic highway capacity (Fig. C.2) the empirical metastability of free flow with respect to an F→S transition at a bottleneck has no sense.

> The standard understanding of stochastic highway capacity contradicts the empirical nucleation nature of traffic breakdown at a bottleneck.

To understand why the standard understanding of stochastic highway capacity contradicts the empirical nucleation nature of traffic breakdown at the bottleneck, we assume that for any given time instant there is a particular value of stochastic highway capacity. The particular value of stochastic highway capacity of free flow at a bottleneck cannot depend on whether there is a congested pattern, which has occurred outside the bottleneck, or not. This means that if at a time instant the flow rate at the bottleneck is less than the particular value of stochastic highway capacity at this time instant, no traffic breakdown can be induced at the bottleneck, when the congested pattern reaches the bottleneck location. This contradicts observations of the empirical induced traffic breakdown at the bottleneck that occurs when the congested pattern reaches the bottleneck location (see Figs. 6.1b, 6.2a, 7.1, and 7.2).

> The use of the standard understanding of stochastic highway capacity does lead to invalid conclusions for traffic applications.

C.2 Lighthill–Whitham–Richards (LWR) Kinematics Wave Theory

C.2.1 Shock Waves

The classical understanding of highway capacity related to Formulas (C.1) and (C.2) is consistent with classical Lighthill–Whitham–Richards (LWR) kinematics wave (shock-wave) theory developed for traffic flow in 1955–1956 by Lighthill, Whitham, and Richards.[30] The basic feature of the LWR theory is as follows: It is assumed that there is a fundamental diagram for traffic flow at a highway bottleneck that determines a single value of the flow rate for each given value of the vehicle density. The maximum flow rate at the fundamental diagram is equal to highway capacity C (Fig. C.1).[31] Condition (C.2) determines the occurrence of traffic breakdown.

As in the standard understanding of highway capacity (Sect. C.1), in the classical LWR theory, it is assumed that when the flow rate at the bottleneck is less than highway capacity, then traffic breakdown cannot occur at the bottleneck, i.e., free flow persists. Contrarily, only when the flow rate at the bottleneck exceeds the capacity given by the maximum flow rate at the fundamental diagram of traffic flow, then traffic breakdown must occur at the bottleneck with the subsequent formation and upstream propagation of shock wave upstream of the bottleneck.

---

[30] See Refs. [85, 115].

[31] In the LWR theory, it is assumed that the fundamental diagram consists of a branch for free flow shown in Fig. C.1 and the branch for congested traffic that is not shown in Fig. C.1; both branches merge at the maximum flow rate on the fundamental diagram related to highway capacity C. A more detailed consideration of the LWR theory can be found, for example, in the books [34, 93, 100].

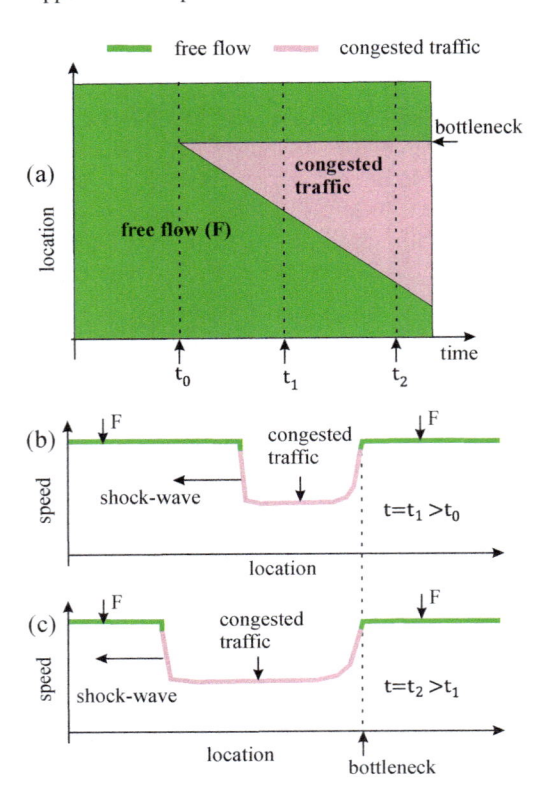

Fig. C.3 Qualitative explanations of kinematic waves (shock waves) in LWR model: **a** Free flow (green) and congested traffic (pink) in the time–road location plane; at $t < t_0$, free flow is at the bottleneck because the flow rate $q \leq C$ and, therefore, condition (C.1) is satisfied; contrarily, at $t \geq t_0$, the flow rate $q > C$ and, therefore, condition (C.2) is satisfied resulting in the occurrence of congested traffic (pink) propagating upstream. **b, c** Qualitative time dependencies of the average vehicle speed for two consequent time instants t_1 (**b**) and t_2 (**c**) shown in (**a**) at time $t > t_0$, when condition (C.2) is satisfied

To illustrate predictions of the LWR theory, we consider a qualitative example of the application of the LWR model shown in Fig. C.3. We assume that at $t < t_0$, the flow rate in free flow at the bottleneck $q \leq C$, i.e., condition (C.1) is satisfied; therefore, no traffic breakdown can occur in free flow (labeled by "free flow (F)" in Fig. C.3a). We assume further that at $t \geq t_0$ the flow rate in free flow at the bottleneck becomes larger than highway capacity C, i.e., condition $q > C$ (C.2) is satisfied: Traffic breakdown does occur at the bottleneck. The LWR theory predicts that traffic breakdown results in the formation of a kinematic wave (shock wave) (labeled by "shock wave" in Fig. C.3b, c). The shock-wave propagation upstream of the bottleneck causes congested traffic at the bottleneck (labeled by "congested traffic" (pink color) in Fig. C.3).

Appendix C: Empirical Induced Traffic Breakdown—Failure of Standard ...

It must be emphasized that the LWR theory *does not contradict* some features of the empirical spontaneous traffic congestion at a bottleneck (Sect. 5.4). Indeed, we might assume that empirical spontaneous traffic breakdown at off-ramp bottleneck B1 shown in Fig. 5.5b, d has occurred due to the satisfaction of condition (C.4) at time about 16:19. Moreover, from a comparison of Figs. 5.6 and 5.7 with Fig. C.3 we might make a conclusion that the development of traffic congestion resulting from traffic breakdown is qualitatively the same in the empirical spontaneous traffic breakdown (Fig. 5.6) and the theoretical traffic breakdown following from the LWR theory (Fig. C.3). Therefore, the following questions arise:

- Why do we state that the LWR theory failed by the explanation of real traffic breakdown?
- Is there an empirical proof for the distinct decision between different theories of traffic breakdown at a highway bottleneck?

This empirical proof is *the empirical induced traffic breakdown* at the bottleneck (Sects. 6.4 and 7.3):

1. The empirical induced traffic breakdown at the bottleneck found in real field traffic data proves the metastability of free flow with respect to traffic breakdown (F→S transition) at the bottleneck within a capacity range (Chap. 7).
2. The empirical induced traffic breakdown at the bottleneck has *no sense* for the classical LWR theory.[32] This proves that the LWR theory cannot explain the empirical nucleation nature of traffic breakdown (F→S transition).

The reason why the LWR theory cannot explain the empirical nucleation nature of traffic breakdown at a bottleneck is as follows. In the LWR theory, it is assumed that if the flow rate at the bottleneck is less than the capacity given by the maximum flow rate at the fundamental diagram for traffic flow, no traffic breakdown can be induced at the bottleneck. This feature of the LWR theory contradicts the observation of the empirical induced traffic breakdown at the bottleneck that does occur when a congested pattern reaches the bottleneck location.

Additionally, we should mention that in the LWR model the downstream front of theoretical congested traffic resulting from traffic breakdown is fixed at the bottleneck (Fig. C.3b, c); this feature of congested traffic distinguishes the phase "synchronized flow" from the phase "wide moving jam" in empirical congested traffic made in the three-phase traffic theory (Sect. 2.1). Nevertheless, we do not use the term "synchronized flow" for congested traffic in the LWR model. The reason is that the term "synchronized flow" introduced in Chap. 2 is related to *empirical* congested traffic that downstream front can be fixed at the bottleneck and that ensures the nucleation nature of the F→S transition at the bottleneck (Sect. 6.2.2). Whereas *empirical* synchronized flow exhibits features that cause the empirical nucleation nature of traffic breakdown (F→S transition), features of *theoretical* congested traffic of the LRW

---

[32] This critical conclusion has been proven in simulations made in the framework of the LWR theory [68]. In particular, these simulations show that no traffic breakdown can be induced at the bottleneck, when the flow rate at the bottleneck satisfies condition $q \leq C$ (C.1).

model contradict basically the empirical nucleation nature of traffic breakdown at the bottleneck (for this reason, we have used different colors for synchronized flow in Fig. 5.7 and for congested traffic in the LWR model in Fig. C.3). A theoretical explanation of synchronized flow that shows the empirical nucleation nature of traffic breakdown at the bottleneck has been given in Chap. 8.

C.2.2 Why Is The Term "Shock Wave" Not Used in This Book?

As we have seen in empirical data (see, for example, Figs. 2.2 and 2.3b), there are many different moving traffic fronts separating the three phases (free flow (F), synchronized flow (S), and wide moving jam (J)) of vehicular traffic (Sect. 2.1.2). Within these fronts traffic variables like vehicle speed and density change sharply. The empirical moving fronts between the traffic phases F, S, and J look like *shock waves* observed in many other systems of natural science. As mentioned above, the term *shock wave* has also been used in the LWR theory of vehicular traffic. For this reason, in standard traffic science, the propagation of a moving jam is often called the propagation of shock waves.

However, it turns out that empirical features of the moving fronts between the traffic phases F, S, and J are qualitatively different from theoretical features of shock waves found in the LWR theory. The most important feature of empirical fronts between the traffic phases F, S, and J is as follows. In real vehicular traffic (Chaps. 6 and 7), a wide moving jam does induce traffic breakdown in metastable free flow with respect to an F→S transition at a bottleneck. Contrarily, shock waves of the LWR theory cannot induce traffic breakdown (F→S transition) at the bottleneck.

> The traffic phenomenon "the empirical induced traffic breakdown at a bottleneck" has *no* sense for the LWR shock-wave theory.

> The terms *kinematic wave* and *shock wave* are automatically associated with the LWR theory. ITS applications of the LWR theory are inconsistent with the empirical nucleation nature of traffic breakdown at the bottleneck. Therefore, the ITS applications based on the LWR theory lead to invalid results for real traffic. For this reason, we do not use the terms *kinematic wave* and *shock wave* in the book.

C.3 Myth About Classical Traffic Flow Instability as Origin of Traffic Breakdown

Many traffic engineers explain traffic breakdown in free flow leading to the onset of traffic congestion as follows[33]: When the preceding vehicle decelerates unexpectedly strongly, then, due to the existence of a time delay in driver deceleration caused by the driver reaction time, the following driver decelerates stronger than it needed to avoid collision with the preceding vehicle. In Sect. 10.2.1, this effect of driver deceleration has been considered driver *over-deceleration*. As a result of over-deceleration, the minimum speed of the following driver becomes lower than the minimum speed of the preceding vehicle. When each of the following drivers exhibits over-deceleration, a growing wave of the local speed decrease in free flow propagating upstream should occur. This growing wave of the local speed decrease in free flow is the classical traffic flow instability (Sect. A.10).[34] Thus, one might assume that the classical traffic flow instability should be the reason for traffic breakdown.

However, as found theoretically in 1993–1995,[35] the development of the classical traffic flow instability should cause the formation of a wide moving jam in free flow (F→J transition) at a bottleneck, *not* an F→S transition that explains traffic breakdown observed at the bottleneck. The theoretical result that the classical traffic flow instability leads to an F→J transition at the bottleneck has later been found in a huge number of traffic flow models in which traffic breakdown is explained by the classical traffic flow instability.[36] In Sect. A.10, it has already been explained why the F→J transition is not observed in real traffic:

- In real world, at any given flow rate in free flow, at which either an F→S transition or an F→J transition can occur at a bottleneck, the probability of the occurrence of the F→S transition at the bottleneck is considerably larger than the probability of the occurrence of the F→J transition at the bottleneck.

For this reason, independent of the mathematical approach used for the simulation of the classical traffic flow instability, the traffic flow models explaining traffic break-

---

[33] For simplicity, we assume here that vehicles cannot pass each other.

[34] The classical traffic flow instability was introduced and theoretically studied in 1958–1961 by Herman, Gazis, Rothery, Montroll, Chandler, and Potts [26, 42, 43, 62] as well as Kometani and Sasaki [77–80].

[35] See papers [70–72].

[36] A diverse variety of the traffic flow models, in which traffic breakdown is explained by the classical traffic flow instability, can be found, e.g., in reviews and books [11, 13, 28, 61, 86, 95, 97, 119, 126]. It should be noted that after the classical traffic flow instability caused by the driver reaction time was introduced in 1958–1961 [26, 42, 43, 62], many mathematical traffic flow models have been developed in which the classical traffic flow instability has been modeled through the use of different mathematical approaches without some explicit use of the driver reaction time (see, e.g., reviews [1, 11, 13, 14, 28, 31, 34, 37, 41, 61, 75, 86, 95, 97, 119, 126] and references there).

down through the classical traffic flow instability cannot explain real traffic breakdown (F→S transition) that exhibits the nucleation nature of traffic breakdown.[37]

Thus, all mathematical traffic flow models in which the classical traffic flow instability is incorporated as the basic mechanism of traffic breakdown are invalid for the modeling of traffic breakdown (F→S transition) in metastable free flow with respect to the F→S transition. As already mentioned, the main reason for this critical conclusion is as follows:

- The classical traffic flow instability predicts that traffic breakdown is the moving jam emergence in free flow at the bottleneck (F→J transition). Contrary to this theoretical result, the general conclusion of empirical observations is that empirical traffic breakdown is an F→S transition at the bottleneck, *not* the F→J transition.
- The empirical metastability of free flow with respect to an F→S transition at a bottleneck has no sense for standard traffic flow models in which the classical traffic flow instability is incorporated as the basic mechanism of the occurrence of traffic congestion.

Real traffic breakdown (F→S transition) at a bottleneck cannot be explained by the existence of the driver reaction time.

Real traffic breakdown (F→S transition) at a bottleneck cannot be explained by the classical traffic flow instability.

C.4 Invalid Applications of Empirical Fundamental Diagram for Understanding Real Traffic

Traffic is a process that occurs in space and time. However, in Sect. 5.2 of the main text of the book, we have already emphasized that many *spatiotemporal features* of traffic breakdown are lost in the empirical fundamental diagram (Fig. 5.2a). The latter means that for a correct understanding of empirical traffic data, the empirical fundamental diagram (Fig. 5.2a) can be used *only* together with spatiotemporal empirical data (Fig. 5.1b) that contain the full information about the development of traffic breakdown in space and time. In other words, the empirical fundamental

---

[37] Proof of this critical statement about the traffic flow models that explain traffic breakdown through the classical traffic flow instability (see reviews of these traffic flow models, e.g., in [1, 11, 13, 14, 28, 31, 34, 37, 41, 61, 75, 86, 95, 97, 119, 126]) can be found in the review [68].

Appendix C: Empirical Induced Traffic Breakdown—Failure of Standard ... 225

diagram can be very useful for *illustrations* of some macroscopic traffic characteristics. However, a consideration of the empirical fundamental diagram without relation to the spatiotemporal empirical data can lead and it does very often lead to incorrect conclusions about traffic breakdown. Respectively, in Sect. 5.2, we have already made the conclusion that from a sole study of the empirical fundamental diagram, the understanding of the nature of real traffic breakdown is not possible. To understand real traffic breakdown, traffic data measured in space and time should be used (Chaps. 6 and 7).

Unfortunately, in contrast to the nature of real traffic as a spatiotemporal phenomenon, since the beginning of traffic science and up to now a sole study of the empirical fundamental diagram is often used for the empirical analysis of vehicular traffic. In other words, the empirical fact that real traffic occurs in space and time, i.e., traffic is fundamentally a spatiotemporal phenomenon has been ignored in many empirical studies in which conclusions about traffic phenomena have solely been derived from a study of empirical data in the empirical fundamental diagram. This can explain why most of the applications of the empirical fundamental diagram without involving the associated spatiotemporal phenomena are basically invalid for understanding real traffic. Examples of these invalid applications are

- the development of methods for the estimation of highway capacity,[38]
- the development of methods for the estimation of highway safety based on the empirical fundamental diagram,
- the evaluation of traffic flow models with the empirical fundamental diagram[39] as well as many other applications,
- the use of so-called *macroscopic* fundamental diagram (MFD) for understanding real traffic in traffic networks.[40]

Therefore, in this book, we have used the fundamental diagram *only* for illustrations of some macroscopic traffic features and characteristics, *not* for the analysis of real spatiotemporal traffic flow phenomena.

---

[38] See, e.g., books [34, 93, 116] and Highway Capacity Manual [63, 64].

[39] See, e.g., paper [113].

[40] An MFD (that is also called as a network fundamental diagram (NFD)) presents a relation between macroscopic traffic variables related to an urban network [30, 45, 88–90]. To find the MFD, some average vehicle density in the whole urban network and the total flow rate in the network should be measured. Empirical studies have shown that the MFD can indeed show a similar relationship between the average vehicle density in the whole network and the total inflow rate in the network [30, 45] as the classical fundamental diagram (Sect. 5.2). Thus, studies of the MFD permit to make the conclusion that the larger the total average vehicle density in the network, the lower is the average speed in the network (see, e.g., [30, 32, 45, 76, 83, 88–90, 128]). However, this conclusion as well as other consequences from the MFD cannot be used for understanding real traffic in a traffic network. Indeed, in the MFD all individual empirical features of traffic breakdown at different network bottlenecks in the network have been averaged. Therefore, complex spatiotemporal traffic phenomena occurring at the different network bottlenecks are lost in the MFD. A more detailed criticism of the MFD approach [30, 32, 45, 76, 83, 88–90, 128] for understanding real traffic in traffic networks can be found in Sect. 4.11.3 of the book [67].

Critical conclusions about invalid applications of the fundamental diagram discussed above are also related to a study of empirical traffic hysteresis effects observed in traffic data presented in the fundamental diagram (or in the speed–density and speed–flow planes). Indeed, we have already mentioned in Sect. 5.3 that there can be a diverse variety of empirical traffic hysteresis effects that are caused by qualitatively different reasons. In particular, the empirical hysteresis effect shown in Fig. 5.4c–e is associated with traffic breakdown (F→S transition) and the subsequent return S→F transition at a highway bottleneck. In contrast, Treiterer's hysteresis phenomenon (Sect. B.3) is associated with the propagation of a vehicle platoon through the three traffic phases F, S, and J; in this case, the hysteresis phenomenon has no relation to bottleneck locations as well as other road locations at which the phase transitions have occurred. Already these two examples show that without a study of empirical traffic data in space and time no correct conclusions about the reason of one or another traffic hysteresis effect can be made.

C.5 Strict Belief in Standard Theories as Reason for Defective Analysis of Empirical Traffic Phenomena

It could be assumed that the failure of standard methodologies of traffic and transportation science for reliable traffic management can be explained by a very long time interval between the development of the classical traffic theories made in the 1950s–1960s and the understanding of the empirical nucleation nature of traffic breakdown (F→S transition) at a highway bottleneck made by the author at the end of the 1990s. During this long time interval, several generations of traffic researches developed a huge number of standard traffic flow theories and models.

In the most fields of science, new experimental and/or empirical phenomena initiate the development of theories and models of the phenomena. Unfortunately, it is often different in the field of standard traffic flow theories: Usually, results of classical traffic flow theories and solutions of associated traffic flow models determine the methodologies for analyzing the empirical traffic data. This prevents the understanding of those new real traffic flow phenomena that cannot be explained by the standard traffic theories.[41]

- A strict belief in the standard traffic theories in traffic and transportation science leads to a defective (and sometimes invalid) analysis of empirical traffic phenomena.
- As above emphasized, standard traffic flow theories and models do not show the nucleation nature of traffic breakdown (F→S transition) at a highway bottleneck. This is probably the reason why the evidence of the empirical induced traffic breakdown (F→S transition) at the bottleneck is simply ignored in the state of the art of traffic and transportation research (see also Sect. C.9).

---

[41] A more detailed criticism of the analysis of empirical traffic phenomena made in the standard traffic research can be found in Sect. 4.11 of the book [67].

Appendix C: Empirical Induced Traffic Breakdown—Failure of Standard ... 227

- In general, we can distinguish two different standard invalid methodologies of a study of empirical traffic flow phenomena:

(i) All empirical traffic flow phenomena that contradict solutions of a traffic model appear to be ignored.
(ii) Real empirical spatiotemporal traffic flow phenomena are averaged either in space and/or in time. As a result of this averaging, important empirical spatiotemporal traffic flow phenomena cannot be found in the averaged data any more. In other words, all empirical features of real traffic phenomena, which cannot be explained by the model, are eliminated through the data averaging.

C.6 Invalid Methods for Traffic Control and Evaluation of Intelligent Transportation Systems (ITS)

Traffic modeling for the evaluation of traffic control, traffic management, the effect of autonomous vehicles on traffic flow as well as many other applications of intelligent transportation systems (ITS) (ITS applications for short) should answer the following questions:

(i) Can a method of traffic control increase highway capacity?
(ii) Can autonomous driving vehicles prevent traffic breakdown at highway bottlenecks?
(iii) Can an approach for traffic management decrease the probability of traffic breakdown?

Many other tasks of traffic modeling are also associated with the analysis of the effect of ITS applications on traffic breakdown at a bottleneck in a traffic network. However, we have explained above that *none* of the standard traffic theories and models can explain the empirical nucleation nature of traffic breakdown (F→S transition) at a highway bottleneck. Therefore, the standard traffic models do lead to invalid results of the evaluation of the effect of ITS applications on real traffic. We can make the following conclusions:

- Most of the standard methods for traffic control, management, and organization of a traffic network as well as other ITS applications are based on the standard understanding of stochastic highway capacity or/and on other standard traffic flow theories and models.[42] However, these standard traffic theories cannot explain the empirical nucleation nature of traffic breakdown (F→S transition) at a highway bottleneck.
- This can explain the failure of the standard methods for traffic control, management, and organization of a traffic network as well as other ITS applications.

---

[42] See, e.g., reviews and books [1, 11–14, 27, 29, 31, 34, 37–40, 44, 48, 49, 61, 63–66, 75, 84, 87, 91, 93, 100, 103, 104, 110, 112, 114, 116–118, 123, 124, 126].

- Results of the evaluation of ITS applications with the use of the standard traffic theories, models, and simulation tools, which are currently the basic traffic flow models for the evaluation of ITS applications,[43] are invalid for real traffic: the standard traffic models cannot be applied for reliable evaluation of ITS applications like the evaluation of the effect of self-driving (autonomous driving) vehicles on traffic breakdown in mixed traffic flow consisting of human driving and autonomous driving vehicles.

C.7 Empirical Nucleation Nature of Emergence of Wide Moving Jam as A Difficulty for Understanding Real Traffic

From the author's experience, for many traffic researchers there is the following difficulty in understanding real traffic. We denote the maximum possible flow rate in free flow by $q_{\text{max}}^{(\text{free})}$ (Fig. B.2). From the empirical feature of the line J (Sect. B.1), it follows that within the flow rate range $q_{\text{out}} \leq q \leq q_{\text{max}}^{(\text{free})}$ real free flow is in a metastable state with respect to the jam emergence (F→J transition)[44]: In the metastable free flow, *theoretically* there can be a large enough local speed decrease in free flow that can be a nucleus for the spontaneous formation of a wide moving jam in free flow (see Fig. A.14c, d of Sect. A.10). Almost all standard traffic flow models incorporating the classical traffic flow instability can simulate this metastability of free flow with respect to the F→J transition. Therefore, the question arises:

- Why does the author criticizes the standard traffic flow models?

The response to this very often question of traffic researchers is as follows:

- In *no* empirical traffic data, the spontaneous moving jam emergence (F→J transition) is observed in free flow: The spontaneous F→J transition remains the theoretical result *only*. This theoretical result has *no* relation to the real world.
- Rather than the spontaneous moving jam emergence (F→J transition), the F→S transition (traffic breakdown) is observed in real free flow at a large enough flow rate.
- Because standard traffic flow models incorporating the classical traffic flow instability are not able to show traffic breakdown in metastable free flow with respect to an F→S transition as it is observed in real free flow, the standard models leads to invalid results of the evaluation of ITS applications.

The reason why spontaneous emergence of moving jams is not observed in real free flow has been explained in Sect. A.10.

---

[43] See, e.g., [1, 11, 13, 14, 29, 31, 34, 37, 40, 44, 48, 49, 61, 63–66, 75, 84, 100, 103, 110, 117, 124, 126]. Proof of this criticism of the basic traffic flow models used for the evaluation of ITS applications can be found in [68].

[44] In empirical observations, the value $q_{\text{max}}^{(\text{free})}/q_{\text{out}} \approx 1.5$ [73, 74].

> In some flow-rate range of free flow, the free flow can be in metastable state with respect both to the F→S transition (traffic breakdown) and to the F→J transition. However, in real free flow *only* the F→S transition can spontaneously be observed. This empirical feature of the metastability of real free flow with respect to the F→S transition could be one of the main difficulties for understanding real traffic.

C.8 Criticism of Three-Phase Traffic Theory

Some traffic researchers have made a criticism of the three-phase traffic theory[45] that can be formulated as follows:

- The three-phase traffic theory is not needed. This is because standard traffic flow theories also succeed in simulating the most essential empirical features of traffic flow described by the three-phase traffic theory.

The criticism of the three-phase traffic theory is *incorrect*.

Indeed, the main reason for the three-phase traffic theory is the explanation of the empirical nucleation nature of traffic breakdown (F→S transition) at a highway bottleneck. *None* of the standard traffic flow theories and models can explain the empirical nucleation nature of traffic breakdown (F→S transition) at the bottleneck.[46]

> The standard traffic and transportation theories cannot explain the empirical nucleation nature of traffic breakdown (F→S transition) at a bottleneck. In contrast, the main reason of the three-phase traffic theory is the explanation of the empirical nucleation nature of traffic breakdown (F→S transition) at the bottleneck. For this reason, the criticism of the three-phase traffic theory is incorrect.

---

[45] See, for example, Refs. [120–122, 127].

[46] This critical conclusion is valid for all traffic flow models developed and used by the authors of Refs. [61, 120–122, 125, 127] whose incorrect criticism of the three-phase traffic theory is often used up to now by many other researchers of the traffic and transportation research community. The same critical conclusion is also valid for all other standard traffic and transportation theories and models (see, e.g., the reviews and books [1, 11, 13, 14, 29, 31, 34, 37, 40, 44, 48, 49, 63–66, 75, 84, 100, 103, 110, 117, 124, 126]). A more detailed criticism of the standard traffic flow models has been made in Sects. 1.7 and 1.13 as well as Chap. 4 of the book [67].

C.9 Why Are Invalid Standard Traffic Models Very Attractive for Traffic Researchers Up To Now?

We have explained above that the reason for the failure of applications of standard traffic theories and models in the real world is the ignoring of the empirical nucleation nature of traffic breakdown (F→S transition) at a highway bottleneck. Nevertheless, the following question can arise:

- Why are invalid standard traffic models very attractive for traffic researchers up to now?

A possible response to this question is as follows: Some of the standard traffic models exhibit important achievements that are as follows[47]:

- Many of the standard traffic flow models incorporate the classical traffic flow instability in free flow (Sects. C.3 and C.7) that exhibits the nucleation nature: The classical traffic flow instability can theoretically occur only if a large enough local speed decrease appears in free flow (Sect. A.10).
- As we have shown in Sect. 10.1, the classical traffic flow instability is responsible for the empirical moving jam emergence in synchronized flow.
- There are many mathematical ideas about simulations of driver time delays in different traffic situations introduced in the standard traffic flow models that can also be very useful for the development of mathematical traffic flow models in the framework of the three-phase traffic theory.[48]

It has already been emphasized that standard traffic flow models incorporating the classical traffic instability can simulate many empirical features of moving jams in congested traffic. For this reason, through an appropriate choice of model parameters the models can show a good agreement with the empirical features of moving jams in congested traffic.[49] Therefore, it is not surprising that these models are the basis of almost all traffic simulation tools for the evaluation of ITS applications.

---

[47] Reviews of these standard traffic flow models can be found, e.g., in [11, 13, 28, 61, 68, 86, 95, 97, 119, 126].

[48] Pioneering ideas for the mathematical description of driver behavioral assumptions of the works by Pipes [111], Herman, Gazis, Rothery, Montroll, Chandler, and Potts [26, 42, 43, 62], Newell [99, 101], Gipps [46], Bando et al. [2–4], Nagel and Schreckenberg [96], Krauß et al. [81, 82], Nagel et al. [98] as well as of many other works (see Chap. 1 of [67]) have also been used in mathematical traffic flow models in the framework of the three-phase traffic theory (see references in the book [67]).

[49] Explanations of the fact why the standard traffic flow models incorporating the classical traffic instability can show very good agreement with the empirical features of moving jams in congested traffic, although the models cannot show the feature of empirical synchronized flow associated with the nucleation nature of traffic breakdown (F→S transition), can be found in Chap. 8 of the book [67].

Appendix C: Empirical Induced Traffic Breakdown—Failure of Standard ...   231

> Standard traffic flow models, in which the classical traffic flow instability is the basic mechanism for the occurrence of traffic congestion, are very attractive for traffic researchers because the models can well describe many of real features of moving jams (stop-and-go traffic) in congested traffic.

Therefore, the following question arises:

- Why should these generally accepted standard traffic flow models used in most traffic simulation tools be invalid for the evaluation of ITS applications?

To respond to this question, we recall that in Sect. C.5 we have already mentioned that the empirical evidence of the metastability of free flow with respect to the F→S transition at a highway bottleneck is simply ignored in the state of the art of the traffic and transportation research.[50] However, rather than a study of moving jams (stop-and-go traffic) in congested traffic, the main task of the evaluation of ITS applications is to study the effect of the ITS applications on highway capacity. Highway capacity is limited by traffic breakdown (F→S transition) at the bottleneck. Thus, a traffic flow model required for the evaluation of ITS applications must simulate traffic breakdown in metastable free flow with respect to an F→S transition at the bottleneck. In other words, the models must simulate the empirical nucleation nature of traffic breakdown (F→S transition) at the bottleneck. However, *none* of the standard traffic models can show the empirical nucleation nature of the F→S transition at the bottleneck. This explains our statement that all traffic simulation tools based on the standard traffic flow models are invalid for a reliable evaluation of the effect of ITS applications on traffic breakdown at the bottleneck.[51]

---

[50] An example of the last statement is the empirical evaluation of standard traffic flow models made during last years with the use of empirical vehicle trajectories measured in congested traffic that is available in the datasets of "New Generation Traffic Simulations (NGSIM)" [102]. The NGSIM empirical traffic data measured on USA highways contain data of empirical synchronized flow in which many moving jams emerge. However, in the datasets of NGSIM there are no traffic data that allow us to study the empirical induced traffic breakdown (F→S transition) at a highway bottleneck.

[51] Examples of some of the traffic simulation tools that are invalid for the evaluation of the effect of ITS applications on traffic breakdown at a highway bottleneck are "VISSIM", "SUMO", "MIT-SIMLab", "DRACULA", "Dynameq", "METANET", "DynaMIT", "AVENUE", "Paramics", and "Aimsun" reviewed in the book [11]. This critical conclusion is also related to many other traffic simulation approaches based on standard traffic flow models (e.g., Newell's model [101], Treiber's intelligent driver model (IDM) [125], and Bando et al. model [2–4]) reviewed, e.g., in [1, 13, 14, 29, 31, 34, 37, 40, 44, 48, 49, 61, 63–66, 75, 84, 100, 103, 110, 117, 124, 126]. Proof of this criticism can be found in the review [68].

A common mistake of the most traffic researchers made by the evaluation of ITS applications is that standard traffic flow models used in simulation tools are applied outside the region of the model applicability: The standard traffic flow models that can describe empirical data for moving jams in congested traffic are not able to simulate the empirical nucleation nature of traffic breakdown (F→S transition) at a highway bottleneck.

References

1. W.D. Ashton, *The theory of traffic flow*, (Methuen & Co. London, Wiley, New York, 1966)
2. M. Bando, K. Hasebe, A. Nakayama, A. Shibata, Y. Sugiyama, Jpn. J. Appl. Math. **11**, 203–223 (1994)
3. M. Bando, K. Hasebe, A. Nakayama, A. Shibata, Y. Sugiyama, Phys. Rev. E **51**, 1035–1042 (1995)
4. M. Bando, K. Hasebe, A. Nakayama, A. Shibata, Y. Sugiyama, J. Phys. I France **5**, 1389–1399 (1995)
5. J.H. Banks, Transp. Res. Rec. **1320**, 83–90 (1991)
6. J.H. Banks, Transp. Res. Rec. **1394**, 17–25 (1993)
7. J.H. Banks, Transp. Res. Rec. **1510**, 1–10 (1995)
8. J.H. Banks, Transp. Res. Rec. **1678**, 128–134 (1999)
9. J.H. Banks, Transp. Res. Rec. **1802**, 225–232 (2002)
10. J.H. Banks, Transp. Res. Rec. **2099**, 14–21 (2009)
11. J. Barceló (ed.), *Fundamentals of Traffic Simulation* (Springer, Berlin, 2010)
12. M.G.H. Bell, Y. Iida, *Transportation Network Analysis* (Wiley, Hoboken, 1997)
13. N. Bellomo, V. Coscia, M. Delitala, Math. Model. Methods Appl. Sci. **12**, 1801–1843 (2002)
14. M. Brackstone, M. McDonald, Transp. Res. F **2**, 181 (1998)
15. W. Brilon, Transp. Res. Rec. **2483**, 57–65 (2015)
16. W. Brilon, J. Geistefeldt, M. Regler, in *Traffic and Transportation Theory*, ed. by H.S. Mahmassani (Elsevier Science, Amsterdam, 2005), pp. 125–144
17. W. Brilon, J. Geistefeld, H. Zurlinden, Transp. Res. Rec. **2027**, 1–8 (2007)
18. W. Brilon, M. Regler, J. Geistefeld, Straßenverkehrstechnik, Heft 3, 136 (2005)
19. W. Brilon, H. Zurlinden, Straßenverkehrstechnik, Heft 4, 164 (2004)
20. M.J. Cassidy, in *Proceedings of the IEEE Intelligent Transportation Systems* (IEEE, Oakland, 2001), pp. 513–535
21. M.J. Cassidy, S. Ahn, Transp. Res. Rec. **1934**, 140–147 (2005)
22. M.J. Cassidy, R.L. Bertini, Transp. Res. B **33**, 25–42 (1999)
23. M.J. Cassidy, B. Coifman, Transp. Res. Rec. **1591**, 1–6 (1997)
24. M.J. Cassidy, A. Skabardonis, A.D. May, Transp. Res. Rec. **1225**, 61–72 (1989)
25. M.J. Cassidy, J.R. Windover, Transp. Res. Rec. **1484**, 73–79 (1995)
26. R.E. Chandler, R. Herman, E.W. Montroll, Oper. Res. **6**, 165–184 (1958)

27. Y.-C. Chiu, J. Bottom, M. Mahut, A. Paz, R. Balakrishna, T. Waller, J. Hicks, Dynamic Traffic Assignment, A Primer. Transp. Res. Circ. **E-C153** (2011), http://onlinepubs.trb.org/onlinepubs/circulars/ec153.pdf
28. D. Chowdhury, L. Santen, A. Schadschneider, Phys. Rep. **329**, 199 (2000)
29. C.F. Daganzo, *Fundamentals of Transportation and Traffic Operations* (Elsevier Science Inc., New York, 1997)
30. C.F. Daganzo, Transp. Res. B **41**, 49–62 (2007)
31. D.R. Drew, *Traffic Flow Theory and Control* (McGraw Hill, New York, 1968)
32. J. Du, H. Rakha, V.V. Gayah, Transp. Res. C **66**, 136–149 (2016)
33. S.M. Easa, A.D. May, Transp. Res. Rec. **772**, 24–37 (1980)
34. L. Elefteriadou, *An Introduction to Traffic Flow Theory* (Springer, Berlin, 2014)
35. L. Elefteriadou, A. Kondyli, W. Brilon, F.L. Hall, B. Persaud, S. Washburn, J. Transp. Eng. **140**, 04014003 (2014)
36. L. Elefteriadou, R.P. Roess, W.R. McShane, Transp. Res. Rec. **1484**, 80–89 (1995)
37. A. Ferrara, S. Sacone, S. Siri, *Freeway Traffic Modelling and Control* (Springer, Berlin, 2018)
38. T.L. Friesz, D. Bernstein, *Foundations of Network Optimization and Games*, Complex Networks and Dynamic Systems, vol. 3 (Springer, New York, 2016)
39. N.H. Gartner, C.J. Messer, A.K. Rathi (eds.), *Special Report 165: Revised Monograph on Traffic Flow Theory* (Transportation Research Board, Washington DC, 1997)
40. N.H. Gartner, C.J. Messer, A.K. Rathi (eds.), *Traffic Flow Theory: A State-of-the-Art Report* (Transportation Research Board, Washington DC, 2001)
41. D.C. Gazis, *Traffic Theory* (Springer, Berlin, 2002)
42. D.C. Gazis, R. Herman, R.B. Potts, Oper. Res. **7**, 499–505 (1959)
43. D.C. Gazis, R. Herman, R.W. Rothery, Oper. Res. **9**, 545–567 (1961)
44. D.L. Gerlough, M.J. Huber, *Traffic Flow Theory Special Report 165* (Transportation Research Board, Washington DC, 1975)
45. N. Geroliminis, C.F. Daganzo, Transp. Res. B **42** 759–770 (2008)
46. P.G. Gipps, Transp. Res. B **15**, 105–111 (1981)
47. B.D. Greenshields, J.R. Bibbins, W.S. Channing, H.H. Miller, in *Proceedings of the Highway Research Board*, vol. 14 (1935), pp. 448–477
48. M. Guerrieri, R. Mauro, *A Concise Introduction to Traffic Engineering* (Springer, Berlin, 2021)
49. F.A. Haight, *Mathematical Theories of Traffic Flow* (Academic Press, New York, 1963)
50. F.L. Hall, Transp. Res. A **21**, 191–201 (1987)
51. F.L. Hall, in [40], pp. 2-1–2-36
52. F.L. Hall, K. Agyemang-Duah, Transp. Res. Rec. **1320**, 91–98 (1991)
53. F.L. Hall, B.L. Allen, M.A. Gunter, Transp. Res. A **20**, 197–210 (1986)
54. F.L. Hall, D. Barrow, Transp. Res. Rec. **1194**, 55–65 (1988)
55. F.L. Hall, W. Brilon, Transp. Res. Rec. **1457**, 35–42 (1994)
56. F.L. Hall, M.A. Gunter, Transp. Res. Rec. **1091**, 1–9 (1986)
57. F.L. Hall, L.M. Hall, Transp. Res. Rec. **1287**, 108–118 (1990)
58. F.L. Hall, V.F. Hurdle, J.H. Banks, Transp. Res. Rec. **1365**, 12–18 (1992)

59. F.L. Hall, B.N. Persaud, Transp. Res. Rec. **1232**, 9–16 (1989)
60. F.L. Hall, A. Pushkar, Y. Shi, Transp. Res. Rec. **1398**, 24–30 (1993)
61. D. Helbing, Rev. Mod. Phys. **73**, 1067–1141 (2001)
62. R. Herman, E.W. Montroll, R.B. Potts, R.W. Rothery, Oper. Res. **7**, 86–106 (1959)
63. *Highway Capacity Manual 2000* (National Research Council, Transportation Research Board, Washington, DC, 2000)
64. *Highway Capacity Manual 2016*, 6th edn. (National Research Council, Transportation Research Board, Washington, DC, 2016)
65. A. Horni, K. Nagel, K.W. Axhausen (eds.), *The Multi-Agent Transport Simulation MATSim* (Ubiquity, London, 2016), http://matsim.org/the-book. https://doi.org/10.5334/baw
66. P. Kachroo, K.M.A. Özbay, *Feedback Control Theory for Dynamic Traffic Assignment* (Springer, Berlin, 2018)
67. B.S. Kerner, *Breakdown in Traffic Networks* (Springer, Berlin, 2017)
68. B.S. Kerner, in *Complex Dynamics of Traffic Management*, ed. by B.S. Kerner. Encyclopedia of Complexity and Systems Science Series (Springer, New York, 2019), pp. 195–283
69. B.S. Kerner, in *Complex Dynamics of Traffic Management*, ed. by B.S. Kerner. Encyclopedia of Complexity and Systems Science Series (Springer, New York, 2019), pp. 387–500
70. B.S. Kerner, P. Konhäuser, Phys. Rev. E **48**, 2335–2338 (1993)
71. B.S. Kerner, P. Konhäuser, Phys. Rev. E **50**, 54–83 (1994)
72. B.S. Kerner, P. Konhäuser, M. Schilke, Phys. Rev. E **51**, 6243–6246 (1995)
73. B.S. Kerner, H. Rehborn, Phys. Rev. E **53**, R1297–R1300 (1996)
74. B.S. Kerner, H. Rehborn, Phys. Rev. E **53**, R4275–R4278 (1996)
75. F. Kessels, *Traffic Flow Modelling* (Springer, Berlin, 2019)
76. V.L. Knoop, H. van Lint, S.P. Hoogendoorn, Phys. A **438**, 236–250 (2015)
77. E. Kometani, T. Sasaki, J. Oper. Res. Soc. Jpn. **2**, 11–26 (1958)
78. E. Kometani, T. Sasaki, Oper. Res. **7**, 704–720 (1959)
79. E. Kometani, T. Sasaki, Oper. Res. Soc. Jpn. **3**, 176–190 (1961)
80. E. Kometani, T. Sasaki, in *Theory of Traffic Flow*, ed. by R. Herman (Elsevier, Amsterdam, 1961), pp. 105–119
81. S. Krauß, Ph.D thesis, (DRL-Forschungsbericht 98-08, 1998), http://www.zaik.de/~paper
82. S. Krauß, P. Wagner, C. Gawron, Phys. Rev. E **54**, 3707–3712 (1996)
83. L. Leclercq, N. Chiabaut, B. Trinquier, Transp. Res. B **62**, 1–12 (2014)
84. W. Leutzbach, *Introduction to the Theory of Traffic Flow* (Springer, Berlin, 1988)
85. M.J. Lighthill, G.B. Whitham, Proc. R. Soc. A **229**, 281–345 (1955)
86. S. Maerivoet, B. De Moor, Phys. Rep. **419**, 1–64 (2005)
87. H.S. Mahmassani, Netw. Spat. Econ. **1**, 267–292 (2001)
88. H.S. Mahmassani, M. Saberi, A. Zockaie, Transp. Res. C **36**, 480–497 (2013)
89. H.S. Mahmassani, J.C. Williams, R. Herman, Transp. Res. Rec. **971**, 121–130 (1984)

90. H.S. Mahmassani, J.C. Williams, R. Herman, in *Proceedings of the 10th International Symposium on Transportation and Traffic Theory*, ed. by N.H. Gartner (Elsevier, New York, 1987), pp. 1–20
91. F.L. Mannering, W.P. Kilareski, *Principles of Highway Engineering and Traffic Analysis*, 2nd edn. (Wiley, New York, 1998)
92. A.D. May, Highw. Res. Rec. **59**, 9–38 (1964)
93. A.D. May, *Traffic Flow Fundamentals* (Prentice-Hall Inc., New Jersey, 1990)
94. A.D. May, P. Athol, W. Parker, J.B. Rudden, Highw. Res. Rec. **21**, 48–70 (1963)
95. T. Nagatani, Rep. Prog. Phys. **65**, 1331–1386 (2002)
96. K. Nagel, M. Schreckenberg, J. Phys. (France) I **2**, 2221–2229 (1992)
97. K. Nagel, P. Wagner, R. Woesler, Oper. Res. **51**, 681–716 (2003)
98. K. Nagel, D.E. Wolf, P. Wagner, P. Simon, Phys. Rev. E **58**, 1425–1437 (1998)
99. G.F. Newell, Oper. Res. **9**, 209–229 (1961)
100. G.F. Newell, *Applications of Queuing Theory* (Chapman Hall, London, 1982)
101. G. F. Newell, Transp. Res. B **36**, 195–205 (2002)
102. Next Generation Simulation Programs (NGSIM), http://ops.fhwa.dot.gov/traffic analysistools/ngsim.htm
103. M. Papageorgiou, I. Papamichail, Transp. Res. Rec. **2047**, 28–36 (2008)
104. S. Peeta, A.K. Ziliaskopoulos, Netw. Spat. Econ. **1**, 233–265 (2001)
105. B.N. Persaud, F.L. Hall, Transp. Res. A **23**, 103–113 (1989)
106. B.N. Persaud, F.L. Hall, L.M. Hall, Transp. Res. Rec. **1287**, 167–175 (1990)
107. B.N. Persaud, V.F. Hurdle, Transp. Res. Rec. **1194**, 191–198 (1988)
108. B.N. Persaud, S. Yagar, R. Brownlee, Transp. Res. Rec. **1634**, 64–69 (1998)
109. B.N. Persaud, S. Yagar, D. Tsui, H. Look, Transp. Res. Rec. **1748**, 110–115 (2001)
110. B. Piccoli, A. Tosin, in *Encyclopedia of Complexity and System Science*, ed. by R.A. Meyers (Springer, Berlin, 2009), pp. 9727–9749
111. L.A. Pipes, J. Appl. Phys. **24**, 274–287 (1953)
112. I. Prigogine, R. Herman, *Kinetic Theory of Vehicular Traffic*, (American Elsevier, New York, 1971)
113. X. Qua, Sh. Wang, J. Zhang, Transp. Res. B **73**, 91–102 (2015)
114. B. Ran, D. Boyce, *Modeling Dynamic Transportation Networks* (Springer, Berlin, 1996)
115. P.I. Richards, Oper. Res. **4**, 42–51 (1956)
116. R.P. Roess, E.S. Prassas, *The Highway Capacity Manual: A Conceptual and Research History* (Springer, Berlin, 2014)
117. M. Saifuzzaman, Z. Zheng, Transp. Res. C **48**, 379–403 (2014)
118. T. Seo, A.M. Bayen, T. Kusakabe, Y. Asakura, Annu. Rev. Control **43**, 128–151 (2017)
119. A. Schadschneider, D. Chowdhury, K. Nishinari, *Stochastic Transport in Complex Systems* (Elsevier Science Inc., New York, 2011)
120. M. Schönhof, D. Helbing, Transp. Sci. **41**, 135–166 (2007)
121. M. Schönhof, D. Helbing, Transp. Res. B **43**, 784–797 (2009)
122. M. Schönhof, M. Treiber, A. Kesting, D. Helbing, Transp. Res. Rec. **1999**, 3–12 (2007)

123. Y. Sheffi, *Urban Transportation Networks: Equilibrium Analysis with Mathematical Programming Methods* (Prentice-Hall, New Jersey, 1984)
124. TRANSIMS (U.S. Department of Transportation, Federal Highway Administration, Washington, DC, 2017), https://www.fhwa.dot.gov/planning/tmip/resources/transims/
125. M. Treiber, A. Hennecke, D. Helbing, Phys. Rev. E **62**, 1805–1824 (2000)
126. M. Treiber, A. Kesting, *Traffic Flow Dynamics* (Springer, Berlin, 2013)
127. M. Treiber, A. Kesting, D. Helbing, Transp. Res. B **44**, 983–1000 (2010)
128. M. Yildirimoglu, M. Ramezani, N. Geroliminis, Transp. Res. C **59**, 404–420 (2015)

Glossary

Bottleneck Traffic breakdown occurs mostly at a bottleneck. There are road bottlenecks and moving bottlenecks. A road bottleneck can be a result of road works, on- and off-ramps, a decrease in the number of road lanes, road curves and road gradients, traffic signal, etc. A moving bottleneck (MB) results from a single vehicle or a group of vehicles that move slower than other vehicles in free flow.

Congested Traffic Congested traffic is considered an empirical state of traffic in which the average speed is lower than the minimum average speed that is still possible in free flow.

Empirical Traffic Data Empirical traffic data (real field traffic data) are solely obtained through measurements in real traffic based on road detectors, video cameras, global navigation satellite system in vehicles, aerial observations, etc.

Free Flow Free traffic flow (free flow for short) is usually observed, when the vehicle density in traffic is small enough. The flow rate increases in free flow with the increase in the vehicle density, whereas the average vehicle speed is a decreasing density function. There is a limit in the density increase in free flow. At this limiting state of free flow, the flow rate and density reach their maximum values while the average speed reaches a minimum value that is still possible in free flow.

F→S transition Traffic breakdown at a highway bottleneck is a phase transition from the free flow traffic phase (F) to synchronized flow traffic phase (S) called an F→S transition. The terms *traffic breakdown* and an *F→S transition* are synonyms.

Highway Capacity Highway capacity is limited by traffic breakdown in free flow at a highway bottleneck. Any flow rate in free flow at the bottleneck, at which traffic breakdown (F→S transition) can occur spontaneously or can be induced, is equal to one of the highway capacities. There are the infinite number of such flow rates: *At any time instant*, there is a range of highway capacities that boundaries are related to a minimum highway capacity and a maximum highway capacity.

Induced Traffic Breakdown (Induced F→S Transition) at Bottleneck An induced traffic breakdown (induced F→S transition) at a bottleneck is traffic breakdown that is induced at the bottleneck through the propagation of a moving congested traffic pattern to the location of the bottleneck.

Moving Jam A moving traffic jam (moving jam for short) is a congested traffic pattern propagating upstream within which the density is large and the vehicle speed is low. The jam is spatially limited by two jam fronts; within the downstream jam front, vehicles accelerate escaping from the jam; within the upstream jam front, vehicles slow down approaching the jam.

Nucleation Nature of Traffic Breakdown (F→S Transition) The empirical nucleation nature of an F→S transition (traffic breakdown) at a highway bottleneck is as follows. A small enough local speed decrease in free flow at the bottleneck does not cause traffic breakdown. Contrarily, a large enough local speed decrease in free flow at the bottleneck leads to the F→S transition at the bottleneck. A local speed decrease leading to the F→S transition is called a *nucleus* for traffic breakdown at the bottleneck.

Standard Traffic and Transportation Science The term "standard" for traffic and transportation science (in particular, traffic theories and models) is related to the traffic theories and models whose fundamentals were developed before the three-phase traffic theory was introduced.

Stochastic Highway Capacity The existence of the infinite number of highway capacities at any time instant means that highway capacity is stochastic.

Synchronized Flow Synchronized flow is a phase of congested traffic. The synchronized flow traffic phase is defined as follows: The downstream front of synchronized flow can be localized at a highway bottleneck. The synchronized flow traffic phase ensures the nucleation nature of the F→S transition at the bottleneck.

Three-Phase Traffic Theory The three-phase traffic theory is a framework for understanding real traffic in three traffic phases: (i) free flow (F), (ii) synchronized flow (S), and (iii) wide moving jam (J). The synchronized flow and wide moving jam traffic phases belong to congested traffic. The main reason for the three-phase traffic theory is the explanation of the empirical nucleation nature of traffic breakdown (F→S transition) at a highway bottleneck.

Traffic Breakdown Traffic breakdown at a highway bottleneck is a transition from free flow to congested traffic at a bottleneck. In all observations, traffic breakdown is a transition from free flow to synchronized flow (F→S transition) at the bottleneck. The terms *F→S transition*, *breakdown phenomenon*, *traffic breakdown*, and *speed breakdown* are synonyms.

Wide Moving Jam Traffic Phase A wide moving jam is a phase of congested traffic. The wide moving jam traffic phase is defined as follows: The wide moving jam is a moving jam that propagates through highway bottlenecks and any traffic states while maintaining the mean velocity of the downstream jam front.

Index

A

Alternations of flow interruption intervals and moving blanks within wide moving jam, 160
Asymmetric deceleration–acceleration driver behavior, 180
Automated driving, 170
Automatic driving, 170
Autonomous driving, 170
Average local speed decrease in free flow at bottleneck, 51

B

Boomerang effect, 150, 156
Bottleneck, 3, 237
Boundaries of 2D states of synchronized flow, 178
Breakdown phenomenon, 1, 238

C

Car-following, 102, 180
Characteristic parameters of propagation of wide moving jam, 206
Classical traffic flow instability, 136, 223
Classification of MSPs, 146
Classification of SPs, 146
Competition between over-acceleration and speed adaptation, 189
Competition between speed adaptation and over-deceleration, 139
Competition of S→F and S→J instabilities, 154, 197
Congested pattern control approach, 114
Congested traffic, 1, 54, 237
Congested traffic definition, 54
Congested traffic pattern, 20
Control of traffic breakdown at bottleneck, 114
Cooperative automated systems, 170
Cooperative driving, 170
Crisis in traffic science, 165
Critical nucleus required for traffic breakdown (F→S transition) at bottleneck, 81, 91, 114
Critical speed for S→F instability, 189
Critical speed for S→J instability, 139
Critical speed for traffic breakdown (F→S transition) at bottleneck, 81, 114
Criticism of three-phase traffic theory, 229

D

Decay of local speed increase in synchronized flow, 188
Defective analysis of empirical traffic phenomena, 226
Definition of MSP, 146
Definition of synchronized flow, 20, 79, 238
Definition of wide moving jam, 20, 238
Density, 1
Density estimation, 51
Different levels of automation in vehicle, 170
Discontinuous character of mean time delay in over-acceleration, 104
Discontinuous character of over-acceleration, 104
Dissolving speed wave of local speed increase in synchronized flow, 188
Dissolving synchronized flow, 129
Distinguishing between free and congested traffic, 54

Distinguishing between synchronized flow and wide moving jam, 20
Double Z-characteristic for phase transitions in traffic flow, 195
Downstream front of synchronized flow, 18
Downstream jam front, 1, 19
Driver over-acceleration, 102
Driver over-acceleration within indifferent zone, 186
Driver over-deceleration, 136, 223
Driver reaction time, 136
Driver speed adaptation, 102
Driver speed adaptation within indifferent zone, 185

E

Empirical anomaly in traffic science, 166
Empirical emergence of moving jams in synchronized flow, 23
Empirical fundamental diagram of traffic flow, 51
Empirical fundamental of transportation science, 5, 98
Empirical induced F→S transition at bottleneck, 5, 22, 71, 89
Empirical induced traffic breakdown at bottleneck, 5, 22, 71, 89
Empirical local speed decrease in free flow at bottleneck, 42, 127
Empirical nucleation nature of F→S transition at bottleneck, 5, 72
Empirical nucleation nature of traffic breakdown at bottleneck, 5, 72
Empirical proof of asymmetric deceleration–acceleration driver behavior, 210
Empirical proof of time delay in over-acceleration, 132
Empirical spontaneous F→S transition at bottleneck, 21, 57
Empirical spontaneous traffic breakdown at bottleneck, 21, 57
Empirical traffic data, 4, 237
Empirical traffic phenomena, 4

F

F→J instability, 200
F→J transition, 200
F→S transition, 21, 237
F→S→F transitions, 129
F→S→J transitions, 23, 135
Flow–density plane, 51
Flow interruption interval within wide moving jam, 159
Flow rate, 34
Free flow, 17, 237
Front between two traffic phases, 18
Fundamental change in meaning of stochastic highway capacity, 167
Fundamental diagram of traffic flow, 51
Fundamental problem for understanding real traffic, 40
Fundamental requirement for traffic flow model, 6, 101, 215
Fundamental requirement for traffic flow theory, 5, 101, 215
Fundamentals of standard traffic and transportation science, 2

G

General congested traffic pattern (GP), 154
Growing moving jam, 138
Growing speed wave of local speed decrease in synchronized flow, 136
Growing speed wave of local speed increase in synchronized flow, 189

H

Heavy bottleneck, 160
Heterogeneous traffic flow, 119
Highway bottleneck, 42
Highway capacity in standard traffic and transportation science, 48, 216
Highway capacity in three-phase traffic theory, 92, 237
Hysteresis effect, 55
Hysteresis loop, 55

I

Incommensurability of standard traffic theories with three-phase traffic theory, 167
Indifferent zone for car-following, 180
Induced F→S transition at bottleneck, 22, 71, 89, 238
Induced traffic breakdown at bottleneck, 22, 71, 89, 238
Infinite number of highway capacities at any time instant, 94
Intelligent transportation systems (ITS), 3, 64, 227
ITS applications, 3, 64, 227

J
Jam absorption effect, 206
Jam feature [J], 20, 205
Jam loop, 210

K
Kinematic wave, 222

L
Largest nucleus for traffic breakdown (F→S transition) at bottleneck, 81, 91
Lighthill–Whitham–Richards (LWR) theory, 219
Line J, 206
Localization of downstream front of synchronized flow at bottleneck, 59
Localized SP (LSP), 71, 147
Localized synchronized flow pattern (LSP), 71, 147
Local speed decrease in free flow at bottleneck, 39
Local speed decrease in free flow at moving bottleneck (MB), 44
Local speed decrease in free flow at off-ramp bottleneck, 41
Local speed decrease in free flow at on-ramp bottleneck, 41
Local speed decrease in synchronized flow, 136
Local speed increase in synchronized flow, 188

M
Macroscopic fundamental diagram (MFD), 225
Main prediction of three-phase traffic theory, 191
Main reason for three-phase traffic theory, 6, 166
Maximum highway capacity, 92, 237
Mean time delay in over-acceleration, 104
Measurement of traffic variables in real traffic, 4
Mega-jam, 160
Metastable free flow with respect to traffic breakdown (F→S transition) at bottleneck, 77, 114
Metastable synchronized flow with respect to S→J instability, 141
Minimum average speed that is still possible in free flow, 54, 237
Minimum highway capacity, 90, 237
Minimum speed within average local speed decrease in free flow at bottleneck, 51
Mixed traffic flow, 170
Moving blank within wide moving jam, 160
Moving bottleneck (MB), 44, 120, 237
Moving congested traffic pattern, 23
Moving general congested traffic pattern (MGP), 154
Moving jam, 1, 135, 238
Moving spatiotemporal congested traffic pattern, 23
Moving synchronized flow pattern (MSP), 146, 147

N
Net distance between two vehicles following each other, 177
Net time gap, 181
Network fundamental diagram (NFD), 225
Non-predictive behavior of synchronized flow, 197
Normal traffic and transportation science, 164
Nucleation-interruption effect of moving jam emergence in synchronized flow, 141
Nucleation-interruption phenomenon at bottleneck, 129
Nucleation nature of F→S transition, 77, 238
Nucleation nature of S→J instability, 141
Nucleation nature of traffic breakdown, 77, 238
Nucleus for emergence of growing moving jam in synchronized flow, 141
Nucleus for F→S transition, 72, 77, 238
Nucleus for S→F instability, 189
Nucleus for traffic breakdown, 72, 77, 238

O
Off-ramp bottleneck, 41
Off-ramp merging region, 41
On-ramp bottleneck, 40
On-ramp merging region, 40
Onset of congestion in free flow, 1
Origin of localization of downstream front of synchronized flow at bottleneck, 59
Over-acceleration, 102
Over-acceleration effect, 102
Over-acceleration within indifferent zone, 186

P

Paradigm shift in traffic and transportation science, 167
Perception of highway capacity in standard traffic and transportation science, 48
Perception of highway capacity resulting from empirical induced traffic breakdown at bottleneck, 98
Perception of vehicular traffic in three-phase traffic theory, 99
Phase transition points on vehicle trajectory, 17, 32
Probabilistic behavior of traffic breakdown, 49
Probe vehicle, 27
Probe vehicle data, 27

Q

Queuing models of vehicular traffic, 2, 49
Queuing theory, 2, 49

R

Range of highway capacities, 92, 168, 237
Range of stochastic highway capacities, 94, 168, 237
Real field traffic data, 4, 237
Region of dissolving synchronized flow, 129
Response of traffic research community to three-phase traffic theory, 169
Response to crisis in traffic science, 166
Road bottleneck, 3, 40, 237

S

S→F instability, 189
S→F transition, 129, 146, 191
S→F→S→J transitions, 154
S→J instability in driver experiment on circular road, 141
S→J transition, 23
S→J→S→F transitions, 154
Safe space gap, 180
Safe time headway, 182
Safety driving condition, 180
Self-maintaining of synchronized flow at bottleneck, 59, 75
Sequence of F→S→F transitions, 129
Sequence of F→S→J transitions, 23, 135
Sequence of moving bottlenecks, 121
Shock wave, 222

Smallest nucleus required for traffic breakdown (F→S transition) at bottleneck, 81
Space gap, 177
Spatiotemporal congested traffic pattern, 21
Spatiotemporal microscopic structure within wide moving jam, 158
Spatiotemporal traffic pattern, 20
Speed adaptation, 102
Speed adaptation effect, 102
Speed breakdown, 1, 60, 238
Speed–density plane, 51
Speed–flow plane, 51
Speed limitation, 120
Speed wave of local speed increase in synchronized flow, 188
Spill-back effect, 96
Spill-over effect, 96
Spontaneous F→S transition at bottleneck, 21, 57
Spontaneous traffic breakdown at bottleneck, 21, 57
Spontaneous traffic breakdown versus induced traffic breakdown at bottleneck, 131
Stable free flow with respect to F→S transition at bottleneck, 90
Stable synchronized flow with respect to S→J transition, 206
Standard perception of highway capacity, 48
Standard traffic and transportation science, 2, 238
Standard traffic and transportation theories, 2, 238
Standard understanding of highway capacity, 49, 168, 216
Standard understanding of stochastic highway capacity, 49, 168, 216
Steady state of synchronized flow, 177
Stochastic highway capacity in standard traffic and transportation science, 49, 168, 216
Stochastic highway capacity in three-phase traffic theory, 92, 168, 237
Stop-and-go traffic, 1, 135
Synchronization space gap, 180
Synchronization time headway, 182
Synchronized flow, 17, 79, 238
Synchronized flow pattern (SP), 71, 145, 147

T

Three-phase traffic theory, 17, 238

Index 243

Time delay in driver deceleration, 136, 223
Time delay in driver reaction, 136
Time delay in over-acceleration, 104
Time headway, 181
TPACC, 172
Traffic breakdown, 1, 21, 238
Traffic management, 2
Traffic oscillations of congested traffic, 1, 135
Traffic pattern, 20
Traffic phase definition, 17
Treiterer's hysteresis phenomenon, 210
Trucks as moving bottlenecks, 120
Two-dimensional (2D) region of states of synchronized flow, 177
2Z-characteristic for phase transitions in traffic flow, 195

U
Upstream front of synchronized flow, 19
Upstream jam front, 1, 19

V
Vehicle acceleration, 181
Vehicle deceleration, 181
Vehicle density, 1
Vehicle density estimation, 51
Vehicle length, 177
Vehicle platoon, 211
Vehicle trajectory, 27

W
Waves in heterogeneous free flow, 120
Wide moving jam, 17, 238
Widening SP (WSP), 147

Z
Z-characteristic for S→J transition, 194
Z-characteristic for traffic breakdown (F→S transition) at bottleneck, 192

Printed in Great Britain
by Amazon